Light Emission in Silicon:
From Physics to Devices

SEMICONDUCTORS
AND SEMIMETALS
Volume 49

Semiconductors and Semimetals

A Treatise

Edited by Robert K. Willardson
    CONSULTING PHYSICIST
    SPOKANE, WASHINGTON

Eicke R. Weber
DEPARTMENT OF MATERIALS SCIENCE
AND MINERAL ENGINEERING
UNIVERSITY OF CALIFORNIA
AT BERKELEY

*In memory of Dr. Albert C. Beer, Founding Co-Editor in 1966
and Editor Emeritus of Semiconductors and Semimetals.
Died January 19, 1997, Columbus, OH.*

Light Emission in Silicon: From Physics to Devices

SEMICONDUCTORS
AND SEMIMETALS

Volume 49

Volume Editor

DAVID J. LOCKWOOD

NATIONAL RESEARCH COUNCIL
INSTITUTE FOR MICROSTRUCTURAL SCIENCES
OTTAWA, CANADA

ACADEMIC PRESS
San Diego   London   Boston
New York   Sydney   Tokyo   Toronto

This book is printed on acid-free paper.

COPYRIGHT © 1998 BY ACADEMIC PRESS

ALL RIGHTS RESERVED.
NO PART OF THIS PUBLICATION MAY BE REPRODUCED OR TRANSMITTED IN ANY FORM OR BY ANY MEANS, ELECTRONIC OR MECHANICAL, INCLUDING PHOTOCOPY, RECORDING, OR ANY INFORMATION STORAGE AND RETRIEVAL SYSTEM, WITHOUT PERMISSION IN WRITING FROM THE PUBLISHER.

The appearance of the code at the bottom of the first page of a chapter in this book indicates the Publisher's consent that copies of the chapter may be made for personal or internal use, or for the personal or internal use of specific clients. This consent is given on the condition, however, that the copier pay the stated per copy fee through the Copyright Clearance Center, Inc. (222 Rosewood Drive, Danvers, Massachusetts 01923), for copying beyond that permitted by Sections 107 or 108 of the U.S. Copyright Law. This consent does not extend to other kinds of copying, such as copying for general distribution, for advertising or promotional purposes, for creating new collective works, or for resale. Copy fees for pre-1997 chapters are as shown on the chapter title pages; if no fee code appears on the chapter title page, the copy fee is the same as for current chapters. 0080-8784/98 $25.00

ACADEMIC PRESS
525 B Street, Suite 1900, San Diego, CA 92101-4495, USA
1300 Boylston Street, Chestnut Hill, Massachusetts 02167, USA
http://www.apnet.com

ACADEMIC PRESS LIMITED
24-28 Oval Road, London NW1 7DX, UK
http://www.hbuk.co.uk/ap/

International Standard Book Number: 0-12-752157-7

PRINTED IN THE UNITED STATES OF AMERICA
97 98 99 00 01 IC 9 8 7 6 5 4 3 2 1

Contents

LIST OF CONTRIBUTORS ix
ABSTRACT xi
PREFACE xiii

Chapter 1 Light Emission in Silicon

David J. Lockwood

I.   Introduction 1
II. The Optoelectronic Age 2
III. Physical Properties of Si 4
IV. Methods for Overcoming the Indirect Bandgap Limitations in Si . . . 6
    1. Brillouin Zone Folding in Atomic Layer Superlattices 6
    2. Band Structure Engineering via Alloying 9
    3. Luminescence via Impurity Centers 11
    4. Si Nanostructures 14
    5. Polymers and Molecules Containing Si 24
    6. Hybrid Methods for Integrating Direct Gap Materials with Si . . 25
V.   Prospects for Si Based Optoelectronic Devices 26
    References 29

Chapter 2 Band Gaps and Light Emission in Si/SiGe Atomic Layer Structures

Gerhard Abstreiter

I.   Introduction 37
II. Structural Properties 40
III. Bandgaps, Band Offsets, and Brillouin Zone Folding 44
IV. Photoluminescence, Electroluminescence, and Photocurrent Measurements . . 54
    1. $Si_{1-x}Ge_x$ Alloy Layers and Quantum Wells 54
    2. $Si_m Ge_n$ Short Period Superlattices 59

    3. $Ge_nSi_mGe_n$ Atomic Layer Structures and Interfaces with Staggered Band Offsets 64
    4. Laterally Confined QWs and Ge-Rich Self-assembled Dots 67
  V. Concluding Remarks 70
    Acknowledgments 70
    References 71

Chapter 3 Radiative Isoelectronic Impurities in Silicon and Silicon-Germanium Alloys and Superlattices

Thomas G. Brown and Dennis G. Hall

  I. Introductory Concepts 78
    1. Isoelectronic Impurity Atoms and Complexes 78
    2. Exciton Binding 79
    3. Historical Perspective: Isoelectronic Impurities in GaP 82
    4. Isoelectronic Impurities in Si: A Resume 83
  II. Isoelectronic Bound Exciton Emission from c-Si 83
    1. Sample Preparation and Processing 83
    2. Photoluminescence from Si:In, Si:Al, and Si:Be 85
    3. Photoluminescence from Chalcogen-related Centers 91
    4. Electroluminescence at Isoelectronic Centers in c-Si 93
  III. Isoelectronic Bound Exciton Emission in Be-Doped SiGe Alloys: A Case Study 94
    1. Photoluminescence from Thick, Be-Doped SiGe Alloys 95
    2. Photoluminescence from Be-Doped SiGe/Si QWs 97
    3. Beryllium Doping During Epitaxial Growth 99
  IV. Device Considerations 103
  V. Concluding Remarks 106
    References 107

Chapter 4 Erbium in Silicon

J. Michel, L.V.C. Assali, M.T. Morse, and L.C. Kimerling

  I. Introduction 111
  II. Er Doping of Si 113
    1. Ion Implantation 113
    2. Solid Phase Epitaxy (SPE) 117
    3. Molecular Beam Epitaxy (MBE) 118
    4. Chemical Vapor Deposition (CVD) 118
    5. Ion-beam Epitaxy (IBE) 120
  III. Diffusivity and Solubility 121
  IV. Light Emission 127
    1. Physics of Light Emission 127
    2. Ligands 133
    3. Electrical Properties 136
    4. Activation and Deactivation Processes 139
  V. Electronic Structure 142
    1. Electronic Structure of Er-related Impurities in Si 144
    2. Isolated Er Impurity in Si 145
    3. Er-related Complexes in Si 148

VI. Light Emitting Diode Design	150
VII. Summary	153
Acknowledgments	153
References	153

Chapter 5 Silicon and Germanium Nanoparticles

Yoshihiko Kanemitsu

I. Introduction	157
II. Fabrication of Silicon (Si) and Germanium (Ge) Nanoparticles	158
1. Si Nanoclusters: Organic Synthesis	158
2. Isolated Si Nanocrystals: Decomposition of Silane Gas	162
3. Porous Si and Ge: Electrochemical Etching	163
4. Si and Ge Nanocrystals in SiO_2 Matrices: Co-sputtering and Ion Implantation	166
5. Ge Nanocrystals: Chemical Methods	168
III. Photoluminescence Mechanism	170
1. Size Dependence of the PL Peak Energy	171
2. Resonantly Excited Luminescence Spectrum	174
3. Three Region Model	177
4. Photoluminescence Dynamics	185
IV. Unique Optical Phenomena	189
1. Nonlinear Optical Properties of π-Si	189
2. Tuning of Luminescence Wavelength	194
V. Summary	200
Acknowledgments	201
References	202

Chapter 6 Porous Silicon: Photoluminescence and Electroluminescent Devices

Philippe M. Fauchet

I. Introduction	206
1. Si Light Emission	206
2. Porous Si	207
II. Properties of the PL Bands	210
1. The "Red" Band	210
2. The "Blue" Band	212
3. The "Infrared" Bands	213
4. The Extrinsic Luminescence Bands	216
III. Origin of the Intrinsic PL Bands	218
1. Quantum Confinement and the Red PL Band	218
2. Si Oxide and the Blue PL Band	222
3. Recrystallization, Dangling Bonds, and the Infrared PL Bands	224
IV. Pure Quantum Confinement and Surface States: A Critical Discussion	226
V. Nonoptical Properties	233
1. Introduction	233
2. Electrical Properties	233
3. Structural Properties	236
VI. Electroluminescent Devices	238

	1. General Survey	238
	2. LED Lifetime	239
	3. Power Efficiency	240
	4. Response Time	241
	5. Spectral Coverage	243
	6. Compatibility with Microelectronics	244
VII.	Conclusions and Outlook	246
	Acknowledgments	247
	References	247

Chapter 7 Theory of Radiative and Nonradiative Processes in Silicon Nanocrystallites

C. Delerue, G. Allan, and M. Lannoo

I.	Introduction	253
II.	Electronic Properties	254
III.	Optical Transitions and Radiative Lifetime	258
IV.	Exchange Splitting and Symmetry of the Crystallites	262
V.	Atomic Relaxation, Stokes Shift, and Self-trapped Exciton	269
	1. Stokes Shift for the Delocalized States	270
	2. The Existence of Self-trapped Excitons	271
VI.	Nonradiative Recombination	279
	1. Recombination on Surface Dangling Bonds	279
	2. Nonradiative Auger Recombination	286
VII.	Screening in Nanocrystallites and Coulomb Charging Effects	292
	1. Hydrogenic Impurities	292
	2. Coulomb Effects and Effective Dielectric Constant	295
VIII.	Conclusion	298
	References	299

Chapter 8 Silicon Polymers and Nanocrystals

Louis Brus

I.	Introduction	303
II.	Silicon Polymers in One, Two, and Three Dimensions	304
	1. Band Structure	304
	2. Luminescence	307
III.	Passivated Silicon Nanocrystals	308
	1. Theory of Optical Properties	308
	2. Nanocrystal Synthesis, Characterization, and Luminescence	309
	3. Comparison Between Nanocrystals and Macroscopic Crysalline Si	319
	4. Physical Size Regimes for Individual Nanocrystals	321
IV.	Electron Transport in Porous Nanocrystal Materials	322
	Acknowledgments	325
	References	326

INDEX	329
CONTENTS OF VOLUMES IN THIS SERIES	337

List of Contributors

Numbers in parenthesis indicate the pages on which the authors' contribution begins.

GERHARD ABSTREITER (37), *Walter Schottky Institut, Technical University Munchen, Am Coulombwall, D-8046 Garching, Germany*

G. ALLAN (253), *Institut d'Electronique et de Microelectronique du Nord, Departement ISEN-B.P. 69, 59652 Villeneuve d'Ascq Cedex, France*

L. V. C. ASSALI (111), *Instituto de Fisica da Universidade de Sao Paulo, CP 20516, 01452-990 Sao Paulo, SP, Brazil*

THOMAS G. BROWN (78), *The Institute of Optics, University of Rochester, Rochester, NY 14627*

LOUIS BRUS (303), *Chemistry Department, Columbia University, New York, NY, 10027*

C. DELERUE (253), *Institut d'Electronique et de Microelectronique du Nord, Departement ISEN-B.P. 69, 59652 Villeneuve d'Ascq Cedex, France*

PHILIPPE M. FAUCHET (206), *Department of Electrical Engineering, The Institute of Optics, Department of Physics and Astronomy, and Laboratory for Laser Energetics, University of Rochester, Rochester, NY 14627*

DENNIS G. HALL (78), *The Institute of Optics, University of Rochester, Rochester, NY 14627*

YOSHIHIKO KANEMITSU (157), *Institute of Physics, University of Tsukuba, Tsukuba, Ibaraki 305, Japan*

L. C. KIMERLING (111), *Department of Materials Science and Engineering, Massachusetts Institute of Technology, Cambridge, MA 02139*

M. LANNOO (253), *Institut d'Electronique et de Microelectronique du Nord, Departement ISEN-B.P. 69, 59652 Villeneuve d'Ascq Cedex, France*

DAVID J. LOCKWOOD (1), *Institute for Microstructural Sciences, National Research Council Canada, Montreal Road, Bldg. M-36, Room 1144, Ottawa, Ontario K1A OR6, Canada*
J. MICHEL (111), *Department of Materials Science and Engineering, Massachusetts Institute of Technology, Cambridge, MA 02139*
M. T. MORSE (111), *Department of Materials Science and Engineering, Massachusetts Institute of Technology, Cambridge, MA 02139*

Abstract

Interest in obtaining useful light emission from silicon based materials has never been greater. This is primarily because there is a strong demand for optoelectronic devices based on silicon, but also because recently there has been significant progress in materials engineering methods. This book reviews the latest developments in this work, which is aimed at overcoming the indirect bandgap limitations in light emission from silicon. Subjects covered in detail include optical band gap engineering through quantum confinement and Brillouin zone folding in Si/Ge and $Si/Si_{1-x}Ge_x$ superlattices and heterostructures, light emission from isoelectronic and erbium impurity centers in silicon, and luminescence in silicon nanoparticles, porous silicon, and silicon polymers. The incorporation of these different materials into devices is described and future device prospects are assessed.

Preface

Although the optical properties of silicon have been explored over many years and are now well understood, interest in light emission in silicon has never been greater. This mostly is due to two significant factors. First, optoelectronics has emerged as a major industry in its own right and there is a strong and continuing desire to incorporate optical devices such as lasers and detectors, and optical modulators, multiplexers, waveguides, and interconnects within the established silicon microelectronics industry. Much of this work is perforce being carried out with group III–V semiconductors such as GaAs, whose optical properties are well suited to such devices. Nevertheless, because of the heavy investment in silicon based technology, the desire to use silicon material for such devices is overwhelming. The main problem is that silicon is an indirect bandgap semiconductor and, therefore, emits light very feebly. This problem is now being addressed through the second prime factor: materials science engineering. This field of research has shown amazing growth in the last decade over a wide variety of materials systems and techniques. In the semiconductor field, growth techniques such as molecular beam epitaxy are able to provide tailored structures with atomic layer precision. Such novel structures, unknown in nature, allow the electronic and optical properties of prototype bulk materials to be significantly altered through, for example, band structure engineering. These powerful new techniques are now being applied in earnest, in attempts to overcome the inherent deficiency of weak light emission in silicon. Significant advances have been made to such an extent that bright light-emitting devices are now being produced from silicon. Considering this rapid progress from materials engineering to devices, it is an appropriate time to review advances in this field. This book is the outcome of an interest expressed by many in such a review.

The various review chapters in this book have been prepared by well-recognized authorities in their fields of research who have worked for a number of years on these topics. The subjects covered in detail include optical band-gap engineering through quantum confinement in quantum wells and Brillouin-zone folding in Si/Ge and Si/Si$_{1-x}$Ge$_x$ heterostructures and superlattices, light emission from isoelectronic and erbium impurity centers in silicon, and luminescence in silicon nanoparticles, porous silicon, and silicon polymers. The incorporation of these different materials into devices is described and future device prospects are assessed. Other emerging topics such as tunable light emission from silicon quantum wires and dots are covered in the introductory chapter. For space reasons, it was decided to limit this review to crystalline materials and thus only passing references are made to light emission in amorphous silicon, which is a huge field of research of its own with a vast literature that has been reviewed frequently elsewhere (see, for example, *Semiconductors and Semimetals*, Volume 21).

The intention has been to provide an up-to-date overview of the significant progress made in the last decade regarding light emission in silicon and of the prospects for devices that are now emerging from this research. The review chapters are self-contained, but are interrelated, and can be approached in any order. They provide not only a summary of the latest achievements but also a basic underpinning of the fundamental concepts of the subject, with the intention that they can be appreciated by a novice or an expert. It is hoped that they will serve both as a useful reference for current researchers in this field and as a tutorial for the next generation. Certainly, there will be further rapid developments in this field of research over the next decade and, it is hoped, a silicon based laser will soon emerge.

David J. Lockwood
Ottawa
January 1997

Light Emission in Silicon:
From Physics to Devices

SEMICONDUCTORS
AND SEMIMETALS
Volume 49

CHAPTER 1

Light Emission in Silicon

David J. Lockwood

INSTITUTE FOR MICROSTRUCTURAL SCIENCES
NATIONAL RESEARCH COUNCIL OF CANADA
OTTAWA, CANADA

I.	INTRODUCTION	1
II.	THE OPTOELECTRONIC AGE	2
III.	PHYSICAL PROPERTIES OF Si	4
IV.	METHODS FOR OVERCOMING INDIRECT BANDGAP LIMITATIONS IN Si	6
	1. Brillouin Zone Folding in Atomic Layer Superlattices	6
	2. Band Structure Engineering via Alloying	9
	3. Luminescence via Impurity Centers	11
	4. Si Nanostructures	14
	5. Polymers and Molecules Containing Si	24
	6. Hybrid Methods for Integrating Direct Gap Materials with Si	25
V.	PROSPECTS FOR Si-BASED OPTOELECTRONIC DEVICES	26
	References	29

I. Introduction

The many and diverse approaches to materials science problems have greatly enhanced our ability in recent times to engineer the physical properties of semiconductors. Silicon, of all semiconductors, underpins nearly all microelectronics today and will continue to do so for some time to come. However, in optoelectronics, the severe disadvantage of an indirect bandgap has limited the application of elemental Si. This book describes a number of approaches to engineering efficient light emission in Si. Here, these different approaches are placed in context together with other methods of obtaining Si light emission that are not covered in the following chapters.

II. The Optoelectronic Age

The ubiquitous Si microelectronics "chip" is taken for granted in modern society. There has been much research involved in producing these high technology marvels and such research continues unabated at a faster and faster pace. Despite the often stated announcement that "the age of GaAs has arrived", it never quite has, and continued developments in Si and, more recently, $Si_{1-x}Ge_x$ alloy technology (Abstreiter, 1992; Kasper and Schäffler, 1991) continue to advance the frontiers of microminiaturization, complexity, and speed. This continued advance has been driven by application requirements in switching technology (e.g., computers) and high speed electronics (e.g., wireless telecommunications). Gallium arsenide and other compound semiconductors have, however, maintained a significant role in the construction of optoelectronic and purely photonic devices where the medium of switching and communication is light itself (Saleh and Teich, 1991).

The merging of Si based electronics with photonics has largely required the pursuit of hybrid technologies for light emitters and modulators (see, e.g., Fig. 1), which are often both expensive and complicated to produce. The

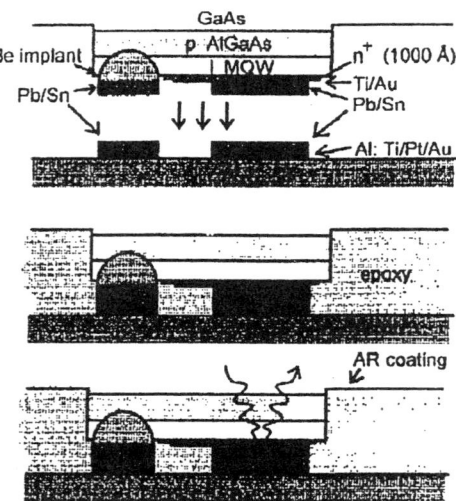

FIG. 1. Three-step hybridization process used to form GaAs multiple quantum well modulators integrated with Si CMOS: (top) fabrication, aligning, and solder bonding of GaAs modulator chip on Si chip; (middle) flowing epoxy etch-protectant between the chips, which is then allowed to harden; (bottom) removal of GaAs substrate using a jet etcher and deposition of antireflection (AR) coating. (From Goossen et al., 1995.)

most satisfactory solution and still mostly a dream as far as light sources are concerned would be optoelectronic devices created entirely from Si based materials, where extensive experience in Si fabrication and processing could be put to best use (Soref, 1993). Already, a wide range of optoelectronic integrated circuits (OEICs) incorporating Si or $Si_{1-x}Ge_x$ as a detector or waveguide have been elaborated (Kasper and Presting, 1990; Kasper and Schäffler, 1991; Soref, 1993, 1996). A recent example of a $Si/Si_{1-x}Ge_x$ photonic device is the rib-waveguide optical modulator shown in Fig. 2, where carrier injection in a $Si_{1-x}Ge_x$ p-i-n heterostructure allows optical intensity modulation in Si at megahertz frequencies (Fernando et al., 1995). Such monolithic modulators operating at much higher (gigahertz) frequencies are required in practice. Nevertheless, the major deficiency in Si based optoelectronic devices remains the lack of suitable light emitters and especially lasers.

The general requirements in Si based light sources are efficient light emitting diodes, lasers, and optical amplifiers for use in optical communications technologies such as fiber optics and displays. Operating wavelengths in the range 0.45–1.6 μm are needed to cover both full color displays and fiber optic operating wavelengths of 1.3 and 1.55 μm. Specific applications for such sources include fiber optic transmitters, optical interconnects within and between computer chips, optical controllers for phased-array microwave antennas, information display screens, printing elements in xerography, and writing and readout of optical compact disc information.

Many quite different approaches to alleviating the miserable light emission in Si ($\sim 10^{-4}$ quantum efficiency at 300 K) have been proposed

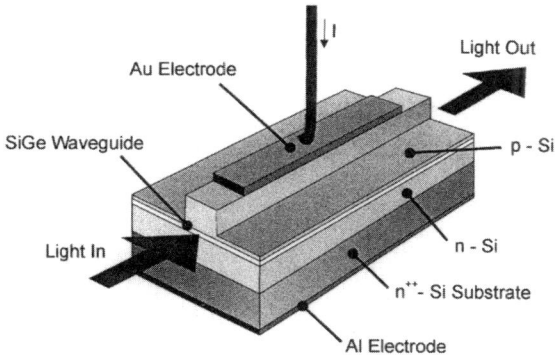

FIG. 2. Schematic representation of a $Si/Si_{0.85}Ge_{0.15}/Si$ p-i-n waveguide modulator. A maximum modulation depth of 66% was obtained at 1.3 μm for a 2 mm long waveguide using a peak pulse current density of 2700 A/cm². (From Fernando et al., 1995.)

and are actively being explored (Iyer and Xie, 1993; Kimerling et al., 1997). Some, such as $Si_{1-x}Ge_x$ quantum well or Si/Ge superlattice structures, rely on band structure engineering, while others rely on quantum confinement effects in low dimensional structures, as typified by quantum dots or porous silicon (π-Si) (Lockwood, 1994). Still another approach is impurity mediated luminescence from, for example, isoelectronic substitution or by the addition of rare earth (RE) ions. An overview of results obtained with these and other methods is given below. However, in order to understand more fully the reasons why such different approaches are necessary, it is appropriate to review first what creates the optical emission problem in crystalline silicon (c-Si).

III. Physical Properties of Si

Silicon crystallizes in the diamond structure (*Properties of Silicon*, 1988), which consists of two interpenetrating face-centered cubic lattices displaced from each other by one quarter of the body diagonal. In zinc-blende semiconductors such as GaAs, the Ga and As atoms lie on separate sublattices, and thus the inversion symmetry of Si is lost in III–V binary compounds. This difference in their crystal structures underlies the disparate electronic properties of Si and GaAs. The energy band structure in semiconductors is derived from the relationship between the energy and momentum of a carrier, which depends not only on the crystal structure but also on the bonding between atoms, their respective bond lengths, and the chemical species. The band structure is often quite complex and can only be calculated empirically. The results of such calculations (Chelikowsky and Cohen, 1976) for Si and GaAs are shown in Fig. 3. The figure shows the dispersion relations for the energy $E(k)$ of an electron (positive energy) or hole (negative energy) for wave vectors k within the first Brillouin zone.

The valence band structure is much the same for many semiconductors and exhibits a maximum at the Brillouin zone center or Γ point (i.e., at $k = 0$). The notable difference between Si and GaAs is that the degeneracy in the $\Gamma_{25'}$ band maximum at $k = 0$ is removed in the case of GaAs, because of the spin–orbit interaction, into Γ_8 and Γ_7 subbands. In general, $E(k)$ has maxima or minima at zone center and zone boundary symmetry points, but additional extrema may occur at other points in the Brillouin zone (see Fig. 3). In the case of Si, the lowest point in the conduction band occurs away from high symmetry points near the X point at the Brillouin zone boundary (along $\langle 001 \rangle$), whereas in GaAs it occurs at the Γ point. The energy gap in a semiconductor is defined as the separation between this

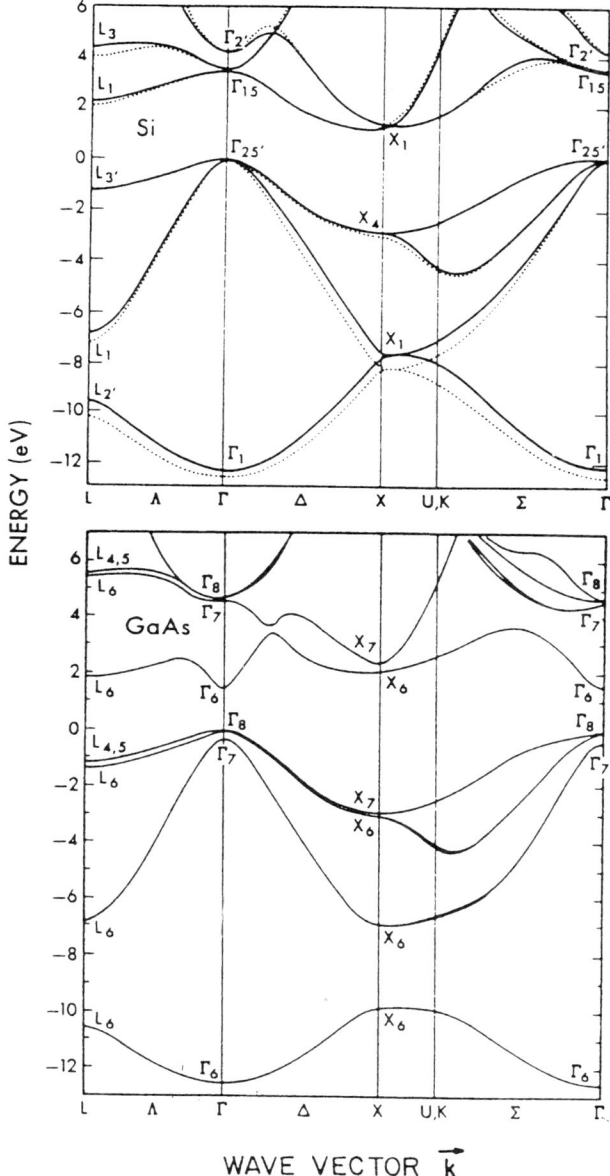

FIG. 3. Theoretical band structures of Si and GaAs. In the case of Si, results are shown for nonlocal (solid line) and local (dashed line) pseudopotential calculations. (From Chelikowsky and Cohen, 1976.)

conduction band minimum and the valence band maximum at the Γ point. For GaAs, the energy gap is classified as direct, because a transition can occur directly at $k = 0$ between initial and final states having the same wave vector. Correspondingly, Si is termed an indirect gap semiconductor, because the initial and final states have different wave vectors.

In direct gap GaAs, an excited electron at the bottom of the conduction band can relax spontaneously back into the valence band by emitting a photon at the bandgap energy. This electron–hole radiative recombination process can only occur in Si if momentum is conserved, that is, the excited electron wave vector must be reduced to zero. This, in pure Si, occurs via the transfer of momentum to a phonon that is created with equal and opposite wave vector to that of the initial state in the conduction band. Such a three-body process is quite inefficient compared with direct gap recombination (Kimerling et al., 1997; Pankove, 1971). This is why Si is such a poor light emitter.

Electron–hole pairs may bind to each other to form excitons, which can be either free or tied to impurities or defects (Kimerling et al., 1997; Pankove, 1971). The decay of such excitons can lead to light emission that may be tunable by, for example, quantum confinement. Such excitonic emission is thus under active investigation in quantum well (QW), wire, and dot structures (Yoffe, 1993).

IV. Methods for Overcoming Indirect Bandgap Limitations in Si

Materials engineering, a relatively new phenomenon in materials science, is now being actively applied to Si in an attempt to overcome indirect bandgap limitations in light emission from Si. In these various attempts, the aim is: (1) to increase the efficiency of luminescence by increasing the overlap of the electron and hole wavefunctions via, for example, confinement and band structure engineering; (2) to tune the wavelength of the emission by forming alloys and molecules; or (3) to induce recombination at impurity centers. Such attempts can often involve several of these factors. Alternatively, hybrid methods are being explored where, for example, direct gap GaAs is joined with Si. Each of these methods is outlined briefly below and many of them are explored in detail in subsequent chapters.

1. BRILLOUIN ZONE FOLDING IN ATOMIC LAYER SUPERLATTICES

In the mid-1970s it was conjectured theoretically by Gnutzmann and Clausecker (1974) that Brillouin zone folding in thin layer superlattices where the layer thicknesses were of the order of the unit cell dimensions could result in a direct (or quasi-direct, as it is now termed) bandgap structure. Growth in the 1980s of high quality $(Si_m Ge_n)_p$ atomic layer superlattices (m and n are the number of monolayers of Si and Ge in each period and p is the number of periods) by molecular beam epitaxy (Kasper and Schäffler, 1991) led impetus to this concept, which was revisited by Jackson and People (1986) and, subsequently, by a number of other theoreticians. The essence of the idea is conveyed in Fig. 4. The new superlattice periodicity d along the growth direction results in a smaller Brillouin zone of size $\pm \pi/d$ compared with that of the original lattice ($\pm 2\pi/a$, where a is the lattice constant). The electronic band structure is then folded back into this new reduced Brillouin zone. For this simple model, it is apparent that the minimum in the conduction band in bulk Si is folded into the Brillouin zone center for $d \approx 5a/2$, which corresponds to 10 monolayers of Si, and a direct gap is evident. In practice, strains within a $Si_m Ge_n$ superlattice together with band offsets at heterointerfaces compromise this naive picture (see, e.g., Brey and Tejedor (1987); Froyen, Wood,

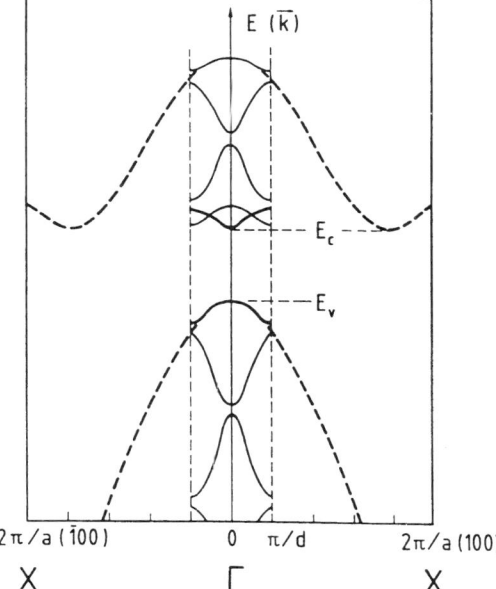

FIG. 4. Schematic representation of the Brillouin zone folding concept in Si resulting from the new superlattice periodicity in the growth direction. Here, the conduction band minimum along the X direction is folded back into the Γ point when the superlattice period is about 10 monolayers of Si. (From Kasper and Schäffler, 1991.)

and Zunger (1987); Hybertsen and Schlüter (1987); Kasper and Schäffler (1991); People and Jackson (1987)). Theory has shown that for certain superlattice periods and when the Si layers are strained a direct energy gap is expected in Si_mGe_n superlattices, but the transition probability is still several orders of magnitude below that of GaAs.

The first experimental evidence of modifications to the Si and Ge band structures in such superlattices was obtained from electroreflectance measurements of Si_4Ge_4 superlattices grown on (001) Si (Pearsall et al., 1987). However, it was not until later on when strain-symmetrized Si_mGe_n superlattices were grown on strain-relaxed thick $Si_{1-x}Ge_x$ alloy buffer layers on Si that first indications of the predicted photoluminescence (PL) intensity enhancement and reduced energy gap were obtained (Kasper and Schäffler, 1991; Zachai et al., 1990). Improvements in crystal growth conditions subsequently led to a positive identification of these new features of Si_mGe_n superlattices (Menczigar et al., 1993). As shown in Fig. 5, the PL no-phonon (NP) peak clearly shifts to lower energy and increases in intensity with increasing superlattice periodicity, as compared with a $Si_{0.6}Ge_{0.4}$ alloy layer of the same average composition as the $Si_{3i}Ge_{2i}$ ($i = 1, 2, 3$) superlattices.

From a device point of view, although infrared emission can readily be obtained at low temperature from such Si_mGe_n structures at energies useful for fiber optic transmission work, the PL and electroluminescence (EL) is

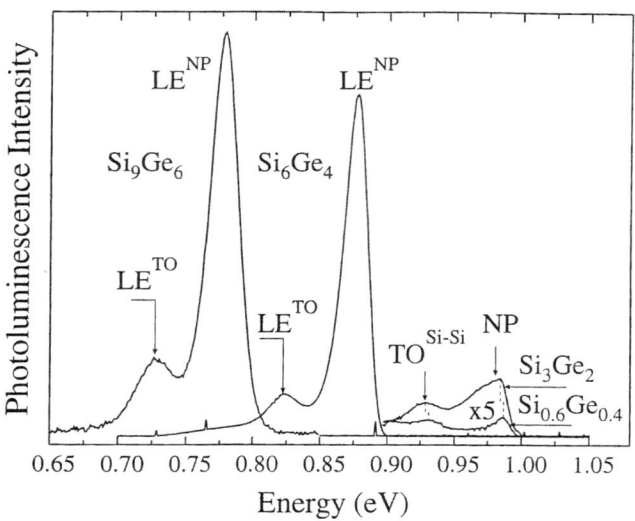

FIG. 5. Low temperature PL spectra of strain-symmetrized Si_mGe_n superlattices and of a $Si_{0.6}Ge_{0.4}$ alloy layer grown on a step-graded SiGe buffer layer on Si. (From Menczigar et al., 1993.)

essentially quenched at room temperature (Menczigar *et al.*, 1992; Presting *et al.*, 1992). Unless there are further major improvements in material quality, it is more likely that these atomic layer superlattices will find eventual use as infrared detectors rather than as emitters (Pearsall, 1994; Presting *et al.*, 1992). However, some promising steps towards room temperature EL structures have been reported recently (Engvall *et al.*, 1993, 1995).

2. Band Structure Engineering via Alloying

Alloying of Ge or C with Si allows engineering of the electronic band structure, where the energy gap may be varied with alloy composition and strain (Pearsall, 1994; People and Jackson, 1990). This is shown, for example, in Fig. 6 for strained $Si_{1-x}Ge_x$ on Si, where the tunability range is appropriate for fiber optic communications. Unfortunately, because of heterostructure stability limitations, the $Si_{1-x}Ge_x$ layer thickness must be kept below the critical thickness, which decreases rapidly with increasing x

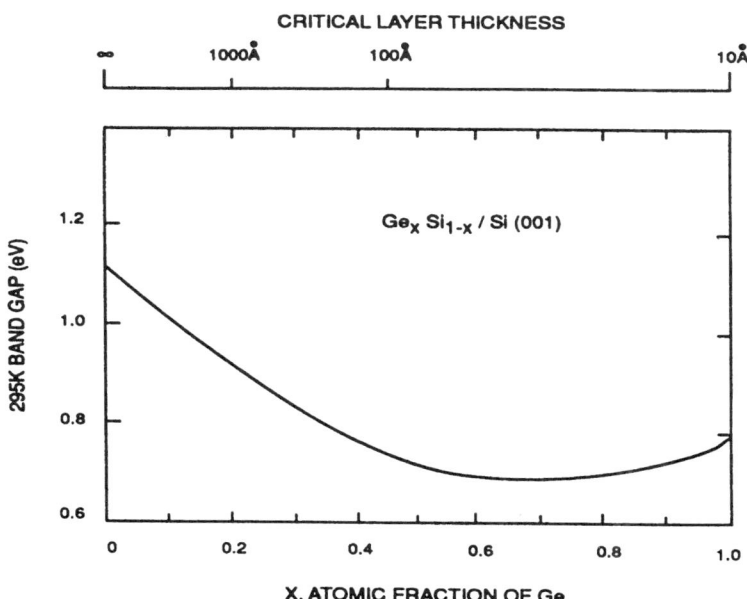

FIG. 6. The bandgap at room temperature of strained $Si_{1-x}Ge_x$ on Si. Also shown is the critical layer thickness as a function of x. (From Pearsall, 1994.)

(see Fig. 6). Thus absorbing/emitting regions in infrared detectors/emitters are necessarily small. Also, the bandgap remains indirect. Despite these severe limitations much research has been carried out on the optical properties of $Si/Si_{1-x}Ge_x$ heterostructures (Pearsall, 1994), which exhibit type I band alignment (Houghton et al., 1995; People and Jackson, 1990), and, to a lesser extent, on $Si/Si_{1-x}C_x$ or even $Si/Si_{1-x-y}Ge_xC_y$ (Orner et al., 1996; St. Amour et al., 1995; Soref et al., 1996), and the properties of infrared emitting devices are being explored.

Electroluminescence and PL have been observed from $Si_{1-x}Ge_x$ in both single layer and superlattice form with increased intensity compared with Si, as shown, for example, in Fig. 7. The strong broad peak seen in PL and EL ($\sim 0.5\%$ internal quantum efficiency) at 0.89 eV in Fig. 7 is typical of results obtained from related studies (see, e.g., Lenchyshyn et al., 1992; Noël et al., 1990; Sturm et al., 1991). The luminescence energy tracks the alloy composition dependence shown in Fig. 6, but is at a lower energy. The recombina-

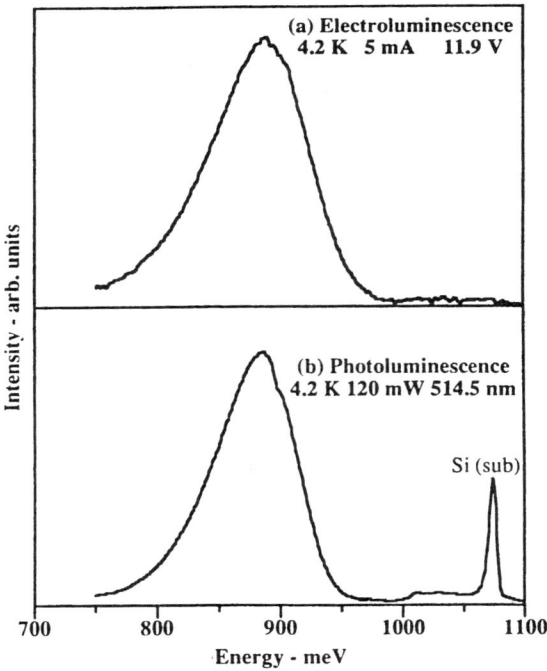

FIG. 7. Broad EL and PL from a $Si_{0.82}Ge_{0.18}$ p-n heterostructure at 4.2 K. A sharper emission line from the Si substrate is also evident. (From Rowell et al., 1990.)

tion mechanism varies depending on the alloy layer thickness and perfection resulting in near band edge and/or excitonic luminescence (Lenchyshyn et al., 1993; Noël et al., 1992; Rowell et al., 1993).

In earlier work, EL from $Si_{1-x}Ge_x$/Si p-n diodes was quenched by increasing the temperature above 80 K (Rowell et al., 1990), but EL was soon reported at temperatures up to 220 K in p-i-n diode structures (Robbins, Calcott, and Leong, 1991). Progress in materials quality and device design has continued to improve EL device performance (see, e.g., Förster et al., 1996; Fukatsu et al., 1992; Kato, Fukatsu, and Shiraki, 1995; Mi et al., 1992; Presting et al., 1996) such that room temperature EL has now been reported at wavelengths near 1.3 μm (Mi et al., 1992; Presting et al. 1996). The major problem with such devices for practical purposes at present is their low efficiency at room temperature (Mi et al., 1992; Presting et al., 1996).

3. LUMINESCENCE VIA IMPURITY CENTERS

Another approach to increasing the EL efficiency of an indirect bandgap semiconductor is to introduce an impurity that localizes the electron and hole, as pioneered in GaP (Thomas, Gershenzon, and Hopfield, 1963). This has been done in Si EL diodes by using, for example, RE impurities (Ennen et al., 1985), carbon complexes (Canham, Barraclough, and Robbins, 1987), and sulfur–oxygen complexes (Bradfield, Brown, and Hall, 1989) as localization centers for electron–hole recombination. Extrinsic luminescence in Si can arise from a variety of sources (Davies, 1989; Kimerling et al., 1997). Here, we concentrate on isoelectronic and RE extrinsic centers, as these are presently the most promising for device applications. Further details may be found in Chapters 3 and 4 of this book.

Isoelectronic centers are created by doping Si with electrically neutral impurities such as the isovalent elements C, Ge, and Sn or a multiple-atom complex with no dangling bonds. Isoelectronic impurities bind free excitons in Si, which can increase the probability of electron–hole recombination due to spatial confinement of the particles. The resultant recombination energy may appear as light or disappear through phonon generation and other nonradiative decay channels (Kimerling et al., 1997). An example of isoelectronic bound exciton emission is shown in Fig. 8 for Si implanted with In. The characteristic sharp NP excitonic emission in PL and EL occurs at 1.11 μm (1.12 eV), which is just below the Si indirect bandgap of 1.17 eV at 14 K. The optical emission intensity decreases with increasing temperature (Brown and Hall, 1986a).

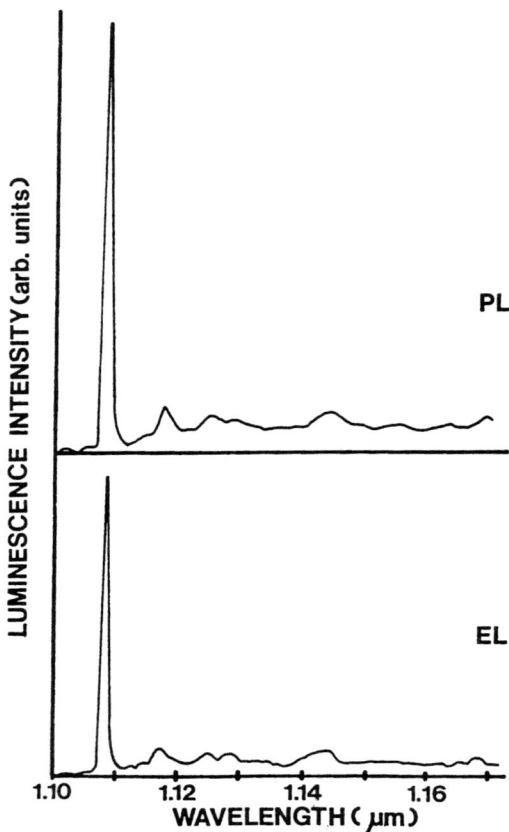

FIG. 8. Sharp PL and EL at 1.11 μm from a quenched Si:In sample at 14 K. (From Brown and Hall, 1986a.)

The optical properties of a variety of such isoelectronic impurity centers including In, Al-N, Be, S, and Se have been studied both in Si and $Si_{1-x}Ge_x$ alloys (Davies, 1989; Kimerling *et al.*, 1997). A luminescence external quantum efficiency of 5% and a lifetime greater than 1 ms have been reported for the S complex emission at 1.32 μm in Si at low temperatures (Brown and Hall, 1986b), but the PL intensity and lifetime decrease sharply with increasing temperature. This variation with increasing temperature is due to exciton dissociation and competing nonradiative recombination processes. The low bound-exciton emission intensity at room temperature militates against isoelectronic-impurity based EL devices at present.

Optical properties of RE ions in solids have been investigated in great

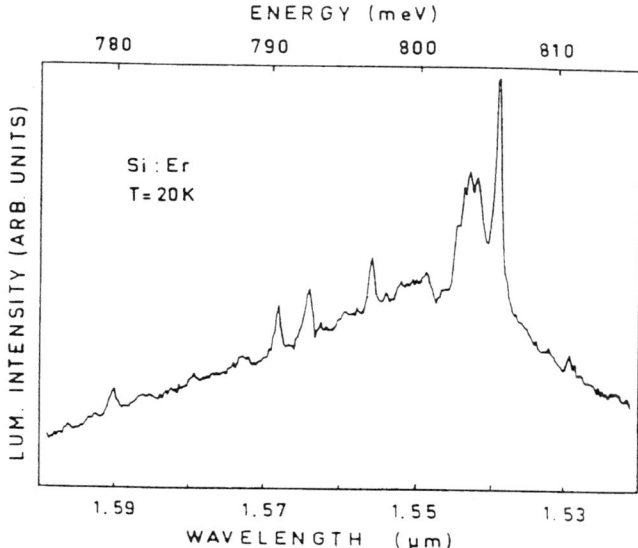

FIG. 9. The low-temperature PL spectrum of Er implanted and annealed Si. (From Ennen et al., 1983.)

detail and are generally well understood (Dieke, 1968). Optical emission of the Er^{3+} ion is of particular interest for semiconductor device applications, because it occurs near 1.5 μm. The Er^{3+} ion emits photons at 1.54 μm in Si (see Fig. 9) by intracenter transitions between Er^{3+}-ion discrete states ($I_{13/2} \rightarrow I_{15/2}$ transition within the 4f electron shell). The excitation of the Er^{3+} ions is a complicated process (Kimerling et al., 1997; Palm et al., 1996) involving first electron–hole carrier generation in Si, then exciton formation, and finally Er excitation by an intracenter Auger process, with a number of competing pathways in the excitation process. Excited state relaxation then occurs via photon emission or, with increasing temperature, via nonradiative backtransfer processes (Michel et al., 1996; Palm et al., 1996). This results in a low quantum efficiency and a marked quenching of luminescence for temperatures above approximately 150 K (Michel et al., 1991; Palm et al., 1996). Nevertheless, research continues with attempts to overcome the Si:Er materials system constraints such as the low solid solubility of Er in Si and the low optical efficiency at room temperature, and room temperature EL devices with improved performance through use of an oxygen codopant are now emerging (see, e.g., Coffa, Franzò, and Priolo, 1996; Michel et al., 1996; Stimmer et al., 1996).

4. SILICON NANOSTRUCTURES

Research on quantum confinement of carriers in Si based nanostructures including π-Si, nanoclusters, and QWs, wires, and dots forms a large part of the work on light emission in Si, and Chapters 5–8 in this book are devoted to different aspects of this topic. Much of this work was stimulated by the discovery in 1990 of bright visible light emission at room temperature in π-Si (Canham 1990). The number of papers published each year on π-Si alone has been approximately 500 recent years. Interest in nanostructures of Si stems from the effects of confinement on carrier wavefunctions when the crystallite diameter is less than the size of the free exciton Bohr radius of 4.3 nm (Yoffe, 1993) in bulk c-Si. Quantum confinement increases the electron–hole wavefunction overlap, resulting in increased light emission efficiency, and shifts the emission peak to higher energy (Brus, 1991; Kimerling et al., 1997).

a. *Porous Si*

Porous Si was discovered in the 1950s by Uhlir (1956). The porous material is created by electrochemical dissolution in HF based electrolytes. Hydrofluoric acid, on its own, etches single-crystal Si extremely slowly, at a rate of only nanometers per hour. However, passing an electric current between the acid electrolyte and the Si sample speeds up the process considerably, leaving an array of deep narrow pores that generally run perpendicular to the Si surface. Pores measuring only nanometers across, but micrometers deep, have been achieved under specific etching conditions.

In 1989, Canham conceived the idea of fabricating Si quantum wires in π-Si by reverting to the much slower chemical HF etch after electrochemically etching c-Si. In this way Canham proposed to join up the pores leaving behind an irregular array of undulating free standing pillars of c-Si only nanometers wide. Canham (1990) observed intense visible PL at room temperature (see Fig. 10) from π-Si that had been etched under carefully controlled conditions. Visible luminescence ranging from green to red in color was soon reported by Cullis and Canham (1991) for other π-Si samples and ascribed to quantum size effects in wires of width ∼3 nm (Canham, 1990). Independently, Lehmann and Gösele (1991) reported on the optical absorption properties of π-Si. They observed a shift in the bulk Si absorption edge to values as high as 1.76 eV that they also attributed to quantum wire formation. Visible PL in π-Si at room temperature was also reported by Bsiesy et al. (1991), Koshida and Koyama (1991), Gardelis et al. (1991), while visible EL was observed by Halimaoui et al. (1991) during

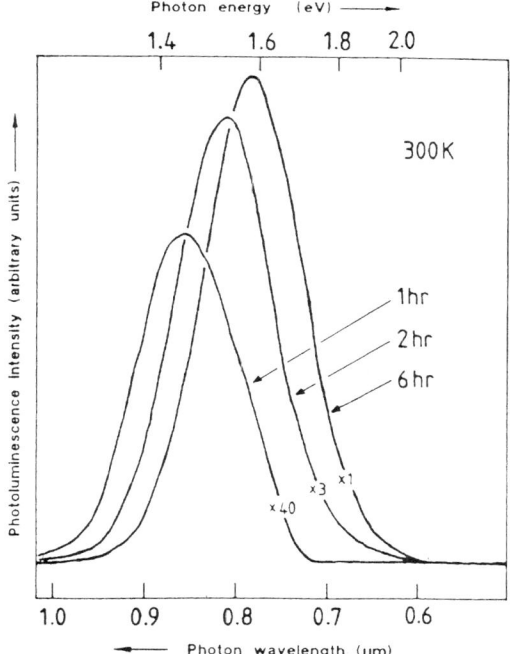

FIG. 10. Room-temperature PL from anodized Si after immersion in 40% aqueous HF for the times indicated. (From Canham, 1990.)

anodic oxidation of π-Si and, later, by Koshida and Koyama (1992) with a diode cell.

Tremendous activity on research into the physical and associated chemical characteristics of π-Si has ensued from these early reports with, unfortunately, considerable duplication of effort. It is impossible to mention all of this work here and interested readers are directed to recent reviews and books (Bensahel, Canham, and Ossicini, 1993; Fauchet et al., 1993; Fauchet, 1996; Feng and Tsu, 1994; Hamilton, 1995; Hérino and Lang, 1995; Iyer, Collins, and Canham, 1992; Kamimura, 1994; Kanemitsu, 1995; Lockwood, 1994; Lockwood et al., 1996; Tischler et al., 1993; Vial, Canham, and Lang, 1993; Vial and Derrien, 1995) for further aspects of this work.

A strong PL signal has been observed from π-Si at wavelengths from near infrared through visible to blue depending on sample porosity and surface chemical treatment. It has even been possible, using specialized preparation techniques, to produce "white" light emitting π-Si (Bensahel, Canham, and

Osscicini, 1993). For discussion purposes, it is convenient to divide these wavelength regions into three: (1) near infrared, (2) red-yellow, and (3) blue.

The most widely studied PL is in the far-red to orange-yellow region, which we shall denote simply as the "red" PL. As evident in Fig. 10, this PL shifts to shorter wavelength with increasing chemical dissolution time. It was soon found that much smaller immersion times were required to produce noticeable blue shifts when chemical dissolution was carried out in the presence of light. The spectra also show a blue shift with increasing anodization current density. The porosity of π-Si increases with increasing anodization current density. Therefore, the behavior of red PL spectra qualitatively reflects the differences in sample porosity and hence in the dimensions of Si nanocrystallites within π-Si. The blue shift of the PL and optical absorption with increasing porosity provided the first important evidence that quantum confinement effects could be playing a role. Nevertheless, after much research, the controversy over the origin(s) of red PL in π-Si persists. This is because the PL peak wavelength and intensity are sensitive to the surface chemistry of π-Si, particularly with regard to relative amounts of hydrogen and oxygen on the surface. Thus, besides the quantum confinement mechanism, various surface state models have been invoked to explain the various results (Lockwood, 1994). Although evidence of quantum confinement effects in π-Si has been obtained via optical absorption measurements (Lockwood, Wang, and Bryskiewicz, 1994), the problems in explaining the PL in such a way are amply demonstrated by the data of Fig. 11. The π-Si samples in this case had a sphere-like morphology (spherites) and the optical gap is seen to be in good agreement with theoretical predictions for quantum dots, but there is a substantial and, as yet, unexplained energy difference between the absorption and emission data.

Oxidation of the π-Si surface has been shown to produce blue PL (Lockwood, 1994). Blue PL is quite weak in as-prepared π-Si. It becomes intense only after strong oxidation and has a much faster decay than red PL. Its origin is of some debate at present. Models currently under consideration include band-to-band recombination in Si nanocrystals, emission from oxide, and emission due to surface states. Present indications are that while the red PL possibly originates from the near-surface region of Si crystallites, the blue PL may emanate from the small c-Si core region.

Near-infrared PL (Lockwood, 1994) at $\sim 0.8\,\text{eV}$ (below the bulk Si bandgap) exhibits complex nonexponential dynamics, with a wide distribution of decay times, and has been assigned to deep level transitions associated with dangling bonds on the surface of Si nanocrystallites.

From these considerations it is apparent that PL in π-Si is very sensitive to the chemistry of π-Si production and treatment. Crystalline Si wires, c-Si

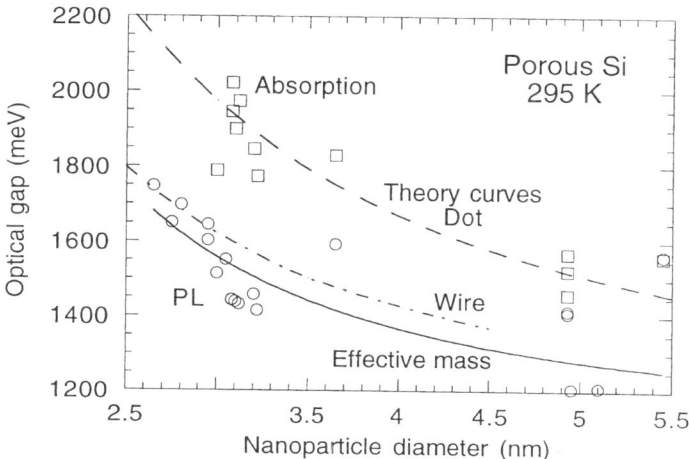

FIG. 11. Dependence of the optical absorption energy gap and PL peak energy on spherite diameter in π-Si samples at room temperature. The solid line is the effective mass model prediction for the optical gap in c-Si spheres, while the broken and short dash-dot lines are theoretical predictions based on a linear combination of atomic orbitals framework for quantum dots and wires, respectively. (From Lockwood and Wang, 1996.)

spherites, and amorphous silicon (a-Si) material, or any combination of them, may be formed in a given sample. Porous Si layers thus formed may be far from uniform, which adds to the difficulties in analyzing their optical properties. Other light emitting species may also be formed on the surface of anodized and otherwise chemically treated Si.

Despite all these disadvantages, the ease of production of π-Si and the facts that the room temperature PL is very efficient (1–10% quantum efficiency) and that it is tunable through blue to near infrared wavelengths have led to impressive efforts to produce practical room-temperature devices. The latest generation of red light emitting diodes (LEDs) have external quantum efficiencies of 0.1% and lifetimes of the order of months (Lockwood et al., 1996). Recently, π-Si LEDs have been integrated into Si microelectronic circuits to provide an addressable LED display (Hirschman et al., 1996). However, improvements in efficiency and power dissipation are necessary for display applications, while an increased modulation frequency (presently ∼1 kHz) is required for optical interconnects. One way to improve the EL efficiency, narrow the band width, improve the directionality, and increase the long-term stability is to insert the LED into a π-Si resonant cavity (Pavesi, Guardini, and Mazzoleni, 1996). Long switching times (up to milliseconds) observed in present π-Si LEDs may yet prove to be an Achilles' heel in optoelectronic applications.

The nonlinear optical properties of π-Si may also prove useful for devices. Second harmonic generation has been observed from π-Si with a magnitude two orders greater than that from c-Si owing to the large surface-to-volume ratio of π-Si (Lo and Lue, 1993). A large optically induced absorption change has been observed in π-Si (Matsumoto et al., 1994). This phenomenon has been used to form all-optical logic gates in π-Si and raises the possibility of fabricating all-optical integrated circuits on Si (Kanemitsu, 1995; Matsumoto et al., 1994, 1995).

b. Silicon Nanoclusters

Rather than produce nanometer size Si crystallites by etching, as in π-Si, there have been numerous attempts at growing them either directly from a gas phase or indirectly by recrystallization within a matrix (Kamimura, 1994; Kimerling et al., 1997; Ogawa and Kanemitsu, 1995; Yoffe, 1993). In fact, the observation of a nanoparticle size dependence of the PL energy in very small Si crystallites passivated with hydrogen (Takagi et al., 1990) predates the similar finding in π-Si (Canham, 1990). Takagi et al. (1990) found that the PL peak energy varied as $1/d^2$ ($3 < d < 5$ nm), where d is the Si nanoparticle diameter, in accordance with quantum confinement effects predicted by a simple effective mass model. As for π-Si, however, the emitted light energy falls below that expected from calculations of the energy gap for Si spheres (Lockwood, 1994). Also, the confinement effect is seen (Schuppler et al., 1994; Takagi et al., 1990) or not seen (Kanemitsu et al., 1993) in emission depending on sample preparation. Interpretation of the nanoparticle PL spectra suffers from the same ambiguities as π-Si, that is, nanoparticle size distribution effects and surface chemistry effects. In addition, the nanoparticle crystal structure may deviate from the cubic diamond structure for very small Si nanoclusters (Kimerling et al., 1997). Recent calculations (Allan, Delerue, and Lannoo, 1996) have shown that luminescence in Si nanocrystallites can be due to excitons trapped at the surface, which is passivated by hydrogen or silicon oxide, while the optical absorption is characteristic of quantum confinement effects. In recent definitive experiments (Brus et al., 1995), the indirect nature of the Si bandgap has been seen from PL and absorption spectra for small (1–2 nm in diameter) surface-oxidized nanocrystals. The red PL quantum efficiency and lifetime is similar to that found for π-Si (Brus et al., 1995), indicating a similar light emission mechanism involving quantum-confined nanocrystal states.

The controversy surrounding the interpretation of PL in Si nanoparticles and in π-Si is displayed throughout the literature and is also evident from the viewpoints expressed in other chapters in this book. The vagaries and complexities of the nanocrystal-interface-surface system are proving difficult to unravel in the short term.

Nanocrystals of Si trapped in some matrix form an attractive system for device fabrication when compared with π-Si, because of increased surface stability and material rigidity. Recently, visible EL has been observed, for example, from Si nanocrystallites embedded in films of a-Si:H (Tong et al., 1996) and from an electrochemically-formed nanocrystalline Si thin film deposited on SnO_2 (Toyama et al., 1996). In the latter case, the p-i-n LED at room temperature emitted orange-red light (1.8 eV) that was readily visible to the eye. The light emission is ascribed variously to near surface states (Tong et al., 1996) and the quantum size effect (Toyama et al., 1996). Substantial progress in the development of such EL structures can be expected over the next few years.

c. *Quantum Wells, Wires, and Dots*

One of the major problems involved in π-Si and Si-nanocluster research and development work is the inhomogeneity of the material. Such inhomogenous broadening effects in PL and EL can be minimized by preparing *uniform* Si structures in the form of QWs, wires, or dots. Such structures can readily be produced directly by modern epitaxial growth techniques such as planar epitaxy, quantum wire formation along wafer steps, and dot self assembly, or indirectly by etching appropriate planar structures in the case of wires and dots. The predicted Si transition energies (Lockwood et al., 1992) due to different degrees of quantum confinement are shown in Fig. 12 (more sophisticated pseudopotential calculations (Zunger and Wang, 1996) give qualitatively similar results), where it can be seen that appreciable confinement effects are seen only for diameters less than 3 nm. Etched structures of this size have been difficult to produce in Si until very recently.

Wells. The simplest approach is to grow thin QWs of Si separated by wide bandgap barriers. Suitable barrier candidates are SiO_2, CaF_2, and Al_2O_3 (Tsu, 1993), and although a number of Si/barrier superlattices have been produced in the past (Lockwood, 1997) none has produced convincing evidence for quantum confinement induced emission until recently. Lu, Lockwood, and Baribeau (1995) reported visible light emission at room temperature from ultrathin-layer Si/SiO_2 superlattices grown by molecular beam epitaxy that exhibited a clear quantum confinement shift with Si layer thickness, as shown in Fig. 13. According to effective mass theory and assuming infinite potential barriers, which is a reasonable approximation since wide gap (9 eV) SiO_2 barriers are used, the energy gap E for

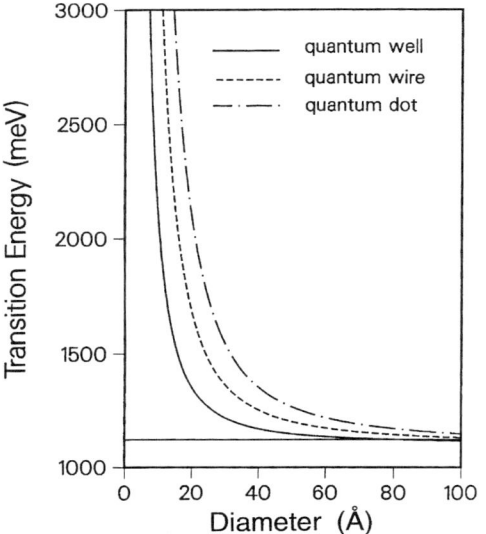

FIG. 12. Optical gap in Si quantum wells, wires, and dots versus system diameter. The transition energy is calculated for the lowest electron and heavy hole eigenergies for infinite confining potentials. The horizontal line is the bulk Si bandgap at room temperature. (From Lockwood et al., 1992.)

one-dimensionally confined Si should vary as

$$E = E_g + \frac{\pi \hbar^2}{2d^2}\left(\frac{1}{m_e^*} + \frac{1}{m_h^*}\right) \quad (1)$$

where E_g is the bulk material bandgap and m_e^* and m_h^* are the electron and hole effective masses (Lockwood, Lu, and Baribeau, 1996). This simple model is a reasonable first approximation to compare with experiment for QWs (Zunger and Wang, 1996). The shift in PL peak energy with Si well thickness d is well represented by Eq. (1), as can be seen in Fig. 13, with

$$E(\text{eV}) = 1.60 + 0.72 d^{-2} \quad (2)$$

The very thin layers of Si ($1 < d < 3$ nm) are amorphous, but nearly crystalline, owing to the growth conditions and the huge strain at Si–SiO$_2$ interfaces (Lu, Lockwood, and Baribeau, 1996). The fitted E_g of 1.60 eV is larger than that expected for c-Si (1.12 eV at 295 K), but is in excellent agreement with that of bulk a-Si (1.5–1.6 eV at 295 K). The indications of

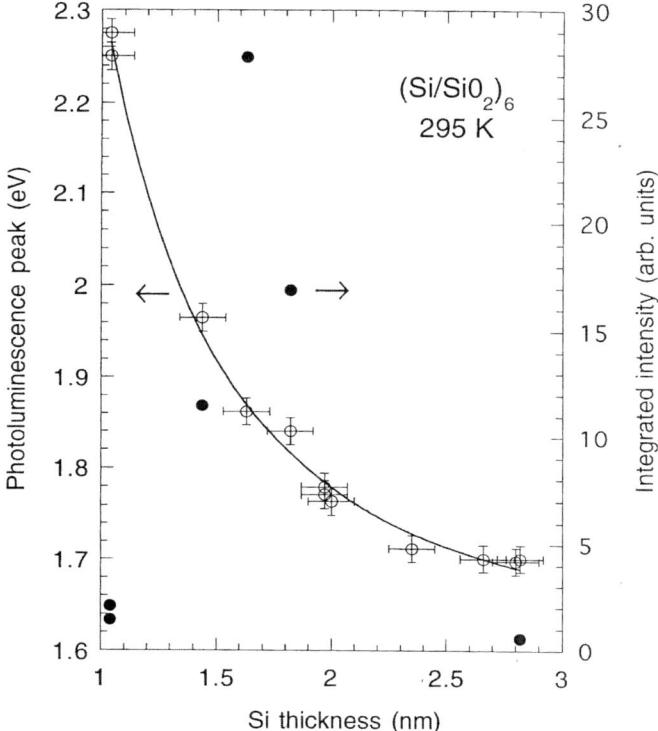

FIG. 13. The PL peak energy (○) and integrated intensity (●) at room temperature in $(Si/SiO_2)_6$ superlattices as a function of Si layer thickness. The solid line is the fit with effective mass theory. (From Lockwood, Lu, and Baribeau, 1996.)

direct band-to-band recombination were confirmed by measurements via X-ray techniques of the conduction and valence band shifts with layer thickness (Lockwood, Baribeau, and Lu, 1996; Lu, Lockwood, and Baribeau, 1995). The fitted confinement parameter of $0.72\,eV/nm^2$ indicates $m_e^* \approx m_h^* \approx 1$, comparable to the effective masses of c-Si at room temperature. The integrated intensity at first rises sharply with decreasing Si thickness until $d \approx 1.5$ nm and then decreases again, which is consistent with QW exciton emission (Brum and Bastard, 1985). The PL intensity is enhanced by factors of up to 100 on annealing and is also selectively enhanced and bandwidth narrowed by incorporation into an optical microcavity (Sullivan et al., 1996), as shown in Fig. 14. The bright PL obtained from as-grown and annealed a-Si/SiO$_2$ superlattices offers interesting prospects for the fabrication of a Si based light emitter that can be tuned from

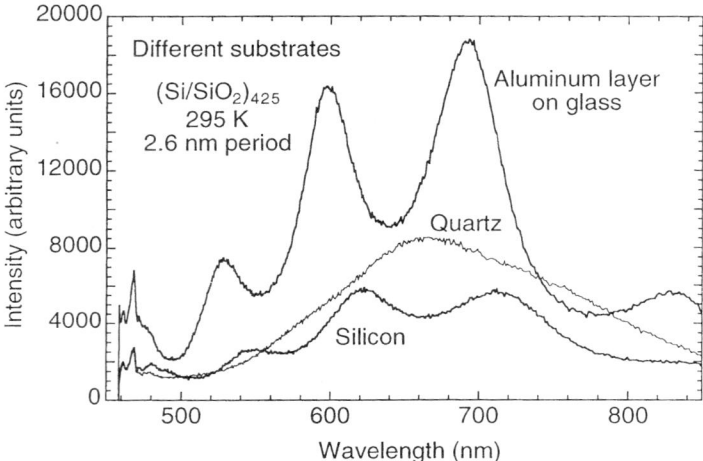

FIG. 14. Room temperature PL of a $(Si/SiO_2)_{425}$ superlattice with a 2.6 nm periodicity deposited on Si, quartz, and Al-coated glass. (From Sullivan et al., 1996.)

500 to beyond 800 nm by varying the a-Si layer thickness and/or the annealing conditions, all using available vacuum deposition technology and standard Si wafer processing techniques. The next important step is to develop LEDs based on such superlattices.

Wires. Quantum wires obtained by etching $Si/Si_{1-x}Ge_x$ heterostructures have been investigated by several groups (see, e.g., Lee et al. (1994); Tang et al. (1993). In PL measurements, wires defined by electron beam lithography and reactive ion etching have shown small blue shifts of up to 30 meV in the $Si_{1-x}Ge_x$ alloy peak at ~ 1.1 eV due to a combination of strain and confinement (Lee et al., 1994; Tang et al., 1993). Alternatively, $Si_{1-x}Ge_x$ wires have been grown on V-groove patterned Si substrates (Usami et al., 1993); the infrared emission (PL and EL) in this case exhibits a large optical anisotropy (Usami et al., 1994). No significant intensity enhancements compared with PL from QW transitions have been realized in these wire structures.

It has not yet been possible to produce thin enough freestanding wires of c-Si by etching techniques to observe quantum confinement effects, although room temperature PL at wavelengths from 400–850 nm is found for pillars with diameters ~ 10 nm (see, e.g., Nassiopoulos, Grigoropoulos, and Papadimitriou (1996b); Zaidi, Chu, and Brueck (1996) and references therein). Recently, an EL device based on Si nanopillars has been produced that emitted red light that was visible to the naked eye (Nassiopoulos, Grigoropoulos, and Papadimitriou, 1996a).

Dots. Attention has now turned to the production of $Si_{1-x}Ge_x$ quantum dots, as these produce the strongest confinement effects for a given diameter or can achieve desired confinements with smaller diameters than for wires (see Fig. 12). Quantum dots fabricated by etching $Si/Si_{1-x}Ge_x$ superlattices have produced 4 K PL at 0.97 eV that is 200 times brighter in 60 nm dots compared with the unetched superlattice PL (Tang *et al.*, 1995). Similar studies of $Si_{1-x}Ge_x$ dots fabricated by self-assembling island growth on Si have shown an increased luminescence efficiency due to the localization of excitons in the dots (Apetz *et al.*, 1995). In the latter case, the dots were buried in Si, which has the advantage of minimizing surface defect recombination. In both cases, EL has been observed from diode structures at low temperatures (Apetz *et al.*, 1995; Tang *et al.*, 1995) and at room temperature (Tang *et al.*, 1995), as shown in Fig. 15. The infrared EL at 4.2 K in the dot

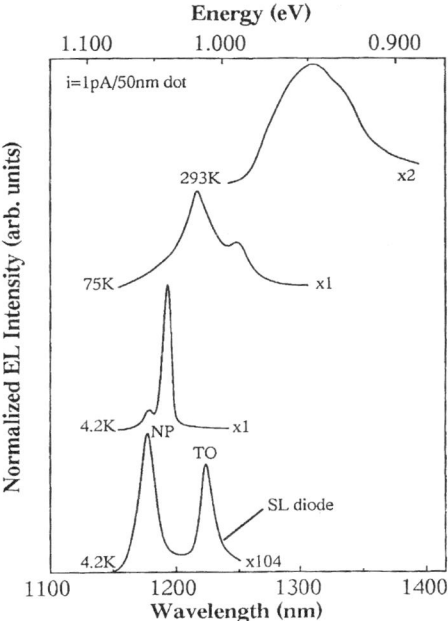

FIG. 15. Temperature dependence of EL spectra of a 50 nm $Si/Si_{0.7}Ge_{0.3}$ quantum dot diode under reverse bias of 0.5 V and an injection current of 1 pA/dot. A reference spectrum from a superlattice (SL) diode is also shown. (From Tang *et al.*, 1995.)

is two orders of magnitude higher in intensity than in the as-grown superlattice (see Fig. 15). At room temperature, the dot EL at 1.3 µm is only 50% less efficient, with a threshold injection current of ~ 0.1 pA/dot and an electrical-input to optical-output power conversion efficiency of 0.14% (Tang et al., 1995). Thus it is conceivable that this work could lead to a new generation of $Si/Si_{1-x}Ge_x$ optoelectronic devices at the optical fiber communication wavelength of 1.3 µm.

5. Polymers and Molecules Containing Si

Bright visible luminescence has been found in a number of Si polymer and molecular compounds (Kamimura, 1994; Kanemitsu, 1995). The most prominent of these is siloxene, $Si_6O_3H_6$ (see, e.g., Stutzmann et al. (1993)) and polysilane and its polymer derivatives (see, e.g., Takeda, 1994).

Siloxene produces bright red PL with characteristics similar to that of the red PL in π-Si, and, at first, it was thought that siloxene or siloxene derivatives formed during the electrochemical etching of Si were responsible for the PL in π-Si (Brandt et al., 1992). Although it is now accepted that siloxene is an unlikely explanation for the optical emission properties of π-Si (Lockwood, 1994), its optical properties have been further investigated with the hope of producing siloxene based devices. Siloxene has a direct bandgap (Deák et al., 1992) and the PL can be wavelength tuned across the visible region (Stutzmann et al., 1993), but its chemical instability at higher temperatures limits its practical usefulness in devices requiring thermal processing.

Polysilane compounds are polymers based on a Si backbone with H atom termination of Si dangling bonds (Miller and Michl, 1989). Derivatives of polysilane are obtained by modification of the Si skeleton structure and by the bonding of various atomic and molecular species to the backbone. Many such derivatives are possible, including siloxene (Miller and Michl, 1989; Takeda, 1994). The electronic structure and optical properties of many of the simpler compounds are now understood (Kamimura, 1994). Linear trans polysilane, $(SiH_2)_n$, comprises a zigzag one-dimensional backbone of Si atoms with each Si atom bonded to two other Si atoms and to two H atoms above and below the Si atom. As such, it can be considered the one-dimensional limit of c-Si. This form of polysilane exhibits a direct bandgap of 3.9 eV and efficiently emits ultraviolet light (Kanemitsu, 1995; Takeda, 1994). The PL characteristics of a variety of Si polymers are shown in Fig. 16. In the case of the chain, sharp ultraviolet emission and absorption peaks are observed, which are attributed to one-dimensional excitons delocalized on the backbone chain. In the branch and ladder structures, broad PL occurs with low quantum efficiency ($10^{-3}-10^{-5}$) at visible

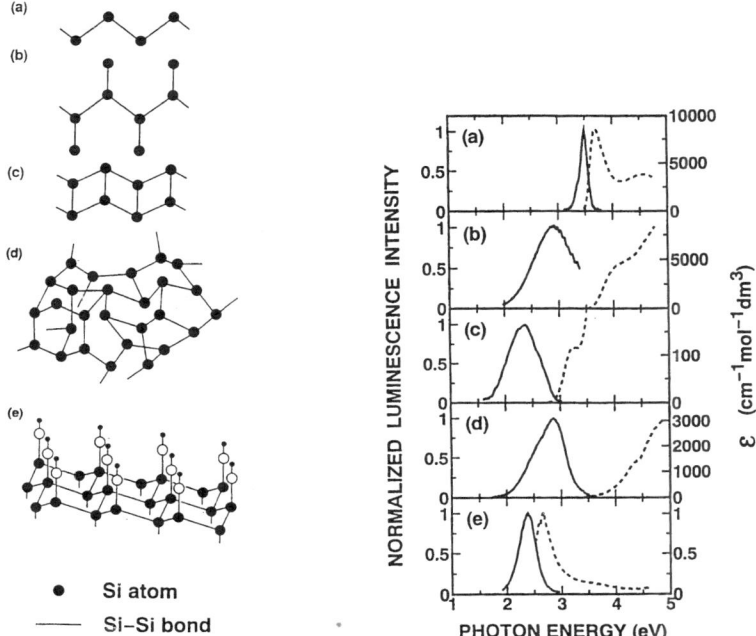

FIG. 16. Schematics of Si polymers with different backbones and their corresponding optical absorption and normalized PL spectra: (a) chain, (b) branch, (c) ladder, (d) network, and (e) planar siloxene. (From Kanemitsu, 1995.)

wavelengths and the excitons are strongly localized. The electronic properties of two-dimensional Si backbone polymers are intermediate between those of $(SiH_2)_n$ and three-dimensional bulk Si; a direct or an indirect bandgap can be obtained depending on the particular configuration (Takeda, 1994). The PL properties of broad-band visible wavelength emission and relatively long lifetime (~ 1 ns) possessed by the branch, ladder, and network Si polymers resemble those of a-Si, because real polymers exhibit structural disorders. Although chain-like Si polymers possess the desired characteristics of an ultraviolet light emitter, much more development work is required before they can be considered for ultraviolet devices based on Si.

6. Hybrid Methods for Integrating Direct Gap Materials with Si

Given that Si light emitters and, in particular, lasers are not yet available for on-chip optoelectronic applications, considerable effort is being placed on marrying dissimilar materials to fashion hybrid devices utilizing Si microelectronics (Iyer and Xie, 1993; Soref, 1993). For example, monolithically integrated GaAs LED arrays with Si driver circuits have been created (Dingle et al., 1993). However, the main interest is in combining III–V semiconductor laser diodes with Si integrated circuits for optical fiber communications or optical interconnects. This requires the growth of III–V materials such as GaAs or InP on Si followed by processing, or the direct bonding of preconstructed III–V laser devices detached from their substrates via an epitaxial lift-off process (Iyer and Xie, 1993; Soref, 1993). Both methods have their disadvantages.

The lattice mismatch of 4% between GaAs and Si creates severe difficulties in maintaining the required low defect density in GaAs for laser production. Dislocations produced by relaxation of the GaAs epitaxial layer are detrimental to device performance and life. Various methods are being tried to alleviate this problem including Ge, superlattice, or graded buffer layers between Si and GaAs, but other difficulties arise from the high GaAs growth temperatures and the differing thermal expansion coefficients of Si and GaAs. Other problems are associated with the change in crystal structure from nonpolar Si to polar GaAs (see Section III); a nonplanar Si growth surface can lead to stacking faults in the GaAs and there is charge build-up at the Si-GaAs interface (Iyer and Xie, 1993). The 8% lattice mismatch for InP on Si is even worse than for GaAs leading to even more severe heteroepitaxy growth problems. Even so, by use of a thick GaAs buffer layer, a 1.54 μm wavelength InGaAs/InGaAsP multiple QW laser operating continuously at room temperature has been produced on a Si substrate (Sugo et al., 1992).

The epitaxial lift-off technique is more straightforward involving wet chemical etching of a release layer, floating off the III–V heterostructure, and transferring it to a planar Si substrate, where it bonds via the van der Waals force (Iyer and Xie, 1993; Soref, 1993). For this bonding technique to work it is essential that the substrate surface be chemically clean and free of particles.

Further significant progress in both techniques can be anticipated within the next few years, and commercial devices will follow once the scaling up of these techniques to the mass production level has been achieved.

V. Prospects for Si-Based Optoelectronic Devices

Although a considerable number of optical detectors and waveguide structures have been created from Si based materials (Bozeat and Loni,

1995; Hall, 1993; Soref, 1993), there is still a paucity of LEDs constructed from Si and, most importantly for many all-Si optoelectronic applications, no lasers. The requirements for an acceptable semiconductor laser for optical fiber applications are rather stringent: 5–10 mW of laser facet power at 1.3 μm, maximum laser threshold less than 70 mA, spectral width less than 10 nm, operation over the temperature range $-40°C$ to $+85°C$, average lifetime of 10^6 h, and low cost (Hall, 1993).

Of the materials systems reviewed here, LEDs made from Si:Er show the most immediate promise for device applications at 1.54 μm. A schematic picture of an optoelectronic device (Michel *et al.*, 1996) comprising an edge emitting Si:Er LED integrated with a Si waveguide on a Si-on-insulator substrate is shown in Fig. 17. The EL linewidth of such LEDs at room temperature is approximately 10 nm (Zheng *et al.*, 1994). This narrow linewidth and fixed emission wavelength augers well for optical fiber communication systems with high bandwidth capacity. Optical gain at 1.54 μm should be obtainable in suitable Si waveguide structures and even laser emission, if the room temperature quantum efficiency can be improved.

Porous Si LEDs emitting at orange-red wavelengths are no longer just a curiosity with the announcement of LEDs having reasonable external quantum efficiencies (0.1%), lifetimes of the order of months, and low driving voltages (2–5 V) in forward bias (Collins, Fauchet, and Tischler, 1997) and also of devices with integrated Si transistor drivers, as shown schematically in Fig. 18. Apart from display applications (Hirschman *et al.*, 1996), however, the long lifetime and broad linewidth of the optical emission will limit optical communications applications of π-Si LEDs, and it is not clear whether current injection lasers will ever be made from π-Si. The need

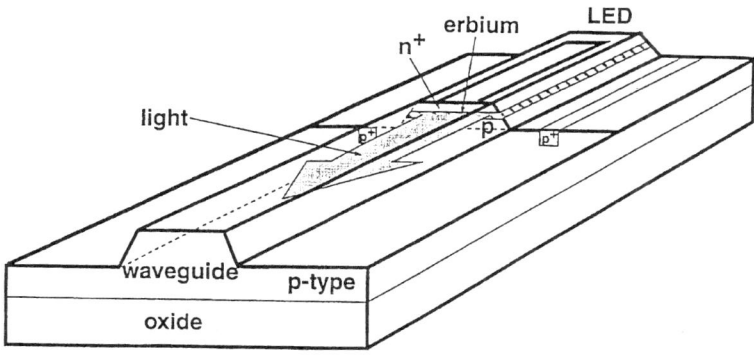

FIG. 17. Schematic representation of a Si:Er edge emitting LED integrated with a Si waveguide. (From Kimerling *et al.*, 1997.)

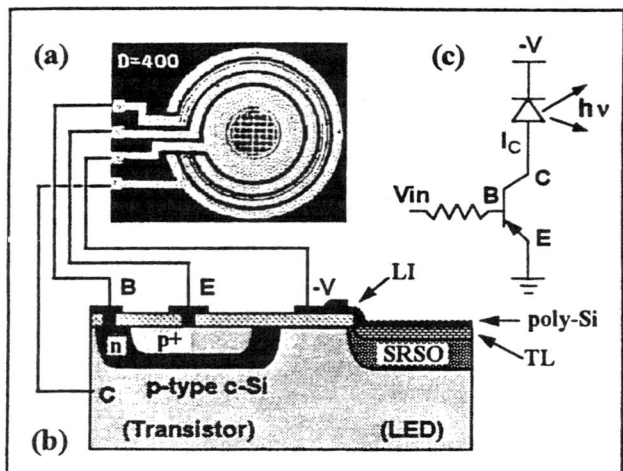

FIG. 18. Integrated π-Si LED/bipolar transistor device operational at room temperature: (a) plan view, (b) cross section, and (c) equivalent circuit. The LED is in the center of the structure and has a 400 μm diameter light-emitting area. Partial oxidation of π-Si in a dilute oxygen ambient has produced Si nanoclusters within an oxide matrix—Si-rich Si oxide (SRSO). (From Hirschman et al., 1996.)

to be compatible with existing large-scale Si processing also leads to difficulties with electrochemically created π-Si. It may be that oxidized Si nanoparticles will eventually prove to be superior to π-Si in this regard and also in device stability, but the long lifetime and wide bandwidth of the emitted light are still going to limit device performance. Nevertheless, π-Si is a versatile material and offers extremely diverse optoelectronic functionality to Si in the areas of infrared and visible waveguiding, photodetection, and photomodulation (Canham et al., 1996).

Light emission from QW and dot structures may yet hold the most promise for producing lasers at wavelengths across the visible into the infrared. The Si/SiO_2 multiple QW structures (Lockwood, 1997) are well suited for visible wavelength lasers at room temperature. Their optical absorption characteristics are ideal for optical pumping in a quantum microcavity, but it is not yet certain if their electrical characteristics are amenable to injection laser design. Quantum dot LEDs made from Si/$Si_{1-x}Ge_x$ (Tang et al., 1995) show considerable potential for laser applications at 1.3 μm. However, much more research and development work on these structures is required before this potential can be realized.

Considerable progress has been made over the last decade on obtaining efficient light emission from a wide variety of Si based materials. This work has led to the development of light emitting devices that are just now reaching useful performance levels. The intensity of research and development on light emission in Si is increasing as a result of these stimulating advances in materials engineering and technology. It is likely that a Si based laser will emerge from this research in the near future, although the actual active laser material could be none of those discussed here, because of the burgeoning diversification (Soref, 1996) in Si-based materials.

REFERENCES

Abstreiter, G. (1992). Engineering the future of electronics. *Physics World* **5**(3), 36–39.
Allan, G., Delerue, C., and Lannoo, M. (1996). Nature of luminescent surface states of semiconductor nanocrystallites. *Phys. Rev. Lett.* **76**, 2961–2964.
Apetz, R., Vescan, L., Hartman, A., Dieker, C., and Lüth, H. (1995). Photoluminescence and electroluminescence of SiGe dots fabricated by island growth. *Appl. Phys. Lett.* **66**, 445–447.
Bensahel, D. C., Canham, L. T., and Osscicini, S. (1993). *Optical Properties of Low Dimensional Structures*. Kluwer, Dordrecht.
Bozeat, R., and Loni, A. (1995). Silicon-based waveguides offer low-cost manufacturing. *Laser Focus World* **31**(4), 97–102.
Bradfield, P. L., Brown, T. G., and Hall, D. G. (1989). Electroluminescence from sulfur impurities in a p-n junction formed in epitaxial Si. *Appl. Phys. Lett.* **55**, 100–102.
Brandt, M. S., Fuchs, H. D., Stutzmann, M., Weber, J., and Cardona, M. (1992). The origin of visible luminescence from "porous silicon": A new interpretation. *Solid State Commun.* **81**, 307–312.
Brey, L., and Tejedor, C. (1987). New optical transitions in Si-Ge strained superlattices. *Phys. Rev. Lett.* **59**, 1022–1025.
Brown, T. G., and Hall, D. G. (1986a). Observation of electroluminescence from excitons bound to isoelectronic impurities in c-Si. *J. Appl. Phys.* **59**, 1399–1401.
Brown, T. G., and Hall, D. G. (1986b). Optical emission at 1.32 μm from sulfur-doped c-Si. *Appl. Phys. Lett.* **49**, 245–247.
Brum, J. A., and Bastard, G. (1985). Excitons formed between excited sub-bands in GaAs-Ga$_{1-x}$Al$_x$As quantum wells. *J. Phys. C: Solid State Phys.* **18**, L789–L794.
Brus, L. (1991). Quantum crystallites and nonlinear optics. *Appl. Phys. A* **53**, 465–474.
Brus, L. E., Szajowski, P. F., Wilson, W. L., Harris, T. D., Schuppler, S., and Citrin, P. H. (1995). Electronic spectroscopy and photophysics of Si nanocrystals: Relationship to bulk c-Si and π-Si. *J. Amer. Chem. Soc.* **117**, 2915–2922.
Bsiesy, A., Vial, J. C., Gaspard, F., Hérino, R., Ligeon, M., Muller, F., Romestain, R., Wasiela, A., Halimaoui, A., and Bomchil, G. (1991). Photoluminescence of high porosity and of electrochemically oxidized π-Si layers. *Surf. Sci.* **254**, 195–200.
Canham, L. T., Barraclough, K. G., and Robbins, D. J. (1987). 1.3-μm light-emitting diode from Si electron irradiated at its damage threshold. *Appl. Phys. Lett.* **51**, 1509–1511.
Canham, L. T. (1990). Silicon quantum wire array fabrication by electrochemical and chemical dissolution of wafers. *Appl. Phys. Lett.* **57**, 1046–1048.
Canham, L. T., Cox, T. I., Loni, A., and Simons, A. J. (1996). Progress towards Si optoelectronics using π-Si technology. *Appl. Surf. Sci.* **102**, 436–441.

Chelikowsky, J. R., and Cohen, M. L. (1976). Nonlocal pseudopotential calculations for the electronic structure of eleven diamond and zinc-blende semiconductors. *Phys. Rev. B* **14**, 556–582.

Coffa, S., Franzò, G., and Priolo, F. (1996). High efficiency and fast modulation of Er-doped light emitting Si diodes. *Appl. Phys. Lett.* **69**, 2077–2079.

Collins, R. T., Fauchet, P. M., and Tischler, M. A. (1997). Porous Si: From luminescence to LEDs. *Physics Today* **50**(1), 24–31.

Cullis, A. G., and Canham, L. T. (1991). Visible light emission due to quantum size effects in highly porous c-Si. *Nature* **353**, 335–338.

Davies, G. (1989). The optical properties of luminescence centers in Si. *Physics Reports* **176**, 83–188.

Deák, P., Rosenbauer, M., Stutzmann, M., Weber, J., and Brandt, M. S. (1992). Siloxene: Chemical quantum confinement due to oxygen in a Si matrix. *Phys. Rev. Lett.* **69**, 2531–2534.

Dieke, G. H. (1968). *Spectra and Energy Levels of Rare Earth Ions in Crystals*. Wiley, New York.

Dingle, B. D., Spitzer, M. B., McClelland, R. W., Fan, J. C. C., and Zavracky, P. M. (1993). Monolithic integration of a light-emitting diode array and a Si circuit using transfer processes. *Appl. Phys. Lett.* **62**, 2760–2762.

Engvall, J., Olajos, J., Grimmeiss, H. G., Presting, H., Kibbel, H., and Kasper, E. (1993). Electroluminescence at room temperature of a Si_nGe_m strained-layer superlattice. *Appl. Phys. Lett.* **63**, 491–493.

Engvall, J., Olajos, J., Grimmeiss, H. G., Kibbel, H., and Presting, H. (1995). Luminescence from monolayer-thick Ge quantum wells embedded in Si. *Phys. Rev. B* **51**, 2001–2004.

Ennen, H., Schneider, J., Pomrenke, G., and Axmann, A. (1983). 1.54-μm luminescence of erbium-implanted III–V semiconductors and Si. *Appl. Phys. Lett.* **43**, 943–945.

Ennen, H., Pomrenke, G., Axmann, A., Eisele, K., Haydl, W., and Schneider, J. (1985). 1.54-μm electroluminescence of erbium-doped Si grown by molecular beam epitaxy. *Appl. Phys. Lett.* **46**, 381–383.

Fauchet, P. M., Tsai, C. C., Canham, L. T., Shimizu, I., and Aoyagi, Y. (1993). *Microcrystalline Semiconductors: Materials Science and Devices*. Materials Research Society, Pittsburgh.

Fauchet, P. M. (1996). Photoluminescence and electroluminescence from π-Si. *J. Lumin.* **70**, 294–309.

Feng, Z. C., and Tsu, R. (1994). *Porous Silicon*. World Scientific, Singapore.

Fernando, C., Janz, S., Tarr, N. G., Normandin, R., Noël, J.-P., and Wight, J. S. (1995). $Si/Si_{0.85}Ge_{0.15}/Si$ p-i-n waveguide optical intensity modulator. *SPIE Proc.* **2402**, 131–142.

Förster, M., Mantz, U., Ramminger, R., Thonke, K., Sauer, R., Kibbel, H., Schäffler, F., and Herzog, H.-J. (1996). Electroluminescence, photoluminescence, and photocurrent studies of Si/SiGe p-i-n heterostructures. *J. Appl. Phys.* **80**, 3017–3023.

Froyen, S., Wood, D. M., and Zunger, A. (1987). New optical transitions in strained Si-Ge superlattices. *Phys. Rev. B* **36**, 4547–4550.

Fukatsu, S., Usami, N., Chinzei, T., Shiraki, Y., Nishida, A., and Nakagawa, K. (1992). Electroluminescence from strained SiGe/Si quantum well structures grown by solid source Si molecular beam epitaxy. *Jpn. J. Appl. Phys.* **31**, L1015–L1017.

Gardelis, S., Rimmer, J. S., Dawson, P., Hamilton, B., Kubiak, R. A., Whall, T. E., and Parker, E. H. C. (1991). Evidence for quantum confinement in the photoluminescence of π-Si and SiGe. *Appl. Phys. Lett.* **59**, 2118–2120.

Gnutzmann, U., and Clausecker, K. (1974). Theory of direct optical transitions in an optical indirect semiconductor with a superlattice structure. *Appl. Phys.* **3**, 9–14.

Goossen, K. W., Walker, J. A., D'Asaro, L. A., Hui, S. P., Tseng, B., Leibenguth, R., Kossives,

D., Bacon, D D., Dahringer, D., Chirkovsky, L. M. F., Lentine, A. L., and Miller, D. A. B. (1995). GaAs MQW modulators integrated with Si CMOS. *IEEE Photonics Tech. Lett.* **7**, 360–362.

Halimaoui, A., Oules, C., Bomchil, G., Bsiesy, A., Gaspard, F., Hérino, R., Ligeon, M., and Muller, F. (1991). Electroluminescence in the visible range during anodic oxidation of π-Si films. *Appl. Phys. Lett.* **59**, 304–306.

Hall, D. G. (1993). The role of Si in optoelectronics. *Mat. Res. Soc. Symp. Proc.* **298**, 367–378.

Hamilton, B. (1995). Porous silicon. *Semicond. Sci. Technol.* **10**, 1187–1207.

Hérino, R., and Lang, W. (1995). *Porous Silicon and Related Materials*. Elsevier, Amsterdam.

Hirschman, K. D., Tsybeskov, L., Duttagupta, S. P., and Fauchet, P. M. (1996). Silicon-based visible light emitting devices integrated into microelectronic circuits. *Nature* **384**, 338–341.

Houghton, D. C., Aers, G. C., Yang, S.-R. E., Wang, E., and Rowell, N. L. (1995). Type I band alignment in $Si_{1-x}Ge_x/Si(001)$ quantum wells: Photoluminescence under applied [110] and [100] uniaxial stress. *Phys. Rev. Lett.* **75**, 866–869.

Hybertsen, M. S., and Schlüter, M. (1987). Theory of optical transitions in Si/Ge(001) strained-layer superlattices. *Phys. Rev. B* **36**, 9683–9693.

Iyer, S. S., Collins, R. T., and Canham, L. T. (1992). *Light Emission from Silicon*. Materials Research Society, Pittsburgh.

Iyer, S. S., and Xie, Y.-H. (1993). Light emission from silicon. *Science* **260**, 40–46.

Jackson, S. A., and People, R. (1986). Optical absorption probability for the zone-folding induced quasi-direct gap in Ge_xSi_{1-x}/Si strained layer superlattices. *Mat. Res. Soc. Symp. Proc.* **56**, 365–370.

Kamimura, H. (1994). *Light Emission from Novel Silicon Materials*. Supplement B to *J. Phys. Soc. Jpn.* **63**. Physical Society of Japan, Tokyo.

Kanemitsu, Y., Ogawa, T., Shiraishi, K., and Takeda, K. (1993). Visible photoluminescence from oxidized Si nanometer-sized spheres: Exciton confinement on a spherical shell. *Phys. Rev. B* **48**, 4883–4886.

Kanemitsu, Y. (1995). Light emission from π-Si and related materials. *Phys. Reports* **263**, 1–91.

Kasper, E., and Presting, H. (1990). Device concepts for SiGe optoelectronics. *SPIE Proc.* **1361**, 302–312.

Kasper, E., and Schäffler, F. (1991). Group-IV compounds In *Strained-Layer Superlattices: Materials Science and Technology* (ed. T. P. Pearsall), Academic Press, Boston, pp. 223–309.

Kato, Y., Fukatsu, S., and Shiraki, Y. (1995). Postgrowth of a Si contact layer on an air-exposed $Si_{1-x}Ge_x/Si$ single quantum well grown by gas-source molecular beam epitaxy, for use in an electroluminescent device. *J. Vac. Sci. Technol. B* **13**, 111–117.

Kimerling, L. C., Kolenbrander, K. D., Michel, J., and Palm, J. (1997). Light emission from silicon. *Solid State Phys.* **50**, 333–381.

Koshida, N., and Koyama, H. (1991). Efficient visible photoluminescence from π-Si. *Jpn. J. Appl. Phys.* **30**, L1221–L1223.

Koshida, N., and Koyama, H. (1992). Visible electroluminescence from π-Si. *Appl. Phys. Lett.* **60**, 347–349.

Lee, J., Li, S. H., Singh, J., and Bhattacharya, P. K. (1994). Low-temperature photoluminescence of SiGe/Si disordered multiple quantum wells and quantum well wires. *J. Electron. Mat.* **23**, 831–833.

Lehmann, V., and Gösele, U. (1991). Porous silicon formation: A quantum wire effect. *Appl. Phys. Lett.* **58**, 856–858.

Lenchyshyn, L. C., Thewalt, M. L. W., Sturm, J. C., Schwartz, P. V., Prince, E. J., Rowell, N. L., Noël, J.-P., and Houghton, D. C. (1992). High quantum efficiency photoluminescence from localized excitons in $Si_{1-x}Ge_x$. *Appl. Phys. Lett.* **60**, 3174–3176.

Lenchyshyn, L. C., Thewalt, M. L. W., Houghton, D. C., Noël, J.-P., Rowell, N. L., Sturm, J. C., and Xiao, X. (1993). Photoluminescence mechanisms in thin $Si_{1-x}Ge_x$ quantum wells. *Phys. Rev. B* **47**, 16655–16658.

Lo, K.-Y., and Lue, J. T. (1993). The optical second-harmonic generation from π-Si. *IEEE Photon. Technol. Lett.* **5**, 651–653.

Lockwood, D. J., Aers, G. C., Allard, L. B., Bryskiewicz, B., Charbonneau, S., Houghton, D. C., McCaffrey, J. P., and Wang, A. (1992). Optical properties of π-Si. *Can. J. Phys.* **70**, 1184–1193.

Lockwood, D. J. (1994). Optical properties of π-Si. *Solid State Commun.* **92**, 101–112.

Lockwood, D. J., Wang, A., and Bryskiewicz, B. (1994). Optical absorption evidence for quantum confinement effects in π-Si. *Solid State Commun.* **89**, 587–589.

Lockwood, D. J., Baribeau, J. M., and Lu, Z. H. (1996). Visible photoluminescence in SiO_2/Si superlattices In *Advanced Luminescent Materials* (eds. D. J. Lockwood, P. M. Fauchet, N. Koshida, and S. R. J. Brueck). The Electrochemical Society, Pennington, pp. 339–347.

Lockwood, D. J., Fauchet, P. M., Koshida, N., and Brueck, S. R. J. (1996). *Advanced Luminescent Materials*. The Electrochemical Society, Pennington.

Lockwood. D. J., Lu, Z. H., and Baribeau, J.-M. (1996). Quantum confined luminescence in Si/SiO_2 superlattices. *Phys Rev. Lett.* **76**, 539–541.

Lockwood, D. J., and Wang, A. G. (1996). Photoluminescence in π-Si due to quantum confinement In *Advanced Luminescent Materials* (eds. D. J. Lockwood, P. M. Fauchet, N. Koshida, and S. R. J. Brueck). The Electrochemical Society, Pennington, pp. 166–172.

Lockwood, D. J. (1997). Quantum confined luminescence in Si/SiO_2 superlattices. *Phase Transitions*, to be published.

Lu, Z. H., Lockwood, D. J., and Baribeau, J.-M. (1995). Quantum confinement and light emission in SiO_2/Si superlattices. *Nature* **378**, 258–260.

Lu, Z. H., Lockwood, D. J., and Baribeau, J.-M. (1996). Visible light emitting Si/SiO_2 superlattices. *Solid-State Electron.* **40**, 197–201.

Matsumoto, T., Hasegawa, N., Tamaki, T., Ueda, K., Futagi, T., Mimura, H., and Kanemitsu, Y. (1994). Large induced absorption change in π-Si and its application to optical logic gates. *Jpn. J. Appl. Phys.* **33**, L35–L36.

Matsumoto, T., Daimon, M., Mimura, H., Kanemitsu, Y., and Koshida, N. (1995). Optically induced absorption in π-Si and its application to logic gates. *J. Electrochem. Soc.* **142**, 3528–3533.

Menczigar, U., Brunner, J., Freiss, E., Gail, M., Abstreiter, G., Kibbel, H., Presting, H., and Kasper, E. (1992). Photoluminescence studies of $Si/Si_{1-x}Ge_x$ quantum wells and Si_mGe_n superlattices. *Thin Solid Films* **222**, 227–233.

Menczigar, U., Abstreiter, G., Olajos, J., Grimmeiss, H. G., Kibbel, H., Presting, H., and Kasper, E. (1993). Enhanced bandgap luminescence in strain-symmetrized $(Si)_m/(Ge)_n$ superlattices. *Phys. Rev. B* **47**, 4099–4102.

Mi, Q., Xiao, X., Sturm, J. C., Lenchyshyn, L. C., and Thewalt, M. L. W. (1992). Room-temperature 1.3 μm electroluminescence from strained $Si_{1-x}Ge_x/Si$ quantum wells. *Appl. Phys. Lett.* **60**, 3177–3179.

Michel, J., Benton, J. L., Ferrante, R. F., Jacobson, D. C., Eaglesham, D. J., Fitzgerald, E. A., Xie, Y.-H., Poate, J. M., and Kimerling, L. C. (1991). Impurity enhancement of the 1.54-μm Er^{3+} luminescence in Si. *J. Appl. Phys.* **70**, 2672–2678.

Michel, J., Zheng, B., Palm, J., Ouellette, E., Gan, F., and Kimerling, L. C. (1996). Erbium doped Si for light emitting devices. *Mat. Res. Soc. Symp. Proc.* **422**, 317–324.

Miller, R. D., and Michl, J. (1989). Polysilane high polymers. *Chem. Rev.* **89**, 1359–1410.

Nassiopoulos, A. G., Grigoropoulos, S., and Papadimitriou, D. (1996a). Electroluminescent device based on Si nanopillars. *Appl. Phys. Lett.* **69**, 2267–2269.

Nassiopoulos, A. G., Grigoropoulos, S., and Papadimitriou, D. (1996b). Light emitting properties of Si nonopillars produced by lithography and etching In *Advanced Luminescent Materials* (eds. D. J. Lockwood, P. M. Fauchet, N. Koshida, and S. R. J. Brueck). The Electrochemical Society, Pennington, pp. 296–306.

Noël, J.-P., Rowell, N. L., Houghton, D. C., and Perovic, D. D. (1990). Intense photoluminescence between 1.3 and 1.8 μm from strained $Si_{1-x}Ge_x$ alloys. *Appl. Phys. Lett.* **57**, 1037–1039.

Noël, J.-P., Rowell, N. L., Houghton, D. C., Wang, A., and Perovic, D. D. (1992). Luminescence origins in molecular beam epitaxial $Si_{1-x}Ge_x$. *Appl. Phys. Lett.* **61**, 690–692.

Ogawa, T., and Kanemitsu, Y. (1995). *Optical Properties of Low-Dimensional Materials*. World Scientific, Singapore.

Orner, B. A., Olowolafe, J., Roe, K., Kolodzey, J., Laursen, T., Mayer, J. W., and Spear, J. (1996). Bandgap of Ge rich $Si_{1-x-y}Ge_xC_y$ alloys. *Appl. Phys. Lett.* **69**, 2557–2559.

Palm, J., Gan, F., Zheng, B., Michel, J., and Kimerling, L. C. (1996). Electroluminescence of erbium-doped Si. *Phys. Rev. B* **54**, 17603–17615.

Pankove, J. I. (1971). *Optical Processes in Semiconductors*. Dover, New York.

Pavesi, L., Guardini, R., and Mazzoleni, C. (1996). Porous Si resonant cavity light emitting diodes. *Solid State Commun.* **97**, 1051–1053.

Pearsall, T. P., Bevk, J., Feldman, L. C., Bonar, J. M., Mannaerts, J. P., and Ourmarzd, A. (1987). Structurally induced optical transitions in Ge-Si superlattices. *Phys. Rev. Lett.* **58**, 729–732.

Pearsall, T. P. (1994). Electronic and optical properties of Ge-Si superlattices. *Prog. Quant. Opt.* **18**, 97–152.

People, R., and Jackson, S. A. (1987). Indirect, quasidirect, and optical transitions in the pseudomorphic (4 × 4)-monolayer Si-Ge strained-layer superlattice on Si(001). *Phys. Rev. B* **36**, 1310–1313.

People, R., and Jackson, S. A. (1990). Structurally induced states from strain and confinement In *Strained Layer Superlattices: Physics* (ed. T. P. Pearsall). Academic Press, Boston, pp. 119–174.

Presting, H., Kibbel, H., Jaros, M., Turton, R. M., Menczigar, U., Abstreiter, G., and Grimmeiss, H. G. (1992). Ultrathin Si_mGe_n strained layer superlattices — a step towards Si optoelectronics. *Semicond. Sci. Technol.* **7**, 1127–1148.

Presting, H., Menczigar, U., Abstreiter, G., Kibbel, H., and Kasper, E. (1992). Electro- and photoluminescence from ultrathin Si_mGe_n superlattices. *Mater. Res. Soc. Symp. Proc.* **256**, 83–88.

Presting, H., Zinke, T., Splett, A., Kibbel, H., and Jaros, M. (1996). Room-temperature electroluminescence from $Si/Ge/Si_{1-x}Ge_x$ quantum-well diodes grown by molecular-beam epitaxy. *Appl. Phys. Lett.* **69**, 2376–2378.

Properties of Silicon. INSPEC, London, 1988.

Robbins, D. J., Calcott, P., and Leong, W. Y. (1991). Electroluminescence from a pseudomorphic $Si_{0.8}Ge_{0.2}$ alloy. *Appl. Phys. Lett.* **59**, 1350–1352.

Rowell, N. L., Noël, J.-P., Houghton, D. C., and Buchanan, M. (1990). Electroluminescence and photoluminescence from $Si_{1-x}Ge_x$ alloys. *Appl. Phys. Lett.* **58**, 957–958.

Rowell, N. L., Noël, J.-P., Houghton, D. C., Wang, A., Lenchyshyn, L. C., Thewalt, M. L. W., and Perovic, D. D. (1993). Exciton luminescence in $Si_{1-x}Ge_x/Si$ heterostructures grown by molecular beam epitaxy. *J. Appl. Phys.* **74**, 2790–2805.

St. Amour, A., Liu, C. W., Sturm, J. C., Lacroix, Y., and Thewalt, M. L. W. (1995). Defect-free band edge photoluminescence and bandgap measurement of pseudomorphic $Si_{1-x-y}Ge_xC_y$ alloy layers on Si(100). *Appl. Phys. Lett.* **67**, 3915–3917.

Saleh, B. A., and Teich, M. C. (1991). *Fundamentals of Photonics*. Wiley, New York.

Schuppler, S., Friedman, S. L., Marcus, M. A., Adler, D. L., Xie, Y.-H., Ross, F. M., Harris, T. D., Brown, W. L., Chabal, Y. J., Brus, L. E., and Citrin, P. H. (1994). Dimensions of luminescent oxidized and π-Si structures. *Phys. Rev. Lett.* **72**, 2648–2651.

Soref, R. A. (1993). Silicon-based optoelectronics. *Proc. IEEE* **81**, 1687–1706.

Soref, R. A. (1996). Silicon-based group IV heterostructures for optoelectronic applications. *J. Vac. Sci. Technol. A* **14**, 913–918.

Soref, R. A., Atzman, Z., Shaapur, F., Robinson, M., and Westhoff, R. (1996). Infrared waveguiding in $Si_{1-x-y}Ge_xC_y$ upon Si. *Opt. Lett.* **21**, 345–347.

Stimmer, J., Reittinger, A., Abstreiter, G., Holzbrecher, H., and Buchal, Ch. (1996). Growth conditions of erbium-oxygen-doped Si grown by MBE. *Mat. Res. Soc. Symp. Proc.* **422**, 15–20.

Sturm, J. C., Manoharan, H., Lenchyshyn, L. C., Thewalt, M. L. W., Rowell, N. L., Noël, J.-P., and Houghton, D. C. (1991). Well-resolved band edge photoluminescence of excitons confined in strained $Si_{1-x}Ge_x$ quantum wells. *Phys. Rev. Lett.* **66**, 1362–1365.

Stutzmann, M., Brandt, M. S., Fuchs, H. D., Rosenbauer, M., Kelly, M. K., Deak, P., Weber, J., and Finkbeiner, S. (1993). Optical properties of siloxene and siloxene derivates In *Optical Phenomena in Semiconductor Structures of Reduced Dimensions* (eds. D. J. Lockwood and A. Pinczuk). Kluwer, Dordrecht, pp. 427–442.

Stutzmann, M., Brandt, M. S., Rosenbauer, M., Fuchs, H. D., Finkbeiner, S., Weber, J., and Deak, P. (1993). Luminescence and optical properties of siloxene. *J. Lumin.* **57**, 321–330.

Sugo, M., Mori, H., Sakai, Y., and Itoh, Y. (1992). Stable continuous wave operation at room temperature of a 1.5-μm wavelength multiple quantum well laser on a Si substrate. *Appl. Phys. Lett.* **60**, 472–473.

Sullivan, B. T., Lockwood, D. J., Labbé, H. J., and Lu, Z.-H. (1996). Photoluminescence in amorphous Si/SiO_2 superlattices fabricated by magnetron sputtering. *Appl. Phys. Lett.* **69**, 3149–3151.

Takagi, H., Ogawa, H., Yamazaki, Y., Ishizaki, A., and Nakagiri, T. (1990). Quantum size effects on photoluminescence in ultrafine Si partices. *Appl. Phys. Lett.* **56**, 2379–2380.

Takeda, K. (1994). Si skeleton high-polymers: Their electronic structures and characteristics In *Light Emission from Novel Silicon Materials* (ed. H. Kamimura). The Physical Society of Japan, Tokyo, pp. 1–29.

Tang, Y. S., Wilkinson, C. D. W., Sotomayor Torres, C. M., Smith, D. W., Whall, T. E., and Parker, E. H. C. (1993). Optical properties of $Si/Si_{1-x}Ge_x$ heterostructure based wires. *Solid State Commun.* **85**, 199–202.

Tang, Y. S., Ni, W.-X., Sotomayor Torres, C. M., and Hansson, G. V. (1995). Fabrication and characterization of $Si-Si_{0.7}Ge_{0.3}$ quantum dot light emitting diodes. *Electron. Lett.* **31**, 1385–1386.

Tang, Y. S., Sotomayor Torres, C. M., Kubiak, R. A., Smith, D. A., Whall, T. E., Parker, E. H. C., Presting, H., and Kibbel, H. (1995). Optical emission from $Si/Si_{1-x}Ge_x$ quantum wires and dots In *The Physics of Semiconductors* (ed. D. J. Lockwood). World Scientific, Singapore, Vol. 2, pp. 1735–1738.

Thomas, D. G., Gershenzon, M., and Hopfield, J. J. (1963). Bound excitons in GaP. *Phys. Rev.* **131**, 2397–2404.

Tischler, M. A., Collins, R. T., Thewalt, M. L. A., and Abstreiter, G. (1993). *Silicon Based Optoelectronic Materials*. Materials Research Society, Pittsburgh.

Tong, S., Liu, X.-N., Wang, L.-C., Yan, F., and Bao, X.-M. (1996). Visible electroluminescence from nanocrystallites of Si films prepared by plasma enhanced chemical vapor deposition. *Appl. Phys. Lett.* **69**, 596–598.

Toyama, T., Matsui, T., Kurokawa, Y., Okamoto, H., and Hamakawa, Y. (1996). Visible photo- and electroluminescence from electrochemically formed nanocrystalline Si thin film. *Appl.*

Phys. Lett. **69**, 1261–1263.

Tsu, R. (1993). Silicon-based quantum wells. *Nature* **364**, 19.

Uhlir, Jr., A. (1956). Electrolytic shaping of Ge and Si. *Bell Syst. Tech. J.* **35**, 333–347.

Usami, N., Mine, T., Fukatsu, S., and Shiraki, Y. (1993). Realization of crescent-shaped SiGe quantum wire structures on a V-groove patterned Si substrate by gas-source Si molecular beam epitaxy. *Appl. Phys. Lett.* **63**, 2789–2791.

Usami, N., Mine, T., Fukatsu, S., and Shiraki, Y. (1994). Optical anisotropy in wire-geometry SiGe layers grown by gas-source selective epitaxial growth technique. *Appl. Phys. Lett.* **64**, 1126–1128.

Vial, J.-C., Canham, L. T., and Lang, W. (1993). "Light Emission from Silicon". *J. Lumin.* **57**, 1–358.

Vial, J. C., and Derrien, J. (1995). *Porous Silicon Science and Technology*. Springer Verlag, Berlin.

Yoffe, A. D. (1993). Low-dimensional systems: Quantum size effects and electronic properties of semiconductor microcrystallites (zero-dimensional systems) and some quasi-two-dimensional systems. *Advan. Phys.* **42**, 173–262.

Zachai, R., Eberl, K., Abstreiter, G., Kasper, H., and Kibbel, H. (1990). Photoluminescence in short-period Si/Ge strained-layer superlattices. *Phys. Rev. Lett.* **64**, 1055–1058.

Zaidi, S. H., Chu, A.-S., and Brueck, S. R. J. (1996). Room temperature photoluminescence from manufactured 1-D Si grating structures In *Advanced Luminescent Materials* (eds. D. J. Lockwood, P. M. Fauchet, N. Koshida, and S. R. J. Brueck). The Electrochemical Society, Pennington, pp. 307–316.

Zheng, B., Michel, J., Ren, F. Y. G., Kimerlin, L. C., Jacobson, D. C., and Poate, J. M. (1994). Room-temperature sharp line electroluminescence at $\lambda = 1.54\,\mu m$ from an erbium-doped, Si light-emitting diode. *Appl. Phys. Lett.* **64**, 2842–2844.

Zunger, A., and Wang, L.-W. (1996). Theory of Si nanostructures. *Appl. Surf. Sci.* **102**, 350–359.

CHAPTER 2

Band Gaps and Light Emission in Si/SiGe Atomic Layer Structures

Gerhard Abstreiter

WALTER SCHOTTKY INSTITUT, TECH. UNIV. MÜNCHEN
D-85748 GARCHING, GERMANY

I. INTRODUCTION	37
II. STRUCTURAL PROPERTIES	40
III. BANDGAPS, BAND OFFSETS, AND BRILLOUIN ZONE FOLDING	44
IV. PHOTOLUMINESCENCE, ELECTROLUMINESCENCE, AND PHOTOCURRENT MEASUREMENTS	54
1. $Si_{1-x}Ge_x$ Alloy Layers and Quantum Wells	54
2. $Si_m Ge_n$ Short Period Superlattices	59
3. $Ge_n Si_m Ge_n$ Atomic Layer Structures and Interfaces with Staggered Band Offsets	64
4. Laterally Confined QWs and Ge-rich Self-assembled Dots	67
V. CONCLUDING REMARKS	70
Acknowledgments	70
References	71

I. Introduction

It was first proposed by Esaki and Tsu (1969) that the electronic band structure of semiconductors may be changed drastically and designed specifically for certain purposes by changing the composition of the materials on a length scale shorter than the de Broglie wavelength of the charge carriers. Based on these ideas, a wide research area dealing with various kinds of man-made semiconductor heterostructures and superlattices has been developed during the past 25 years. The enormous progress became possible due to the fast development and improvement of epitaxial crystal growth techniques, which nowadays allow a control of layer sequences of different semiconductor materials with a precision on an atomic scale. The most widely studied and used heterostructures and superlattices employ III–V semiconductors, especially GaAs and $Al_x Ga_{1-x} As$. The good lattice matching, the well controlled epitaxial growth techniques, the direct energy gap over a wide range of composition, and the small effective masses of conduction band electrons make this material system ideal for band structure engineering. Apart from numerous novel basic effects that have been demonstrated with such semiconductor structures of lower dimensionality, heterostructure research has led also to new device applications like low-noise, high-frequency transistors and quantum well (QW) lasers. Semiconductor electronics, however, is strongly dominated by Si based technologies and this is not expected to change during the next 20 years. It is, therefore, a challenge to combine the advantages of heterostructures and superlattices with the mature Si microelectronics. The present Si technology does not allow easily the integration of components with optical or optoelectronical functions. The semiconductor Si itself and also Ge suffer in this respect from their indirect fundamental bandgap. A few years after the first proposal that the band structure can be tailored by superimposing a periodic superlattice potential onto the crystal lattice potential (Esaki and Tsu, 1969) it was suggested theoretically by Gnutzmann and Clausecker (1974) that this may open the possibility to create a direct gap semiconductor from materials with an indirect energy gap. A periodic sequence of a few atomic planes of Si and Ge leads to a new larger lattice constant in one direction and, as a consequence, the Brillouin zone is reduced along this axis. A proper choice of the superlattice period length results in a Brillouin zone folding such that initially indirect conduction band minima are shifted back to the Γ-point at the center of the reduced Brillouin zone. Early experimental attempts were made to verify these predictions (Kasper, Herzog, and Kibbel, 1975). However, the quality of the anticipated superlattices was not good enough to be able to demonstrate the zone folding effect. It was about 10 years ago when the first high quality Si/SiGe and Si/Ge heterostructures and superlattices were achieved by molecular beam epitaxy (MBE) (Bean et al., 1984) and electroreflection experiments were reported on pseudomorphic Si/Si_4Ge_4 superlattice structures (Pearsall et al., 1987) showing signals below the Si bandgap, which were interpreted as superlattice induced interband transitions. However, it soon became pretty clear by considering the band offsets of these structures (Abstreiter et al., 1986) that zone-folding is not expected to lead to a direct gap in those samples. This first report on new optical transitions in Si/Ge multilayer structures nevertheless stimulated an enormous amount of theoretical work, calculating the band structure of Si/Ge superlattices (Brey and Tejedor, 1987; Froyen, Wood, and Zunger, 1987, 1988; Gell, 1989a, 1989b; Hybertsen and Schlüter, 1987; Hybertsen et al., 1988; Jaros et al., 1992; Morrison, Jaros, and Wong, 1987; Morrison and Jaros, 1988; Satpathy, Martin, and van de Walle, 1988; Schmid et al., 1990, 1991, 1992; Tserbak, Polatoglou, and Theodorou, 1993; Turton and Jaros, 1990; Walle and Martin, 1986; Wong et al., 1988). The major result of all these calculations is that for certain period lengths and for certain strain distributions one expects indeed a direct fundamental energy gap. The optical transition matrix element at this energy gap was found, however, to be at least three orders of magnitude weaker as compared with GaAs. Therefore one talks of a quasi-direct energy gap. Most of the band structure calculations came to the conclusion that the Si layers have to be under tensile strain in the plane in order to achieve a quasi-direct energy gap. This can be extracted already from very simple arguments taking band offsets and strain induced splittings of the conduction band into account (Abstreiter *et al.*, 1985, 1986; Eberl *et al.*, 1987; Hybertsen *et al.*, 1988; People and Jackson, 1987; Zachai *et al.*, 1988, 1990). Pseudomorphic structures on Si substrates do not contain the required strained Si layers. Therefore strain relaxed $Si_{1-x}Ge_x$ alloy layers on Si have been used as new "virtual" substrates to realize Si_mGe_n superlattices whose period length is given by m atomic planes of Si and n atomic planes of Ge (Kasper *et al.*, 1988; Zachai *et al.*, 1988, 1990). Luminescence studies of such strain symmetrized superlattices indeed showed a strong enhancement of photoluminescence (PL) and a reduced energy gap as compared to the random alloy, in reasonable agreement with theoretical predictions (Zachai *et al.*, 1990). The early experiments suffered, however, from the large number of misfit dislocations present in strain relaxed $Si_{1-x}Ge_x$ alloys. The experimental results obtained have been, therefore, discussed controversially in the literature (Schmid *et al.*, 1991). This has been improved more recently by using graded $Si_{1-x}Ge_x$ buffer layers and new techniques for achieving sharp interfaces even at relatively high growth temperatures. The improved sample quality led to a clear identification of the new fundamental bandgaps in Si_mGe_n superlattices (Menczigar *et al.*, 1993; Olajos *et al.*, 1992, 1994). Enhanced optical transition probabilities have been observed as well. More recent experiments have shown that a similar enhancement of the optical matrix elements may be obtained just by one or a few interfaces that break the symmetry of the crystal sufficiently in order to allow for relatively strong no-phonon assisted interband optical transitions (Brunner, Winter, and Eberl, 1996; Usami, Shiraki, and Fuakatsu, 1996). Also, exciton localization seems to play an important role in the observed optical spectra (Jaros and Beavis, 1993; Menczigar *et al.*, 1993).

In this chapter we review various aspects of Si/Ge heterostructures and superlattices, including thin pure layers embedded in Si, Si_mGe_n short period superlattices, and some newer developments of Ge islands in Si. Such structures alter the crystal symmetry, which may lead to enhanced optical interband matrix elements. The main emphasis is put on the fundamental energy gaps, which lie between 0.7 eV to 1 eV and thus cover the wavelength range between 1.6 μm to 1 μm that is of interest for optical fiber communications. This chapter is organized in the following way: In Section II we present briefly the essential structural properties of the Si/Ge heterosystem, followed by a discussion of the bandgaps and band offsets in Section III. In Section IV we present various experimental results on the fundamental energy gaps as studied by PL, electroluminescence (EL) and photocurrent spectroscopy. The chapter ends with some concluding remarks on perspectives of Brillouin zone folding and exciton localization in Si/Ge with respect to enhanced optical emission.

II. Structural Properties

Both Si and Ge crystallize in the diamond structure, which is a face-centered cubic lattice with two atoms per unit cell. The cubic lattice constant a differs by about 4% and is 5.431 Å for Si and 5.657 Å for Ge. Si and Ge can be mixed to form random alloys in the entire compositional range between the pure constituents. The cubic lattice constant of the alloy is to a good approximation obtained by a linear interpolation between a_{Si} and a_{Ge}. Pseudomorphic or lattice matched epitaxial growth of a $Si_{1-x}Ge_x$ alloy on Si substrates results in strained layers. The lateral in-plane lattice constant $a_\parallel$ is then equal to the substrate lattice constant, if the thickness of the substrate and the lateral extension is much larger than the thickness of the epitaxial layer. The lattice constant in growth direction $a_\perp$ is smaller in this case, which altogether results in a tetragonal distortion of the lattice. The deformation depends on the elastic constants. For growth on (100) surfaces the lattice constant in the growth direction is given by:

$$a_\perp = a_i \left[1 - 2\frac{c_{12}}{c_{11}} \left(\frac{a_\parallel}{a_i} - 1 \right) \right] \tag{1}$$

where a_i is the intrinsic lattice constant of the epitaxial layer and c_{12} and c_{11} are the elastic constants. The relevant strain components are then:

$$\varepsilon_\parallel = \frac{a_\parallel}{a_i} - 1, \; \varepsilon_\perp = \left(\frac{a_\perp}{a_\parallel} - 1 \right) = -2\frac{c_{12}}{c_{11}} \varepsilon_\parallel \tag{2}$$

The built in strain leads to a shift and splitting of the conduction and valence bands, which will be discussed in Section III.

The elastic deformation of the epitaxial overlayer becomes energetically more and more unfavorable with increasing thickness. Beyond a certain critical thickness the strain is relaxed, at least partially, by creation of misfit dislocations. Details of the strain relaxation depend on growth conditions like growth temperature and growth rate and on the lattice mismatch. The creation of misfit dislocations and their gliding is the important relaxation mechanism for small mismatch and large layer thickness. Another possibility to decrease the strain energy is the formation of islands or three-dimensional growth (Stranski–Krastanow growth mode). This occurs mainly for large lattice mismatch and high growth temperatures. A critical thickness can be deduced for the formation of dislocations by defining a balance of the relevant forces, which then depends on the lattice mismatch and the Poisson ratio. This was studied in detail, for example, by Matthews and Blakeslee (Matthews and Blakeslee, 1974). The critical thickness determined in this way is in good agreement with experimental results for high growth temperatures. With lower growth temperatures it is, however, possible to achieve considerably larger pseudomorphic layer thicknesses before relaxation occurs. This can be understood, for example, by the excess-stress concept of Dodson and Tsao (1987, 1988, 1988). In this model it is assumed that excess-stress is necessary to activate the gliding of dislocations. The amount of extra stress depend on growth temperature. For high temperatures the Dodson–Tsao concept approaches the Matthews–Blakelsee model. For pure Ge on Si the critical thickness is only about 4 to 6 monolayers and for growth temperatures above roughly 550°C relaxation occurs via nucleation of three-dimensional islands.

Short period Si/Ge superlattices grown on Si substrates are asymmetrically strained because $a_\parallel$ is fixed by the substrate and only the Ge layers are strained. Each Ge layer adds deformation energy and the total critical thickness is given approximately by the average Ge content in the superlattice. The consequence is that for a Si_4Ge_4 or Si_5Ge_5 periodic structure, only a very limited number of periods can be grown. Therefore, the samples used for the experiments of Pearsall *et al.* (1987, 1989) essentially consisted of Si_nGe_m quantum wells (QWs) embedded in Si, because the total thickness of the Si_nGe_m superlattice was only a few nanometers. This problem can be overcome by the concept of a virtual substrate, which consists of a strain relaxed $Si_{1-x}Ge_x$ buffer layer on Si (Kasper *et al.*, 1987, 1988). An optimized buffer should be defect free close to its surface and the misfit dislocations should all be concentrated close to the epilayer-substrate interface. This has not been achieved ideally, because a finite number of threading dislocations are always penetrating to the surface. The quality of such buffer layers has been improved considerably in the past 10 years, for example, by using the concept of graded buffers, where the Ge content is increased gradually from a low concentration over a thickness of more than a micrometer to the desired value (Fitzgerald *et al.*, 1991; LeGoues *et al.*, 1992; Schäffler *et al.*, 1992). When the growth temperature is kept high enough (typically above

700°C) the relaxation occurs close to the Matthews–Blakeslee critical thickness in several successive steps with the continuously or step-wise increasing Ge content. The lattice mismatch is thus continuously adjusted and kept rather small. With this concept it was possible to reduce the dislocation density in the top layers to the order of 10^5–10^6 cm^{-2}. Strain-symmetrized or so-called free-standing superlattices can be grown on such virtual substrates with a large overall thickness. For example, the in-plane lattice constant of a relaxed $Si_{0.6}Ge_{0.4}$ alloy buffer layer is equal to the average lattice constant of a $Si_m Ge_n$ superlattice with $m/n = 3/2$. In such symmetrically strained superlattices there exists no global critical thickness anymore and also the critical thickness of the individual Si and Ge layers in each period is increased, because of the smaller lattice mismatch to the virtual substrate. The Si and Ge layers are strained in opposite directions, which leads to a strain compensation. The period length of a $Si_m Ge_n$ superlattice is consequently given by:

$$d_\perp^{SLS} = \frac{a_\perp^{Si}}{4} \cdot m + \frac{a_\perp^{Ge}}{4} \cdot n \qquad (3)$$

Strictly speaking this equation is only valid for even $m + n$ values, because the superlattices have different symmetries depending on the number of Si (m) and Ge (n) layers. The covalent bonding between Si and Ge is along the [110] directions, which are not equivalent in the superlattice. For odd numbers of $m + n$ the periodicity is doubled. This has indeed been observed in electron diffraction of a Si_3Ge_2 superlattice, whose diffraction pattern looks similar to that of a Si_6Ge_4 superlattice (Eberl et al., 1991). This microscopic symmetry has, however, no consequence on the bandgaps as it is discussed in this context. Therefore we keep the definition of period length as defined in Eq. (3), which is based on the layer thicknesses.

Another important structural parameter that influences strongly the optical properties of short period superlattices is the interface sharpness. It has been shown that sharp interfaces between Si and Ge are only achieved at growth temperatures below 400°C under typical growth conditions using MBE (Abstreiter et al., 1989). At such low temperatures, however, the crystal quality suffers and the luminescence is often dominated by defects. At higher growth temperature there is a strong tendency of Ge segregation (Nützel and Abstreiter, 1996) and finally also interdiffusion between Si and Ge. This problem has also been reduced a few years ago by using so-called surfactants like Sb to keep sharp interfaces at growth temperatures above 500°C (Jorke, 1988; Usami, Fukatsu, and Shiraki, 1994). The combination of optimized graded buffers together with Sb as a surfactant enabled us to

2 BAND GAPS AND LIGHT EMISSION IN Si/Ge ATOMIC LAYER STRUCTURES

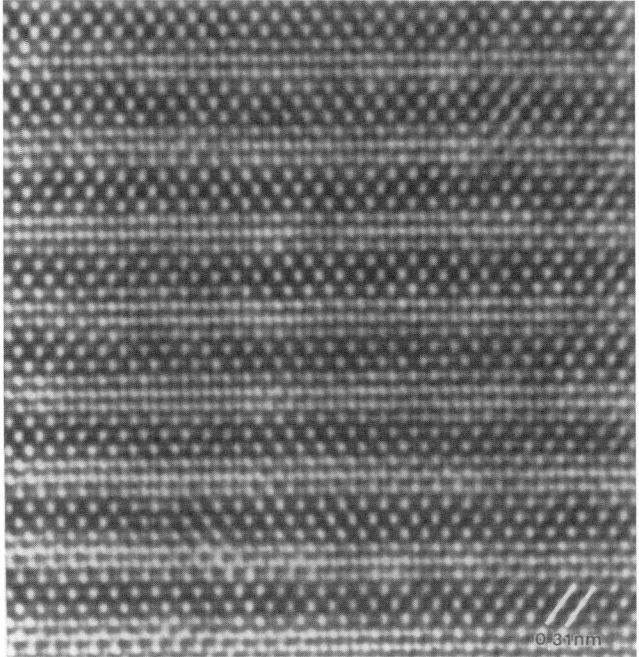

FIG. 1. Lattice image of a Si_5Ge_5 strain symmetrized superlattice. (From Stenkamp and Jäger, 1993.)

grow high-quality short-period Si_mGe_n superlattices. Transmission electron micrographs of some of these structures are shown in Figs. 1 and 2 as examples (Jäger and Mayer, 1995; Stenkamp and Jäger, 1993). Figure 1 is the lattice image of a Si_5Ge_5 strained layer superlattice grown by MBE at low temperatures. For Si, the dark contrast dots correspond to projections of closely spaced atomic columns. Under these imaging conditions the Ge content leads to a half-period contrast. The abruptness of the contrast indicates a gradient in composition of about 1 to 2 monolayers. The interfaces are coherent, but there exist steps with a height of 1 monolayer. The average nominal thickness of 5 monolayers is confirmed. In Fig. 2 a bright field and a lattice image of a short period Si_9Ge_6 superlattice grown by MBE on a step-graded $Si_{0.6}Ge_{0.4}$ buffer layer is shown. The growth temperature was about 500°C and Sb was used as a surfactant to maintain sharp interfaces between Si and Ge. The individual layers of the superlattice appear clearly separated by a reversal in the contrast between the Si and the

FIG. 2. Bright field and lattice image of a Si_6Ge_9 short period superlattice. (From Jäger and Mayer, 1995.)

Ge. Again the layers show coherent interfaces with some 1 or 2 monolayer steps. The first Ge layer is thicker than the following ones. Overall, an excellent quality of the short period superlattice is again demonstrated. The structural results have been confirmed by X-ray analysis (Kopensteiner et al., 1994) and by a detailed evaluation of Raman spectra (Schorer et al., 1994), where also the stability with respect to annealing has been studied.

III. Bandgaps, Band Offsets, and Brillouin Zone Folding

Both Si and Ge are indirect semiconductors with the valence band maximum at the Γ-point in the center of the Brillouin-zone. In Si the 6-fold degenerate conduction band minima are along the [100] directions at about $k = 0.85\, 2\pi/a$, close to the X-point of the Brillouin-zone. These minima are called Δ-minima. For pure Ge the conduction band minima are along the [111] directions at the L-point (Brillouin-zone boundary). In $Si_{1-x}Ge_x$ alloys the band structure remains Si-like up to a Ge concentration of about 85%, where the crossover between Δ and L-minima occurs (Braunstein,

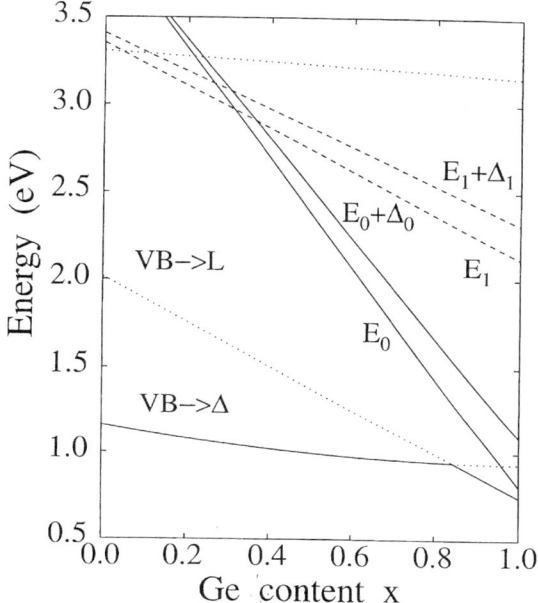

FIG. 3. Energy gaps of unstrained $Si_{1-x}Ge_x$ alloys.

Moore, and Herman, 1958; Braunstein, 1963). The energy gap is slowly decreasing with increasing x as long as the Δ-minima are lowest in energy. The conduction band at the L-point and especially at the Γ-point show a much stronger dependence on the Ge content. The corresponding energy gaps, including also E_0, $E_0 + \Delta_0$, E_1, $E_1 + \Delta_1$ and E_0' are plotted in Fig. 3 versus x (Braunstein, Moore, and Herman, 1958; Kline, Pollak, and Cardona, 1968; Weber and Alonso, 1989). The fundamental energy gap at 4.2 K as determined from interband transitions, is for the Δ-minima approximated by:

$$E_g(\text{eV}) = 1.17 - 0.43x + 0.206x^2 \qquad (4)$$

For $x > 0.85$ the following behavior for the energy gap to the L-point is obtained:

$$E_g(\text{eV}) = 2.01 - 1.270x \qquad (5)$$

For Si/Ge heterosystems and superlattices strain plays an important role. This alters the band structure and the bandgaps drastically. The biaxial stress can be separated into a hydrostatic part and a uniaxial part. The hydrostatic stress leads to a shift of the conduction and valence bands. The uniaxial part in addition causes a splitting of the conduction and valence bands. In linear deformation potential theory the six-fold degenerate conduction band splits for layer growth in the [001] direction into two-fold degenerate Δ-minima in the growth direction and 4-fold degenerate minima in the plane. The splitting is given by:

$$\Delta E_u^{\Delta(2)} = +\frac{2}{3}\Xi_u^\Delta[\varepsilon_\perp - \varepsilon_\|] \tag{6}$$

$$\Delta E_u^{\Delta(4)} = -\frac{1}{3}\Xi_u^\Delta[\varepsilon_\perp - \varepsilon_\|] \tag{7}$$

The hydrostatic part of the stress leads to an identical shift of the Δ(2) and Δ(4) minima:

$$\Delta E_{\text{hyd}}^\Delta = \left[\Xi_d^\Delta + \frac{1}{3}\Xi_u^\Delta\right](2\varepsilon_\| + \varepsilon_\perp) \tag{8}$$

Ξ_d^Δ and Ξ_u^Δ are the linear deformation potentials for the Δ conduction band states. The conduction band minima at the L and Γ points do not show a splitting. They also shift, however, with the hydrostatic part of the stress.

The situation is more complex for the valence bands. The uniaxial part of the stress along the [001] direction leads to an interaction between the light hole and spin–orbit split-off valence band. For this situation the splitting of the heavy hole, light hole and spin–orbit split-off valence bands is given by (Balslev, I., 1966; Bir and Pikus, 1974; Walle, and Martin, 1986):

$$\Delta E^{hh} = -b(\varepsilon_\perp - \varepsilon_\|) \equiv \delta \tag{9}$$

$$\Delta E^{lh} = -\delta + \frac{1}{2}(-(\Delta_0 + \delta) + \sqrt{(\delta + \Delta_0)^2 + 8\delta^2}) \tag{10}$$

$$\Delta E^{soh} = -\frac{1}{2}(-(\Delta_0 + \delta) + \sqrt{(\delta + \Delta_0)^2 + 8\delta^2}) \tag{11}$$

where b is the uniaxial linear deformation potential for the valence bands and Δ_0 is the spin–orbit splitting. The hydrostatic part of the stress causes a shift of the valence band of:

$$\Delta E_{hyd}^{VB} = a_V(2\varepsilon_\| + \varepsilon_\perp) \tag{12}$$

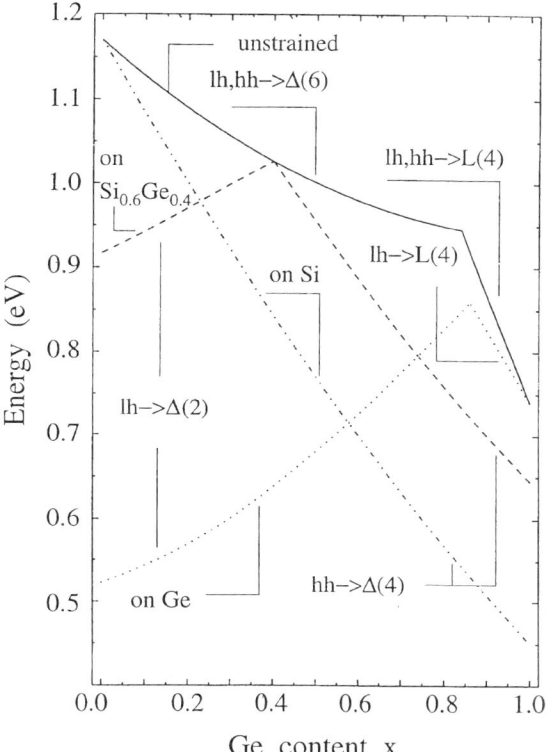

FIG. 4. Fundamental bandgaps for strained and unstrained $Si_{1-x}Ge_x$ alloys.

where a_V is the hydrostatic linear deformation potential of the valence bands. For $Si_{1-x}Ge_x$ alloys the deformation potentials are usually interpolated between those of pure Si and pure Ge. The energy gaps between the conduction and valence band edges are given by

$$E_g^{hh} = E_g^i + \Delta E_{hyd}^i - \Delta E_{hyd}^{VB} + \Delta E_u^i - \Delta E^{hh} \quad (13)$$

$$E_g^{lh} = E_g^i + \Delta E_{hyd}^i - \Delta E_{hyd}^{VB} + \Delta E_u^i - \Delta E^{lh} \quad (14)$$

In Fig. 4 the calculated bandgaps are shown for pseudomorphic $Si_{1-x}Ge_x$ alloys on Si (100), Ge (100) and $Si_{0.6}Ge_{0.4}$ (100) substrates. Also shown is the fundamental energy gap of unstrained alloys (Weber and Alonso, 1989). For pseudomorphic layers on Si (compressive strain) the energy gap is determined by the Δ(4) conduction band minima and the heavy hole valence band. For strained layers on $Si_{0.6}Ge_{0.4}$ the fundamental bandgap is given by the separation of the light hole valence band and the Δ(2) conduction band minima for $0 < x < 0.4$ (tensile strain) and again by the heavy hole valence band and the Δ(4) minima for $0.4 < x < 1$. On Ge substrates the bandgap is determined by Δ(2) and light hole valence band states for $0 < x < 0.86$. For high Ge content a turn-over to the L conduction band minima occurs. Independent of the sign of the built-in strain, the energy gap of the alloy layers is always reduced as compared to the unstrained situation. The general behavior as discussed here is also found in more sophisticated band structure calculations (Rieger and Vogl, 1993). There exist, however, some quantitative differences.

Another important quantity for understanding the optical and electronical properties of Si/Ge based heterostructures and superlattices is the band offset or the band discontinuity at the interfaces. The valence band offset has been studied in detail theoretically by van de Walle and Martin (1986); Colombo, Resta, and Baroni (1991). According to Colombo, Resta, and Baroni (1991) the average valence band offset is given by:

$$\Delta E_{V,av} = (0.47 - 0.06 x_{sub})(x_{epi} - x_{sub})[eV] \qquad (15)$$

where x_{sub} and x_{epi} are the Ge contents in the strain relaxed substrate and in the pseudomorphic epitaxial layer, respectively. For pure Ge on Si this results in an offset of the averaged valence bands of 0.47 eV. Together with the band splittings, as determined above, we are now able to calculate the various band edges. The results are shown in Fig. 5 for $Si_{1-x}Ge_x$ layers on Si (100) substrates and on $Si_{0.6}Ge_{0.4}$ relaxed buffer layer substrates for heavy hole and light hole valence bands and the Δ(2) and Δ(4) conduction band minima. Zero energy was chosen to be at the top of the valence band of the unstrained semiconductor.

In the biaxially compressed alloy layer grown on a Si substrate, the in-plane conduction band minima (Δ(4)) and the heavy hole valence band states form the fundamental energy gap (Fig. 5(a)). The energy of the (Δ(4)) conduction band hardly varies with the Ge concentration of the alloy layer. The major part of the difference in energy gap appears in the valence band. For pure Ge on Si one obtains a valence band offset of between 0.7 and 0.8 eV. The conduction band discontinuity on the other hand is so small that no significant localization of electrons is expected in the SiGe layer. For high Ge concentrations a staggered band offset is expected, which means that the top of the valence band lies in SiGe while the bottom of the conduction band is in Si. This can be seen more clearly in Fig. 6, where just the band

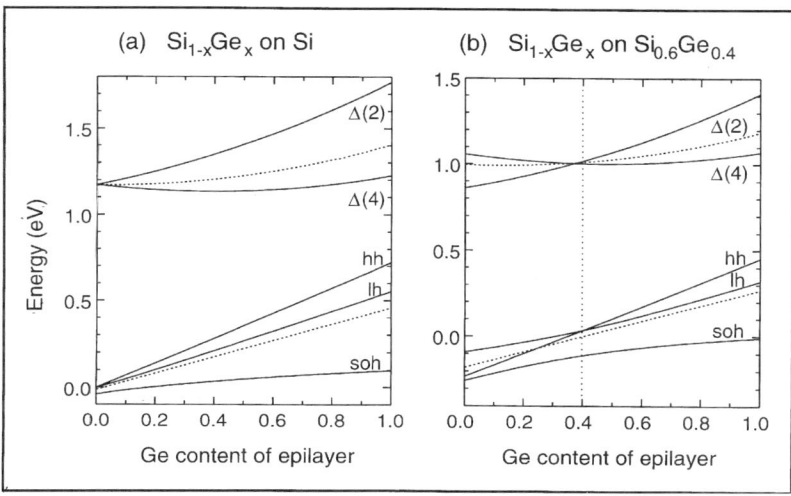

FIG. 5. Energetic behavior of the conduction and valence band edges for $Si_{1-x}Ge_x$ alloy layers on (a) Si substrates and (b) on $Si_{0.6}Ge_{0.4}$ substrates.

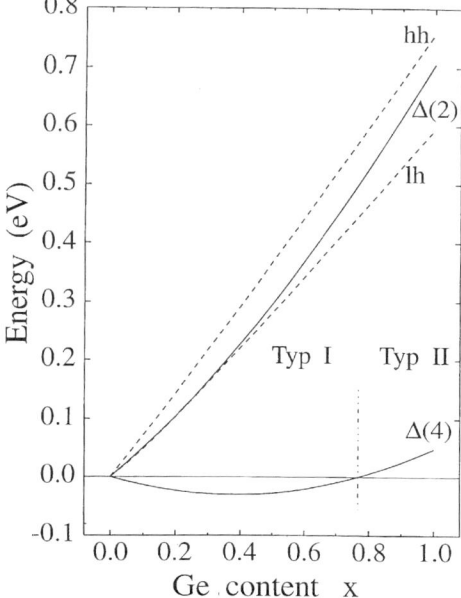

FIG. 6. Band offsets of strained $Si_{1-x}Ge$ alloy layers on Si substrates.

offsets of the ($\Delta(2)$) and ($\Delta(4)$) conduction band minima and the valence band maxima are plotted.

The situation changes drastically if strain relaxed SiGe buffer layers are used as a new substrate. Such substrates lead to tensile strain in epitaxial Si layers with the consequence of a lowering and splitting of the conduction band. The behavior of the valence and conduction bands of SiGe alloy layers on a relaxed $Si_{0.6}Ge_{0.4}$ substrate layer is shown in Fig. 5(b), as an example. In this case the lowest conduction band of strained Si is formed by the two-fold degenerate ($\Delta(2)$) valleys. This leads to a strong confinement of electrons in the Si layers, which was first demonstrated by the electron transfer in modulation doped strain symmetrized Si/SiGe multilayer structures (Abstreiter et al., 1985). This kind of band-offset was later used to achieve high mobility electron channels in strained Si layers on strain relaxed SiGe alloys (Ismail et al., 1994; Mii et al., 1991; Schäffler et al., 1992). The optical properties of such structures are determined by the staggered band alignment at the interface, that is, holes are localized in the Ge rich alloys and electrons in the Si. Interband transitions are, therefore, indirect in real and k-space. There is, however, the hope that symmetry breaking increases the optical transition probability, because the $\Delta(2)$ conduction band minima are oriented normal to the interface. Strong localization or zone-folding in appropriate superlattices may result in quasi-direct optical transitions where no phonon is needed for momentum conservation. In the case that both the electrons and the holes are localized close to the interface, the overlap of the wavefunction can be large enough to produce an appreciable oscillator strength for the optical transition (Brunner, Winter, and Eberl, 1996; Usami, Shiraki, Fukatsu, 1996). Si_mGe_n superlattices on strain relaxed SiGe alloys consist of a repetition of m monolayers of Si and n monolayers of Ge with staggered band-offsets at the interfaces. If the ratio m/n corresponds to the random Si concentration in the alloy, one talks of a strain symmetrized superlattice. In this way the integral of the built-in strain in the superlattice is zero due to the opposite stress in the Si and Ge layers, as discussed already in Section II. If the layers are thin enough one expects the formation of minibands. The superlattice minibands and the resulting energy gaps can be estimated from simple Kronig-Penney type calculations using the periodic potential variations given by the band-offsets and the appropriate effective masses in each band. Periodic band edge variations for a Si_6Ge_4 superlattice on various substrates with different in-plane lattice constants are shown in Fig. 7. Results of miniband calculations using the effective mass approximation and envelope function theory are shown in Fig. 8 for the lowest $\Delta(2)$ and $\Delta(4)$ conduction minibands and the highest hh and lh valence minibands versus superlattice period length assuming strain symmetrized Si and Ge layers with a thickness ratio of 3/2

2 BAND GAPS AND LIGHT EMISSION IN Si/Ge ATOMIC LAYER STRUCTURES 51

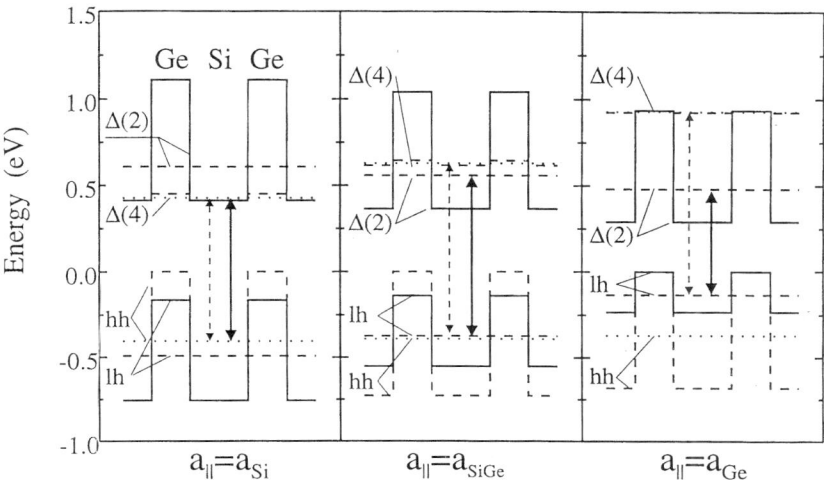

FIG. 7. Band edge modulation for Si_mGe_n superlattices with $m/n = 3/2$ and different strain distributions.

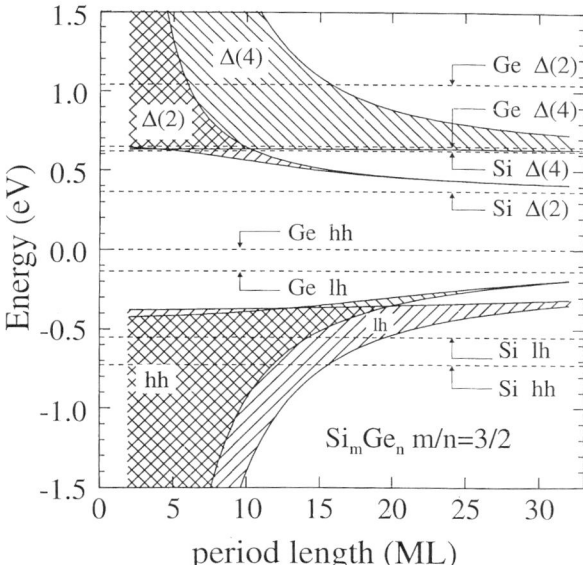

FIG. 8. Lowest conduction and highest valence minibands of strain symmetrized Si_mGe_n short period superlattices. The dashed lines indicate the band edges of the wells and barriers.

(Zachai et al., 1990). The energies of the wells and barriers are indicated by the dashed lines. For small period lengths the miniband widths get larger than the potential modulations. Due to the large effective mass of the $\Delta(2)$ electrons in superlattice directions ($m_l^{Si} = 0.92 m_o$) the miniband is already quite narrow at period lengths as small as 10 monolayers. This means that the electrons are more localized in the Si layers while the heavy holes are in the Ge layers, giving rise to spatially indirect transitions at the interfaces.

This kind of calculation, however, does not give information on the position of the conduction band minimum with respect to the three-dimensional Brillouin zone. As discussed above, a nearly exact backfolding of the $\Delta(2)$ conduction band minima in the growth direction is expected, for example, for a period length of 10 monolayers. In strain symmetrized superlattices this minimum is then also lowest in energy, as required for a quasi-direct energy gap. A complete band structure calculation of a strain symmetrized Si_5Ge_5 superlattice is shown in Fig. 9 (Tserbak, Polatoglou, and Theodorou, 1993) as one example of the many calculations that exist in literature. The expected band folding is observed in the Γ-Z direction. The 10 monolayer period also leads to a conduction band minimum at the Γ-point ($k = 0$). The "unfolded" $\Delta(4)$ minima in Γ-X direction are higher in energy. This is due to the strained Si layers. The quasi-direct energy gap of the superlattice is smaller than the gap of a random SiGe alloy with the same average composition, which is also a consequence of the staggered

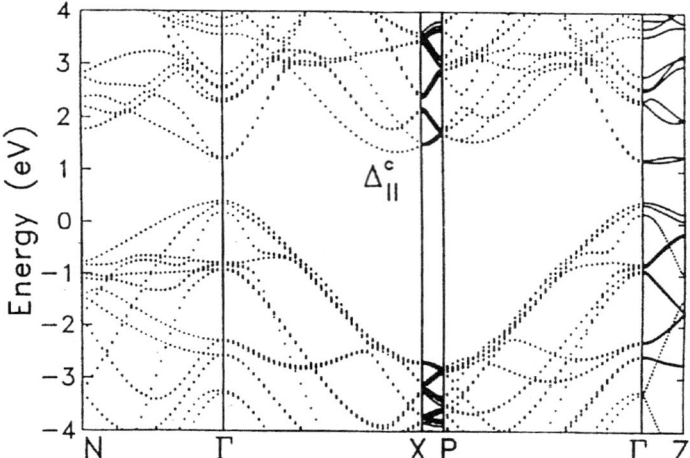

FIG. 9. Electronic band structure of a Si_5Ge_5 strain symmetrized superlattice. (From Tserbak, Polatoglou and Theodorou, 1993.)

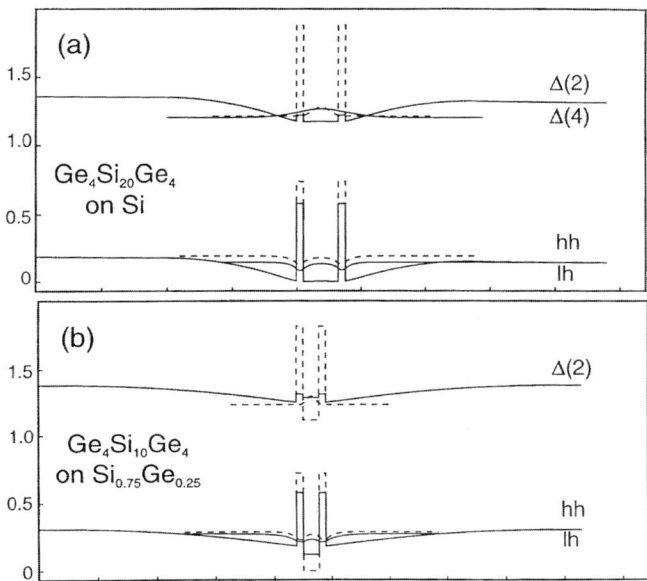

FIG. 10. Band edge behavior of (a) a p-type doped $Ge_4Si_{20}Ge_4$ coupled QW structure embedded in Si and (b) a $Ge_4Si_{10}Ge_4$ coupled QW structure embedded in $Si_{0.75}Ge_{0.25}$.

band lineup. The basic idea of the formation of quasi-direct energy gaps in such superlattices assumes coherent scattering at the periodic interfaces. However, this zone folding does not change the symmetry of the states at the conduction band minima, which means that the transition remains essentially dipole forbidden. The oscillator strength is consequently not much further enhanced than in the single staggered heterointerface. However, one is able to vary the overlap of the wave functions and add up many interfaces in such superlattices.

An additional parameter with which to optimize the overlap of the electron and hole wave function is the combination of heterolayers with the space charge fields of selective doping layers. For example, p-type doping can be used to create an attractive potential for the electrons. Figure 10 shows specially designed monolayer type potential wells that make use of such space charge fields to enhance the localization of electrons in the region of the Ge layers, which form the potential well for the holes. The p-type doping leads to a population of the lowest valence band subbands in the coupled Ge wells of the $Ge_4Si_{20}Ge_4$ layer sequence. Such $Ge_nSi_mGe_n$ structures may be considered as one and a half periods of a Si_mGe_n

superlattice. The doping induces a band bending, which creates a potential well for electrons with its minimum in the 20 monolayer Si layer. If the coupled Ge QWs are grown on Si, which means only Ge layers are strained, then the lowest subbands of the $\Delta(2)$ and $\Delta(4)$ conduction bands are almost at the same energy. The potential well in the conduction band gets deeper with higher doping, which brings the $\Delta(2)$ subband down faster compared to the $\Delta(4)$ subband due to the larger effective mass normal to the interfaces. In Fig. 10(b), a $Ge_4Si_{10}Ge_4$ structure is embedded in a $Si_{0.75}Ge_{0.25}$ strain relaxed alloy. The resulting compressive strain in the Si layer lowers the $\Delta(2)$ conduction band. The confinement of the electrons is consequently enhanced. By varying parameters like the individual layer thickness and the built-in strain or strain distribution, it is possible to optimize the overlap between the electron and hole wave functions. It has been shown theoretically that the oscillator strength can be almost as high as in Si_mGe_n superlattices (Jaros and Beavis, 1993; Jaros et al., 1993; Jaros, 1996; Turton and Jaros, 1995). It was also possible to observe relatively strong PL and EL in such structures up to room temperature, as will be discussed in the next section.

IV. Photoluminescence, Electroluminescence, and Photocurrent Measurements

1. $Si_{1-x}Ge_x$ Alloy Layers and Quantum Wells

Si and $Si_{1-x}Ge_x$ are semiconductors with an indirect energy gap with the valence band maximum at the Γ-point and the conduction band minimum along the Δ-direction close to the X-point within the first Brillouin zone. Direct optical transitions between these states are, therefore, not possible because of momentum conservation. In high-purity single-crystal Si, the bandgap related PL is observed in the form of phonon replica where the momentum conservation is fulfilled by the emission of an appropriate phonon. For low excitation power, which means a low density of electron hole pairs, excitons are formed before recombination. The luminescence spectrum then consists of several sharp and broad lines, which involve free excitons with single and multiphonon emission. The strongest processes are the so called Si-TO^Δ, Si-TA and the Si-$TO^\Delta + O^\Gamma$ transitions (Dean, Haynes, and Flood, 1967). At high exciton densities and for temperatures below 20 K, a condensation of excitons into electron–hole droplets is observed, which leads to broader emission lines about 15 meV below the energy of the TO and TA phonon replica. With impurities like shallow donors or acceptors bound excitons are also observed a few millielectron-volts below the free exciton lines (Dean, Haynes, and Flood, 1967). Their intensities saturate at high excitation power. In Si_xGe_{1-x} one observes a very similar PL spectrum with all the phonon replicas. However, the phonon spectrum is changed, which gives rise to more lines, especially for the optical phonons. The most dominant emission lines consist of SiGe-TA, SiGe-TO_{GeGe}^A, SiGe-TO_{SiGe}^A and SiGe-TO_{SiSi}^A where TO_{GeGe}^A, TO_{SiGe}^A and TO_{SiSi}^A are the optical phonon energies of the Ge-Ge, Si-Ge and Si-Si vibrations in the alloy. Another difference is the appearance of a so-called no phonon (NP) line. Such emissions, which do not involve a phonon, are possible due to perturbations of the periodic crystal that in the present case are mainly due to alloy fluctuations. A detailed analysis of the optical emission spectra of bulk $Si_{1-x}Ge_x$ alloys is given in Weber and Alonso (1989). In such alloys one also often observes defect related luminescence a few hundred millielectron-volts below the bandgap energy. These emission lines can be due to dislocations or other deep traps, which have not been identified in all cases in detail. In narrow $Si_{1-x}Ge_x$ QWs the low temperature PL spectra also exhibit all the features discussed above. The epitaxial SiGe layers studied before 1990, however, showed only broad defect-related luminescence (see Noël et al. (1990)), which may be due either to dislocations or to Ge clusters that were formed during growth. Most of the layers were grown at relatively low temperatures, around 300 to 400°C, at that time in order to avoid segregation or nucleation of three-dimensional growth. The first evidence for bandgap related luminescence from SiGe alloys was reported by Terashima et al. (1990) for layers grown at 650°C. A breakthrough was achieved by Sturm et al. (1991) who observed well resolved excitonic transitions in $Si_{1-x}Ge_x$ multi-QWs (MQWs) grown by chemical vapor deposition (CVD) also at 650°C. Robbins, Calcott, and Leong (1991) used EL to study $Si/Si_{0.8}Ge_{0.2}$ MQWs. They also observed fundamental interband transitions in their samples grown by CVD at 610°C. Shortly after those first observations of excitonic luminescence in SiGe QWs, many groups worldwide were able to grow samples both with MBE and CVD whose qualities were good enough to show excitonic luminescence (Brunner et al., 1992, 1993; Dutartre et al., 1991; Fukatsu et al., 1992, 1993; Fukatsu, 1994; Lenchyshyn et al., 1992; Menczigar et al., 1992; Mi et al., 1992; Northrop et al., 1992; Robbins, Calcott, and Leong, 1991; Spitzer et al., 1992; Vescan et al., 1992; Wachter et al., 1992; Xiao et al., 1992). Most of the samples studied were grown at temperatures above 600°C, which seems to improve the material quality considerably. A typical PL spectrum of a $Si_{0.76}Ge_{0.24}$ QW with a layer thickness of 30 Å is shown in Fig. 11. Around 1.1 eV one observes the free and bound exciton of the Si substrate or the Si buffer layer. the QW related excitonic luminescence covers the

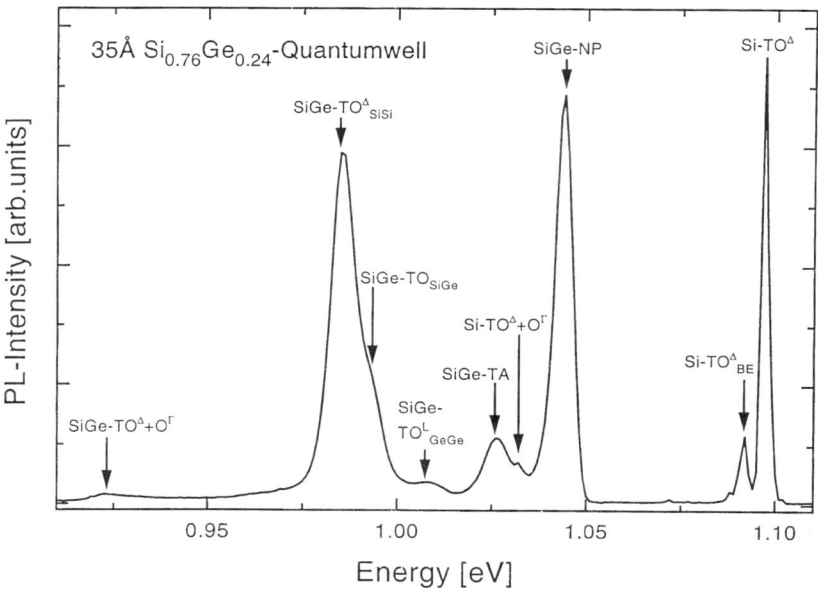

FIG. 11. Photoluminescence of a 30 Å thick $Si_{0.76}Ge_{0.24}$ QW embedded pseudomorphically in Si.

energy range between 0.9 and 1.05 eV and exhibits the NP line and the various phonon replica as discussed above. The energetic positions of these lines depend not only on the Ge concentration but also on the well width, due to the confinement energy of the heavy hole subband and the $\Delta(4)$ subband. The luminescence energy is given by:

$$E_{Lum} = E_{gap} + hh_0 + (\Delta(4))_0 - E_{Ex} - \Delta E_B - E_{Phon} \qquad (16)$$

where E_{gap} is the fundamental energy gap of the strained SiGe alloy (in-plane lattice constant of Si), hh_0 and $(\Delta(4))_0$ are the subband energies in the valence and conduction band respectively, E_{Ex} is the exciton binding energy, ΔE_B is the additional binding energy if the exciton is bound to a shallow impurity, and E_{Phon} is the energy of the phonon emitted in the radiative emission process. E_{gap} depends strongly on the Ge content, while the layer thickness mainly affects the hh_0 subband energy because of the large valence band offset. The effective energy gap can be tuned by these two parameters. For high Ge concentrations, however, the critical thickness limits the shift of the energy gap to lower energies. For pure Ge layers the

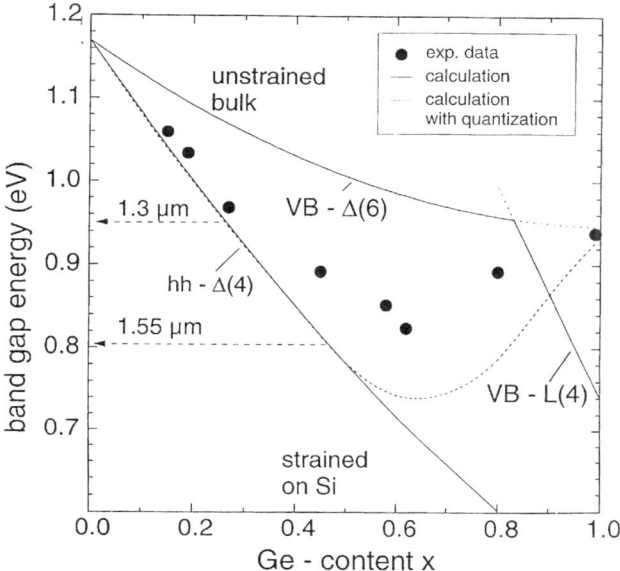

FIG. 12. Measured NP energies of various strained SiGe QWs, and bandgaps of strained and unstrained SiGe alloys.

critical thickness is only about 4 monolayers, which means that the hh_0 subband is very close to the top of the valence band of the surrounding barrier. Experimentally observed luminescence energies for QWs of various Ge concentrations are plotted in Fig. 12 together with the theoretically expected gap of the strained SiGe alloy. Part of the shift of the experimental values to higher energies for high Ge concentrations is due to the quantum mechanical size effect in the narrow layers. It is possible experimentally to reach the energy corresponding to an optical wavelength of 1.3 μm with strained SiGe layers in Si. It is, however, difficult to reach the long wavelength range around 1.55 μm. This is in part due to the critical thickness, but also to interdiffusion, which cannot be neglected at the high growth temperatures used, especially for high Ge concentrations (Gail *et al.*, 1995, 1996). Photoluminescence and EL of Si/SiGe MQW structures have been studied in detail by many authors (see Fukatsu (1994)). Apart from some special samples with high Ge concentration, both the PL and the EL were found to show a strong temperature quenching, with basically no QW related luminescence observable at room temperature. An example is given in Fig. 13, where the EL of a pin diode with five $Si_{0.73}Ge_{0.27}$ QWs embedded in the intrinsic region is plotted for various temperatures. At

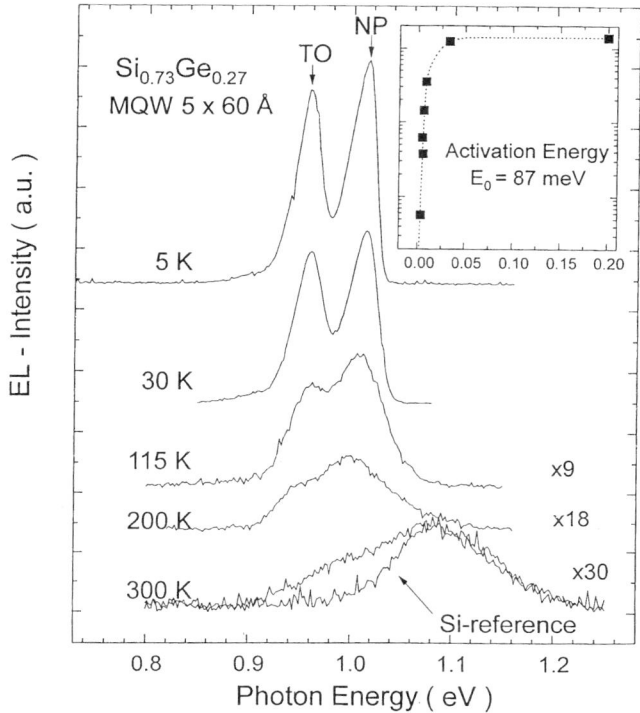

FIG. 13. Temperature dependence of EL of 60 Å thick $Si_{0.73}Ge_{0.27}$ QWs embedded in a Si pin diode.

room temperature only a weak and broad signal from the 60 Å thick QWs is observed on top of the Si emission, while at low temperatures the EL is clearly dominated by the QW emission. Based on these results it was concluded that strained SiGe QWs embedded in Si are not very promising candidates for efficient light emitting devices at room temperature, although room temperature PL is observable in deep SiGe QWs, especially when they are also p-type doped (Fukatsu *et al.*, 1993; Gail *et al.*, 1996; Mi *et al.*, 1992). One is, however, able to tune the energy gap to the wavelength range of 1.3 μm to 1.55 μm, which is the interesting region for optical fiber communication. Such pseudomorphic SiGe QWs, therefore, might be of interest for detection of infrared light. A waveguide geometry is required for this application due to the weak absorption in the thin QWs (Bernhard-Höfer *et al.*, 1995; Splett *et al.*, 1994).

2. Si_mGe_n SHORT PERIOD SUPERLATTICES

Short period superlattices are expected to have stronger optical emission and absorption because of Brillouin zone folding and symmetry breaking mechanisms at the many interfaces. In addition strain symmetrized short period superlattices can be grown to a much larger overall thickness. A prerequisite for the achievement of high quality superlattices is the growth of graded SiGe buffer layers with a low density of threading dislocations and a smooth surface. The best buffer layers have been obtained by a grading of the Ge content from a low value to the desired concentration over a thickness of more than 1 μm. This is exemplified by the PL spectra shown in Fig. 14 for three graded buffer layers with a final Ge-concentration of about 30%. The graded buffers start with 5% Ge followed by a linear increase to 30% over a thickness of 0.5, 1, and 1.5 μm. Each buffer has a cap layer with a thickness of 0.5 μm comprising a constant $Si_{0.7}Ge_{0.3}$ alloy layer. The spectra show clearly that steeper grading of the Ge content leads to strongly enhanced defect related luminescence and only a week excitonic luminescence from the alloy on top.

The sample with the 1.5 μm thick graded buffer on the other hand exhibits very nicely the NP line and the phonon replica of the unstained $Si_{0.7}Ge_{0.3}$ alloy layer and a relatively weak luminescence from the dislocations and defects in the thick graded buffer. Such graded buffer layers were used to fabricate strain symmetrized short period superlattices (see Figs. 1 and 2). In order to have sharp interfaces between the thin Si and Ge layers, a monolayer of Sb was used as a surfactant and growth temperatures were kept as low as 500°C. The first PL of such a Si_6Ge_4 superlattice was studied by Zachai et al. (1988, 1990). It showed already quite strong PL about 110 meV below the NP line of the corresponding $Si_{0.6}Ge_{0.4}$ alloy. For these early experiments a thin buffer with a constant Ge concentration was used, which resulted in a high density of threading dislocations. Therefore, the origin of the PL was discussed controversially. However, the interpretation of the results as bandgap related luminescence was supported later by a systematic study of various superlattices with different period lengths keeping the ratio of Si to Ge layer thickness constant at 3/2. Low temperature luminescence spectra for Si_9Ge_6, Si_6Ge_4, and Si_3Ge_2 superlattices and a $Si_{0.6}Ge_{0.4}$ alloy reference layer are shown in Fig. 15 (Menczigar et al., 1993). The NP line of the 0.5 μm thick strain relaxed alloy on a graded buffer is approximately 970 meV. The superlattices have been grown on top of such structures with a total thickness of 0.2 μm. Adding the 0.2 μm thick Si_3Ge_2 superlattice results in a small downward shift and a slight increase of the intensity of the NP line. This shift and increase is much more pronounced

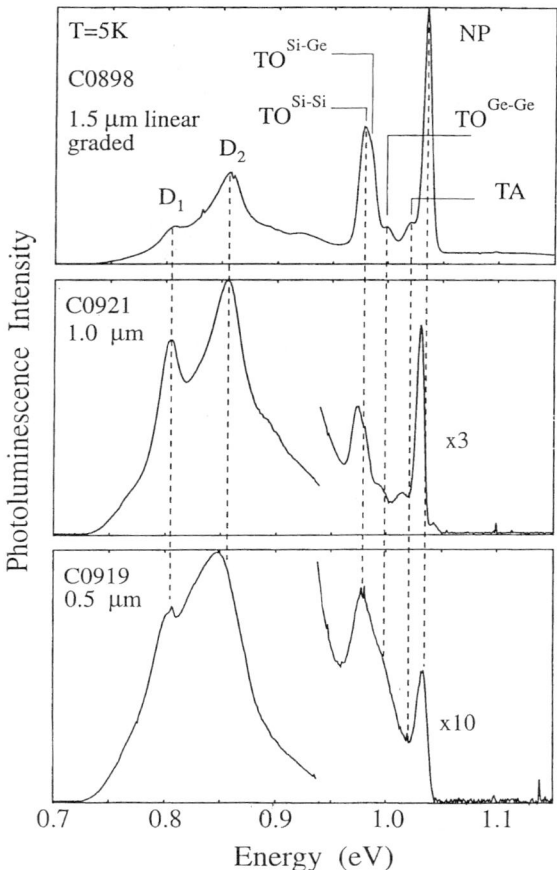

FIG. 14. Photoluminescence of three graded SiGe buffer layers with a linear increase of the Ge content from 5 to 30% over a width of 1.5 μm, 1 μm, and 0.5 μm. Each sample has a 0.5 μm thick top layer with a constant Ge concentration of 30%.

for the Si_6Ge_4 and the Si_9Ge_6 superlattices. The shift to lower energies is expected from theory and is also observed in photocurrent measurements in similar structures, which were fabricated as pin diodes (Olajos et al., 1992, 1994). A superlattice has a smaller energy gap than a random alloy with the same average Ge concentration. This shift to lower energy increases with increasing period length. This is a direct consequence of the band offsets and the staggered band lineup. The enhancement of the NP line might be evidence for the Brillouin-zone-folding effect. However, as will be shown

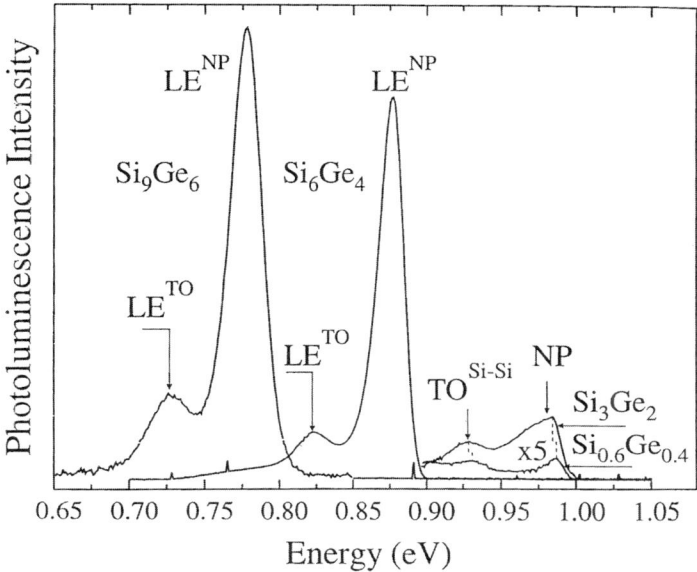

FIG. 15. Low temperature PL spectra of 0.2 μm thick strain symmetrized Si_3Ge_2, Si_6Ge_4, and Si_9Ge_6 superlattices and of a 0.5 μm thick strain relaxed $Si_{0.6}Ge_{0.4}$ alloy on Si substrates with a step graded SiGe buffer layer. (From Menczigar et al., 1993.)

later, a similar enhancement with respect to the TO-phonon replica is observed for just one interface. Therefore, it is believed nowadays that the symmetry breaking at the interface together probably with interface roughness and localization of excitons is responsible for this enhancement (Jaros et al., 1992; Jaros and Beavis, 1993). The pump power and temperature dependence of the superlattice luminescence also indicates very clearly that localized excitons give rise to the strong NP line. A coherent superposition of the electron wave function, as expected for a quasi-direct gap in the sense of Brillouin zone folding, is probably washed out by irregularities of the interfaces and strain fluctuations, although superlattice effects are very clearly observed in phonon Raman spectroscopy, X-ray diffraction and transmission electron microscopy. A comparison between the energies of the NP lines and the onset of photocurrent spectra shows that the NP emission is very close to the fundamental bandgap of the superlattices (Menczigar et al., 1993; Olajos et al., 1992, 1994). Annealing experiments and selective etching of the top layers also show directly that the observed spectra are related to the bandgap of the superlattices. The spectra shift continuously to higher energies with annealing time until they approach the energy gap

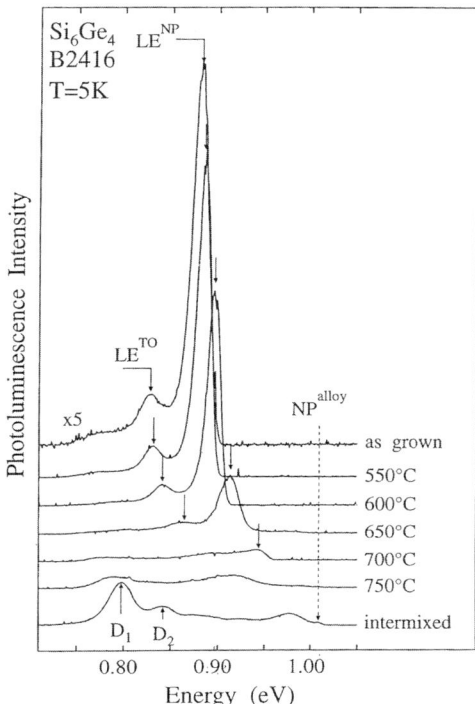

FIG. 16. Photoluminescence spectra of a strain symmetrized Si_6Ge_4 superlattice for various temperature annealing steps. The annealing time was 1 h at the temperature given beside each spectrum. (From Menczigar, 1993.)

of the random alloy. Examples of such annealing experiments are shown in Fig. 16 for the Si_6Ge_4 superlattice. This reflects directly the interdiffusion of the strain symmetrized superlattices, which has also been studied in detail by Raman spectroscopy. Figure 17 shows a series of low temperature luminescence spectra after etching successively into the top layers. This sequence shows very clearly the origin of the various luminescence lines. The as-grown sample exhibits only one strong luminescence line whose origin is the Si_6Ge_4 superlattice. Etching into the superlattice results in a fast reduction of this luminescence peak. A defect line appears at lower energies. At higher energies one observes the NP line and the TO-phonon replica of the $Si_{0.6}Ge_{0.4}$ strain-relaxed alloy. Etching into the graded buffer layer results in the characteristic dislocation related lines D_1 and D_2 (Weber and Alonso, 1989), which are present close to the Si interface. These luminescence measurements are very surface sensitive, because a blue laser that has

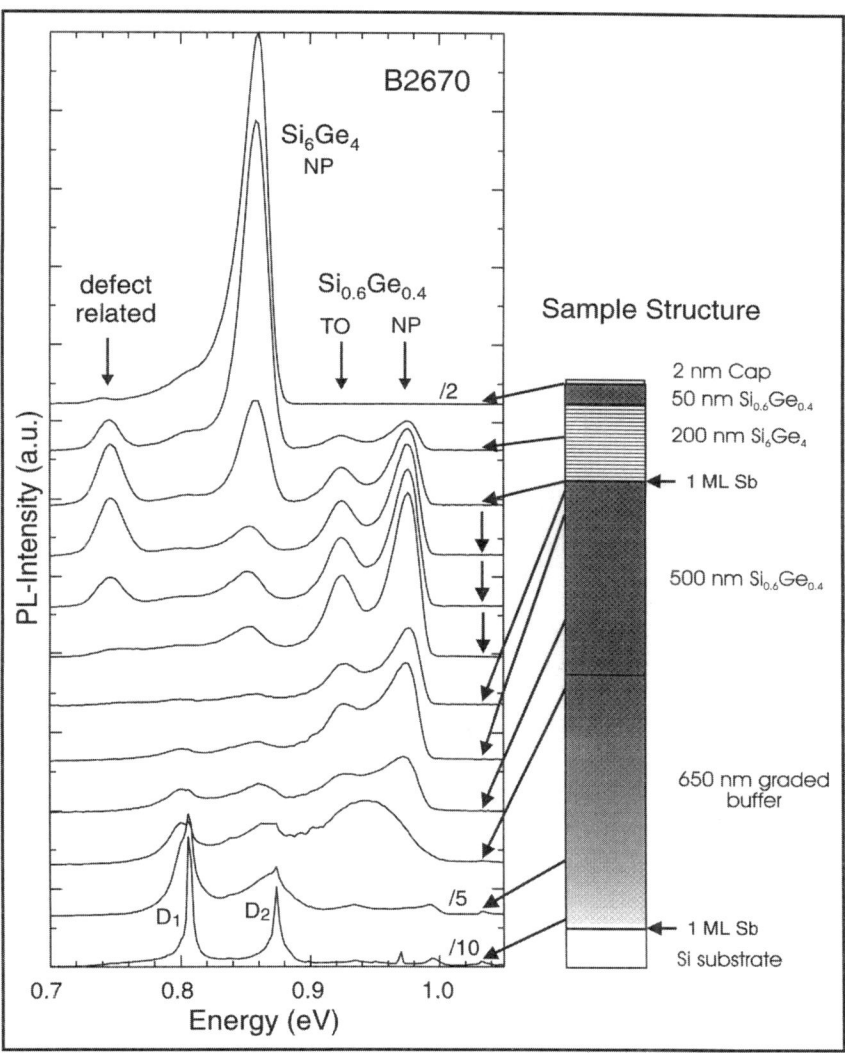

FIG. 17. Photoluminescence of a Si_6Ge_4 superlattice grown on a graded buffer after successive etching of the top layers. The layer sequence is shown schematically. The arrows indicate the correspondence of the spectra with the etching depth.

a small penetration depth was used for excitation. In addition the bandgaps of the epitaxial layers are increasing towards the Si substrate. This means that the excitons stay close to the surface. The PL intensity of such superlattices is strongly enhanced at low temperatures, but there is already a pronounced temperature quenching at temperatures of about 20 K. This indicates that the thermally activated mobility of excitons from their localization sites leads to strong nonradiative recombination. It also shows that the quality of the short period Si_mGe_n superlattices is still too low for application as efficient emitters, although there have been reports for enhanced PL of pin diodes at room temperature with increasing current density (Engvall et al., 1995). The smaller fundamental energy gap of superlattices as compared to random alloys and the unlimited thickness that is, in principle, achievable due to strain symmetrization makes such superlattices interesting candidates for front-on infrared detectors in the 1.3 to 1.55 μm range on Si substrates.

3. $Ge_nSi_mGe_n$ Atomic Layer Structures and Interfaces with Staggered Band Offsets

It was shown theoretically a few years ago that an enhanced oscillator strength is possible by localization of electrons and holes close to a Si/Ge interface (Jaros et al., 1992, 1993; Jaros and Beavis, 1993). To optimize such structures, special $Ge_nSi_mGe_n$ coupled QWs have been designed recently that show strong luminescence up to room temperature (Gail et al., 1995; Olajos et al., 1996). Coupled Ge QWs with a Si layer in between are necessary to obtain confinement of electrons. One Ge well leads only to localization of holes in the narrow Ge layer. The band alignment of such structures has been discussed briefly in Section III. Selected results of the band structure are shown in Fig. 10, as obtained from self consistent calculations. For pseudomorphic growth on Si, the fundamental energy gap of an undoped structure is determined by the confined heavy hole states in the Ge wells and the $\Delta(4)$ in-plane conduction band minima. A p-type doping of the whole sequence induces a band bending, which tends to localize the electrons in the $Ge_nSi_mGe_n$ layers. Such coupled QW have been embedded in the intrinsic region of a p-n junction, which results in an overall band bending. A series of samples has been grown in which the Ge well width was changed between 2 and 4 monolayers. The thickness of the Si layer was kept constant at 20 monolayers and the whole structure was embedded in a 100 Å thick strained $Si_{0.85}Ge_{0.15}$ alloy cladding layer. The PL of these structures is shown in Fig. 18 and the luminescence of the GeSiGe layers consists of a NP line and a TO assisted transition. The spectra shift to smaller energies

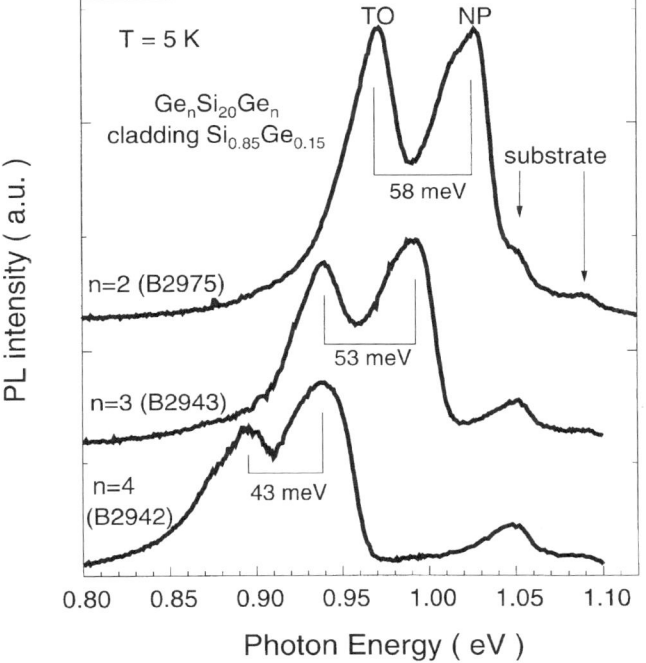

FIG. 18. Low temperature PL of $Ge_nSi_{20}Ge_n$ coupled QW structures with $n = 2$, 3, and 4. (From Gail et al., 1996.)

with increasing Ge layer thickness, as expected and the energetic position of the NP line agrees well with the calculated fundamental bandgap. The luminescence of these coupled well structures persists up to room temperature with approximately the same intensity if the Ge wells are thick enough (Engvall et al., 1995; Gail et al., 1995, 1996). Electroluminescence spectra of samples with Ge layer thickness of 2 and 4 monolayers and an asymetric SiGe cladding layer are shown in Fig. 19. The asymmetric cladding layer leads to a better confinement of the carriers in the p-n junction. At low temperature both samples show the characteristic GeSiGe QW spectrum. The broad and weak room temperature PL of the thinner sample is shifted upwards in energy and is related to Si. There is no indication of the coupled QW anymore. On the other hand, PL of the sample with 4 monolayers of Ge remains strong at roughly the energy of the low temperature peaks. The integrated intensity is even increased compared to the spectrum obtained at low temperature. It was found that the temperature quenching depends

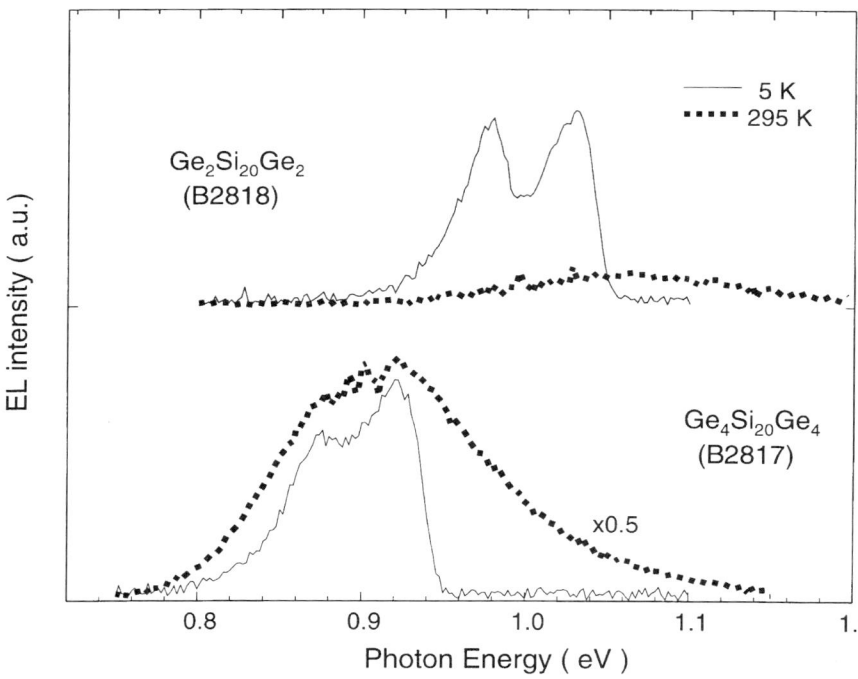

FIG. 19. Low temperature and room temperature EL of Si pin diodes with embedded $Ge_nSi_{20}Ge_n$ coupled QWs with $n = 2$ and 4. A $Si_{0.85}Ge_{0.15}$ cladding layer was added on one side to compensate locally the space charge field of the p-n junction. This leads to a more efficient trapping of the carriers. (From Gail et al., 1996.)

mainly on the confinement of the holes in the Ge coupled QWs (Gail et al., 1995, 1996). The external quantum efficiency of such structures depends on the injection current and is found to be at most 10^{-5} at room temperature, which, however, is similar to the values obtained for short period superlattices at low temperatures.

A better way to obtain a stronger confinement of the electrons and especially to shift the $\Delta(2)$ subband further down is the use of strain relaxed SiGe graded buffer layers. The band alignment of such a structure is shown in Fig. 10(b). Such samples have been grown recently. First results show that the PL is about three times more intense as compared to similar samples grown pseudomorphically on Si substrates. However, with such relaxed buffer layers one has a lower quality of the diodes due to the remaining threading dislocations, which is a similar problem to the one discussed already in the context of short period superlattices.

The enhanced luminescence observed in such monolayer type heterostructures indicates that exciton localization normal to the interfaces, but also in the layer planes, plays an important role in the optical properties, especially in the luminescence efficiency.

Recently, enhanced luminescence was also observed in QWs with staggered band offset where just one interface plays a role (Brunner, Winter, and Eberl, 1996; Usami et al., 1995; Usami, Shiraki, and Fukatsu, 1996). The band edge PL of SiGe/strained-Si/SiGe QWs grown on a strain-relaxed SiGe buffer layer was studied in Usami, Shiraki, and Fukatsu (1996). Spatially indirect luminescence of photoexcited electrons localized in the strained Si layer and holes confined to the neighboring SiGe layer was observed. It was shown that in such neighboring confinement structures, which consist basically of one period of a superlattice, a clear enhancement of the NP transition is observable for certain layer thicknesses. A similar staggered band offset is obtained by using $Si_{1-x}Ge_x/Si_{1-y}C_y$ double QW structures on Si. Brunner, Winter, and Eberl (1996) observed a strong enhancement of the NP line of the spatially indirect transition in this system, where the holes are in the $Si_{1-x}Ge_x$ alloy layer while the electrons are in the $Si_{1-y}C_y$ conduction band potential well. The NP line is especially enhanced for certain layer thicknesses where the electron-hole overlap is optimum. The advantage of the $Si_{1-y}C_y$ alloy is that the lowest conduction subband is formed by the $\Delta(2)$ valleys for pseudomorphic growth on Si. Therefore no graded buffer layer is needed. The fabrication of strain symmetrized short period superlattices grown directly on Si should be possible with this material combination.

4. LATERALLY CONFINED QWs AND Ge-RICH SELF-ASSEMBLED DOTS

As discussed above, lateral confinement of excitons may lead to an enhancement of optical recombination in Si/Ge heterostructures. Apart from the increased oscillator strength, this is probably also due to the fact that the excitons, which are localized spatially, are not mobile anymore. Thus they cannot reach nonradiative recombination centers. There are several approaches in the literature to achieve artificial lateral confinement of excitons in SiGe QWs. Excitonic luminescence was reported for wires grown onto V-grooved substrates (Usami et al., 1994) and for dots and wires into oxide windows (Vescan, 1994) or through shadowing masks (Brunner et al., 1994, 1995). The fabrication of laterally structured QWs by the use of local epitaxy through shadow masks, leads to mesa islands with SiGe QWs embedded completely inside the structure. There is no contact of the QWs with oxide interfaces or surfaces. The smallest mesa size achievable with this method is not determined by the geometrical size of the opening of the shadow mask. It is limited by surface diffusion of Si and Ge during growth. Therefore island sizes with lateral extensions below 1 μm are barely reachable, although shadow masks with openings as small as 0.2 μm have been used. It was found that the absolute QW luminescence intensity increases with decreasing island size, concomitant with an increase in the PL energy. The shift in energy can be explained by a loss of Ge in the SiGe QW due to surface diffusion, which is more and more pronounced for smaller sizes. The enhanced intensity might be due to a reduced probability of having nonradiative recombination centers in the smaller islands.

Another way of achieving lateral confinement directly by epitaxy is the use of the Stranski–Krastanow growth mode for Ge on Si. For high Ge concentrations and for growth temperatures above approximately 550°C, Ge forms three-dimensional islands after deposition of a few monolayers. The formation of such islands has been studied extensively during the past few years (Eaglesham and Carullo, 1990; Hansson *et al.*, 1992, 1994; Mo *et al.*, 1990; Voigtländer and Zinner, 1993). Such self-assembled dots have been overgrown with Si cap layers and studied by PL (Apetz *et al.*, 1995;

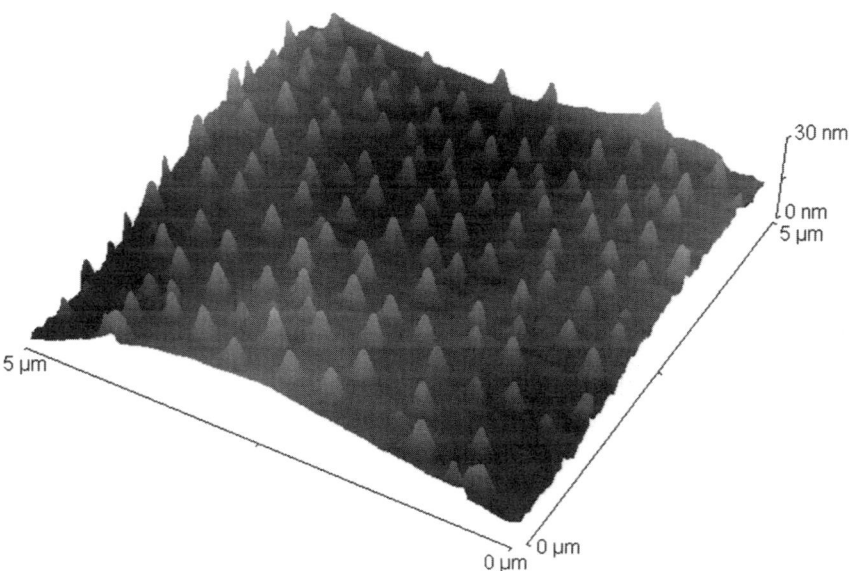

FIG. 20. Atomic force microscopy image of Ge islands formed after deposition of 5.5 monolayers of Ge onto Si at a substrate temperature of 670°C.

2 BAND GAPS AND LIGHT EMISSION IN Si/Ge ATOMIC LAYER STRUCTURES **69**

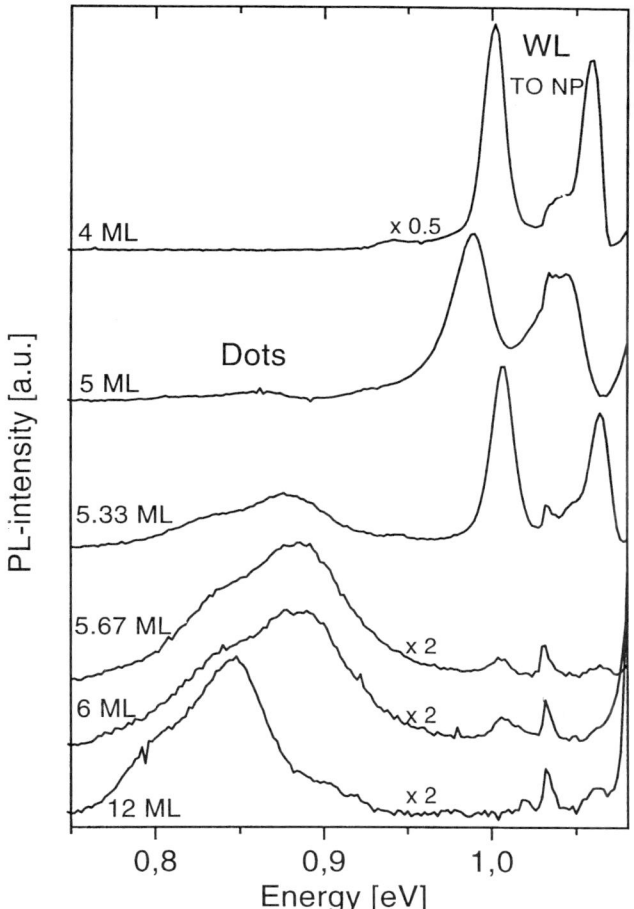

FIG. 21. Low temperature PL of a series of narrow Ge-rich QWs and dots embedded in Si. Between 4 and 5 monolayers (ML) of Ge there appear characteristic features of dots in the spectrum at energies between 0.8 and 0.9 eV, concomitant with an upward shift of the luminescence peaks of the narrow Ge wetting layer (WL). (From Schittenhelm et al., 1995.)

Schittenhelm et al., 1995; Sunamura, Shiraki, and Fukatsu, 1995). A summary of the growth, structural, and optical properties as well as the first devices and possible future device applications of Ge rich SiGe islands in Si has been published recently (Abstreiter et al., 1996). Figure 20 shows a surface image of Ge islands on Si as obtained by atomic force microscopy. For certain growth conditions these islands can be achieved with a rather narrow distribution of their sizes. When overgrown with Si, such islands show PL or EL in the wavelength range between 1.3 and 1.55 μm. The transition from QW to island luminescence is shown in Fig. 21 for a series of samples with different amounts of Ge. The island formation occurs between 4 and 5 monolayers of deposited Ge when a growth temperature of 740°C is used. Details of such PL studies are discussed in Schittenhelm *et al.*, 1995; Sunamura, Shiraki, and Fukatsu, 1995.

The conduction and valence band diagram of Ge islands embedded in Si is rather complicated due to the inhomogeneous strain distribution. We expect a strong confinement for holes in the islands and essentially no confinement for electrons, which makes it again difficult to obtain strong light emission. However, this kind of band offset may lead to other applications like optical writable memories, which make use of the trapping and long lifetime of holes in such islands or dots (Abstreiter *et al.*, 1996).

V. Concluding Remarks

There have been many different attempts during the past ten years to achieve efficient light emission in the near infrared regime using Si/Ge heterostructures. Some of these concepts have been discussed in this chapter. None of these efforts led to a high enough quantum efficiency at room temperature such that an application as an efficient light emitter would appear on the horizon. The Si/Ge heterostructure research, however, resulted in an excellent understanding of the band structures and the ability to control material combinations on an atomic scale. This allows the possibility for precise band structure engineering and various other applications in microelectronics and optoelectronics. For efficient light emitters based on Si in the wavelength range of 1.3 to 1.55 μm, however, other novel and perhaps unconventional concepts should be tried. These may include the use of C in the Si/Ge heterosystem, as mentioned briefly in this chapter, or special quantum dots with localization of both electrons and holes. There are also new promising results emerging from epitaxially grown Er^{3+} doped Si with room temperature PL at 1.54 μm (see Stimmer (1996)). Such new approaches, in combination with the well-known properties of the Si/SiGe heterosystem, are promising and worthwhile investigating and may lead finally to efficient near infrared light emitting devices based on Si.

Acknowledgments

Most of the work discussed in this chapter is based on the theses of Josef Brunner, Markus Gail, Ulrich Menczigar, Peter Schittenhelm, and

Reinhard Zachai. Their work at the Walter Schottky Institute was supported financially by various projects funded by the European Community, the Siemens AG, and the Bayerische Forschungsstiftung. These projects were performed in close collaboration with Erich Kasper, Friedrich Schäffler, Hartmut Presting, and Horst Kibbel from the Daimler research center in Ulm; Janos Olajos, Jesper Engvall, and Hermann G. Grimmeiss from University Lund; Günther Bauer and coworkers from University Linz; Milan Jaros from University of Newcastle; and George Theodorou from University of Thessaloniki. Their many contributions can be identified from the references. I gratefully acknowledge the fruitful collaboration with all partners during the past 10 years. The whole list of references collected in this chapter covers only a small part of all the fascinating work that has been performed worldwide in this area in various laboratories. I apologize for having left out many interesting publications.

REFERENCES

Abstreiter, G., Brugger, H., Wolf, T., Jorke, H., and Herzog, H.-J. (1985). Strain-induced two-dimensional electron gas in selectively doped Si/Si_xGe_{1-x} superlattices. *Phys. Rev. Lett.* **54**, 2441–2444.

Abstreiter, G., Brugger, H., Wolf, T., Zachai, R., and Zeller, Ch. (1986). Optical and electronic properties of Si/SiGe superlattices. *Springer Series in Solid-State Sciences* **67**, 130.

Abstreiter, G., Eberl, K., Friess, E., Wegscheider, W., and Zachai, R. (1989). Si/Ge strained layer superlattices. *J. Cryst. Growth* **95**, 431–438.

Abstreiter, G., Schittenhelm, P., Engel, C., Silveira, E., Zrenner, A., Meertens, D., and Jäger, W. (1996). Growth and characterization of self-assembled Ge-rich islands on Si. *Semicond. Sci. Technol.* **11**, 1521–1528.

Apetz, R., Vescan, L., Hartmann, A., Dieker, C., and Lüth, H. (1995). Photoluminescence and electroluminescence of SiGe dot fabricated by island growth. *Appl. Phys. Lett.* **66**, 445–447.

Balslev, I. (1966). Influence of Uniaxial Stress on the Indirect Absorption Edge in Si and Ge. *Phys. Rev.* **143**, 636–647.

Bean, J. C., Feldman, L. C., Fiory, A. T., and Nakahara, S. (1984). $Si_{1-x}Ge_x$ Strained layer superlattice grown by MBE. *J. Vac. Sci. Technol.* **A2**, 436.

Bernhard-Höfer, K., Zrenner, A., Brunner, J., Abstreiter, G., Wittman, F., and Eisele, I. (1995). Strained $Si_{1-x}Ge_x$ multi-quantum well waveguide structures on (110) Si. *Appl. Phys. Lett.* **66**, 2226–2228.

Bir, G. L., and Pikus, G. E. (1974). *Symmetry and Strain-induced Effects in Semiconductors.* Wiley, New York.

Braunstein, R., Moore, A. R., and Herman, F. (1958). Intrinsic optical absorption in Ge-Si alloys. *Phys. Rev.* **109**, 695–710.

Braunstein, R. (1963). Valence band structure of Ge-Si Alloys. *Phys. Rev.* **130**, 869–879.

Brey, L., and Tejedor, C. (1987). New optical transitions in Si-Ge strained superlattices. *Phys. Rev. Lett.* **59**, 1022–1025.

Brunner, J., Menczigar, U., Gail, M., Friess, E., and Abstreiter, G. (1992). Bandgap luminescence in pseudomorphic SiGe quantum wells grown by MBE. *Thin Solid Films* **222**, 27–29.

Brunner, J., Menczigar, U., Gail, M., Friess, E., and Abstreiter, G. (1993). Influence of growth conditions on the photoluminescence of pseudomorphic MBE grown $Si_{1-x}Ge_x$ quantum wells. *J. Cryst. Growth* **127**, 443–446.

Brunner, J., Rupp, T. S., Gossner, H., Ritter, R., Eisele, I., and Abstreiter, G. (1994). Excitonic luminescence from locally grown SiGe wires and dots. *Appl. Phys. Lett.* **64**, 994–996.

Brunner, J., Jung, W., Schittenhelm, P., Gail, M., Abstreiter, G., Gonderman, J., Hadam, B., Koester, T., Spangenberg, B., Roskos, H. G., Kurz, H., Gossner, H., and Eisele, I. (1995). Local epitaxy of Si/SiGe wires and dots. *J. Cryst. Growth* **157**, 270–275.

Brunner, K., Winter, W., and Eberl, K. (1996). Spatially indirect radiative recombination of carriers localized in $Si_{1-x-y}Ge_xC_y/Si_{1-y}C_y$ double quantum well structure on Si substrates. *Appl. Phys. Lett.* **69**, 1279–1281.

Colombo, L., Resta, R., and Baroni, S. (1991). Valence-band offset at strained Si/Ge-interfaces. *Phys. Rev. B* **44**, 5572–5579.

Dean, P. J., Haynes, J. R., and Flood, W. F. (1967). New radiative recombination processes involving neutral donors and acceptors in Si and Ge. *Phys. Rev.* **161**, 711–729.

Dodson, B. W., and Tsao, J. Y. (1987). Relaxation of strained-layer semiconductor structures via plastic flow. *Appl. Phys. Lett.* **51**, 1325–1327.

Dodson, B. W., and Tsao, J. Y. (1988). Non-Newtonian strain relaxation in highly strained SiGe heterostructures. *Appl. Phys. Lett.* **53**, 2498–2500.

Dodson, B. W., and Tsao, J. Y. (1988). Stress dependence of dislocation glide activation energy in single-crystal SiGe alloys up to 2.6 Gpa. *Phys. Rev. B* **38**, 12383–12387.

Dutartre, D., Brémond, G., Souifi, A., and Benayattou, T. (1991). Excitonic luminescence from Si-capped strained $Si_{1-x}Ge_x$ layers. *Phys. Rev. B* **44**, 11525–11527.

Eaglesham, D. J., and Carullo, M. (1990). Dislocation-free Stranski-Krastanov growth of Ge on Si(100). *Phys. Rev. Lett.* **64**, 1943–1946.

Eberl, K., Krötz, G., Zachai, R., and Abstreiter, G. (1987). Structural, composition and optical properties of ultrathin Si/Ge superlattices. *J. de Phys.* **C5**, 329–332.

Eberl, K., Wegscheider, W., Schorer, R., and Abstreiter, G. (1991). Microscopic symmetry properties of (001) Si/Ge monolayer superlattices. *Phys. Rev. B* **43**, 5188–5191.

Engvall, J., Olajos, J., Grimmeiss, H. G., Kibbel, H., and Presting, H. (1995). Luminescence from monolayer-thick Ge quantum wells embedded in Si. *Phys. Rev. B* **51**, 2001–2004.

Esaki, L., and Tsu, R. (1969). Superlattice and negative conductivity in semiconductors. IBM Research Note RC-2418.

Fitgerald, E. A., Xie, Y. H., Green, M. L., Brasen, D., and Kortan, A. R. (1991). Strain-free $Si_{1-x}Ge_x$ layers with low threading dislocation densities grown on Si substrates. *Appl. Phys. Lett.* **59**, 811–813.

Froyen, S., Wood, D. M., and Zunger, A. (1987). New optical transitions in strained Si-Ge superlattices. *Phys. Rev. B* **36**, 4547–4550.

Froyen, S., Wood, D. M., and Zunger, A. (1988). Structural and electronic properties of epitaxial thin layer Si_nGe_n superlattices. *Phys. Rev. B* **37**, 6893–6907.

Fukatsu, S. Fujita, K., Yaguchi, H., Shiraki, Y., and Ito, R. (1991). Self-limitation in the surface segregation of Ge atoms during Si molecular beam epitaxial growth *Appl. Phys. Lett.* **59**, 2103–2105.

Fukatsu, S., Yoshida, H., Usami, N., Fujiwara, A., Takahashi, Y., Shiraki, Y., and Ito, R. (1992). Systematic blue shift of exciton luminescence in strained $Si_{1-x}Ge_x/Si$ quantum well structures grown by gas source Si molecular beam epitaxy. *Thin Solid Films* **222**, 1–4.

Fukatsu, S., Usami, N., and Shiraki, Y., Nishida, and Nakagawa, K. (1993). High temperature operation of strained $Si_{0.65}Ge_{0.35}/Si(111)$ p-type multiple-quantum-well light-emitting diode grown by solid source Si molecular-beam epitaxy. *Appl. Phys. Lett.* **63**, 967–970.

Fukatsu, S. (1994). Luminescence investigation on strained $Si_{1-x}Ge_x/Si$ modulated quantum wells. *Solid-State Electronics* **4–6**, 817–823.
Gail, M., Brunner, J., Nützel, J. F., Abstreiter, G., Engvall, J., Olajos, J., and Grimmeiss, H. (1995). Optical study of diffusion limitation in MBE growth of SiGe quantum wells. *Semicond. Sci. Technol.* **10**, 319–325.
Gail, M., Abstreiter, G., Olajos, J., Engvall, J., Grimmeiss, H. G., Kibbel, H., and Presting, H. (1995). Room-temperature photoluminescence of $Ge_mSi_nGe_m$ structures. *Appl. Phys. Lett.* **66**, 2978–2980.
Gail, M., Jung, W., Brunner, J., Schittenhelm, P., Nützel, J. F., and Abstreiter, G. (1996). Diffusion effects and luminescence in thin SiGe/Si layers. *Solid State Phenom.* **47–48**, 473–484.
Gell, M. A. (1989a). Direct-gap Si/Ge superlattices. *Phys. Rev. B* **40**, 1966–1968.
Gell, M. A. (1989b). Optical window in strained-layer Si/Ge microstructures. *Appl. Phys. Lett.* **55**, 484–485.
Gnutzmann, U., and Clausecker, K. (1974). Theory of direct optical transitions in an optical indirect semiconductor with a superlattice structure. *Appl. Phys.* **3**, 9–14.
Hansson, P. O., Albrecht, M., Strunk, H. P., Bauser, E., and Werner, J. H. (1992). Dimensionality and critical sizes of GeSi on Si(100). *Thin Solid Films* **216**, 199.
Hansson, P. O., Albrecht, M., Dorsch, W., Strunk, H. P., and Bauser, E. (1994). Interfacial energies providing a driving force for Ge/Si heteroepitaxy. *Phys. Rev. Lett.* **73**, 444–447.
Hybertsen, M. S., and Schlüter, M. (1987). Theory of optical transitions in Si/Ge(001) strained-layer superlattices. *Phys. Rev. B* **36**, 9683–9693.
Hybertsen, M. S., Schlüter, M., People, R., Jackson, S. A., Lang, D. V., Pearsall, T. P., Bean, J. C., Vandenberg, J. M., and Bevk, J. (1988). Origin of the optical transitions in ordered Si/Ge(001) superlattices. *Phys. Rev. B* **37**, 10195–10198.
Ismail, K., Le Goues, F. K., Saenger, K. L., Arafa, M., Chu, J. O., Mooney, P. M., and Meyerson, B. S. (1994). Identification of a mobility-limiting scattering mechanism in modulation-doped Si/SiGe heterostructures. *Phys. Rev. Lett.* **73**, 3447–3450.
Jäger, W., and Mayer, J. (1995). Energy-filtered transmission electron microscopy of Si_mGe_n superlattices and Si-Ge heterostructures. I. Experimental results. *Ultramicroscopy* **59**, 33–45.
Jaros, M., Beavis, A. W., Hagon, P. J., Turton, R. J., Miloszewski, A., and Wong, K. B. (1992). Quantitative theory of optical properties of Si-Ge heterostructures. *Thin Solid Films* **222**, 205–208.
Jaros, M., and Beavis, A. W. (1993). Localization at imperfect interfaces and its role in optical spectra of quantum well structures. *Appl. Phys. Lett.* **63**, 669–671.
Jaros, M., Beavis, A. W., Corbin, E., Hagon, J. P., Turton, R. J., and Wong, K. B. (1993). Effect of ordering interface imperfections and clusters and external electric fields on optical spectra of Si-SiGe heterostructures. *J. Vac. Sci. Technol.* **B11**, 1689.
Jaros, M. (1996). *Solid State Phenom.* **47–48**, 463.
Jorke, H. (1988). Surface segregation of Sb on Si(100) during MBE growth. *Surf. Sci.* **193**, 569–572.
Kasper, E., Herzog, H. J., and Kibbel, H. (1975). A One-dimensional SiGe superlattice grown by UHV epitaxy. *Appl. Phys.* **8**, 199–205.
Kasper, E., Herzog, H.-J., Jorke, H., and Abstreiter, G. (1987). Strained layer Si/SiGe superlattices. *Superlat. and Micros.* **3**, 141–146.
Kasper, E., Kibbel, H., Jorke, H., Brugger, H., Friess, E., and Abstreiter, G. (1988). Symmetrically strained Si/Ge superlattices on Si substrates. *Phys. Rev. B* **38**, 3599–3601.
Kline, J. S., Pollak, F. H., and Cardona, M. (1968). Electroreflectance in Ge-Si alloys. *Helv. Phys. Acta* **41**, 968–977.
Kopensteiner, E., Bauer, G., Kibbel, H., and Kasper, E. (1994). Investigation of strain-

symmetrized and pseudomorphic superlattices by X-ray reciprocal space mapping. *J. Appl. Phys.* **76**, 3489.

LeGoues, L. K., Le, Meyerson, B. S., Morar, J. F., and Kirchner, P. D. J. (1992). *Appl. Phys.* **71**, 4230.

Lenchyshyn, L. C., Thewalt, M. L. W., Sturm, J. C., Schwartz, P. V., Prinz, E. J., Rowell, N. L., Noël, J.-P., and Houghton, D. C. (1992). High quantum efficiency photoluminescence from localized excitons in $Si_{1-x}Ge_x$. *Appl. Phys. Lett.* **60**, 3174–3176.

Matthews, J. W., and Blakeslee, A. E. (1974). Defects in epitaxial multilayers. *J. Cryst. Growth.* **27**, 118–125.

Matthews, J. W., and Blakeslee, A. E. (1975). Defects in epitaxial multilayers. *J. Cryst. Growth* **29**, 273–280.

Menczigar, U., Brunner, J., Friess, E., Gail, M., Abstreiter, G., Kibbel, H., Presting, H., and Kasper, E. (1992). Photoluminescence studies of $Si/Si_{1-x}Ge_x$ quantum wells and Si_mGe_n superlattices. *Thin Solid Films* **222**, 227–233.

Menczigar, U., Abstreiter, G., Olajos, J., Grimmeiss, H. G., Kibbel, H., Presting, H., and Kasper, E. (1993). Enhanced bandgap luminescence in strain-symmetrized $(Si)_m/(Ge)_n$ superlattices. *Phys. Rev. B* **47**, 4099–4102.

Mi, Q., Xiao, X., Sturm, J. C., Lenchyshyn, L. C., and Thewalt, M. L. W. (1992). Room-temperature 1.3 μm electroluminescence from strained $Si_{1-x}Ge_x$/Si quantum well. *Appl. Phys. Lett.* **60**, 3177–3179.

Mii, Y. J., Xie, Y. H., Fitzgerald, E. A., Don Monroe, Thiel, F. A., Weir, B. E., and Feldman, L. C. (1991). Extremely high electron mobility in Si/Ge_xSi_{1-x} structures grown by molecular beam epitaxy. *Appl. Phys. Lett.* **59**, 1611–1614.

Mo, Y. W., Savage, D. E., Swartzentruber, B. S., and Lagally, M. G. (1990). Kinetic pathway in Stranski-Krastanov growth of Ge on Si(001). *Phys. Rev. Lett.* **65**, 1021–1024.

Morrison, J., Jaros, M., and Wong, K. B. (1987). Strain-induced confinement in $Si_{0.75}Ge_{0.25}$(Si/$Si_{0.5}Ge_{0.5}$) (001) superlattice systems. *Phys. Rev. B* **35**, 9693–9707.

Morrison, J., and Jaros, M. (1988). Electronic and optical properties of ultrathin Si/Ge(001) superlattice. *Phys. Rev. B* **37**, 916–921.

Noël, J.-P., Rowell, N. L., Houghton, D. C., and Perovic, D. D. (1990). Intense petoluminescence between 1.3 and 1.8 μm from strained $Si_{1-x}Ge_x$ alloys. *Appl. Phys. Lett.* **57**, 1037–1039.

Northrop, G. A., Morar, J. F., Wolford, D. J., and Bradley, J. A. (1992). Observation of strong $Si/Si_{1-x}Ge_x$ narrow quantum-well near-edge luminescence under applied hydrostatic pressure. *Appl. Phys. Lett.* **61**, 192–194.

Nützel, J. F., and Abstreiter, G. (1996). Segregation and diffusion on semiconductor surfaces. *Phys. Rev. B* **53**, 13551–13558.

Olajos, J., Engvall, J., Grimmeiss, H. G., Menczigar, U., Abstreiter, G., Kibbel, H., Kasper, E., and Presting, H. (1992). Band-gap of strain-symmetrized, short-period Si/Ge superlattices. *Phys. Rev. B* **46**, 12857–12860.

Olajos, J., Engvall, J., Grimmeiss, H. G., Menczigar, U., Gail, M., Abstreiter, G., Kibbel, H., Kasper, E., and Presting, H. (1994). Photo- and electroluminescence in short-period Si/Ge superlattice structures. *Semicond. Sci. Technol.* **9**, 2011–2016.

Olajos, J., Engwall, J., Grimmeiss, H. G., Gail, M., Abstreiter, G., Presting, H., and Kibbel, H. (1996). Confinement effects and polarization dependence of luminescence from monolayer-thick Ge quantum wells. *Phys. Rev. B* **54**, 1922–1927.

Pearsall, T. P., Bevk, J., Feldman, L. C., Bonar, J. M., Mannaerts, J. P., and Ourmazd, A. (1987). Structurally induced optical transitions in Ge-Si superlattices. *Phys. Rev. Lett.* **58**, 729–732.

Pearsall, T. P., Bevk, J., Bean, J. C., Bonar, J., Mannaerts, J. P., and Ourmazd, A. (1989). Electronic structure of Ge/Si monolayer strained-layer superlattices. *Phys. Rev. B* **39**,

3741–3757.
People, R., and Jackson, S. A. (1987). Indirect, quasidirect, and optical transitions in the pseudomorphic (4 × 4)-monolayer Si-Ge strained-layer superlattice on Si(001). *Phys. Rev. B* **36**, 1310–1313.
Rieger, M. M., and Vogl, P. (1993). Electronic-band parameters in strained SiGe alloys on SiGe substrates. *Phys. Rev. B* **48**, 14276–14287.
Robbins, D. J., Calcott, P., and Leong, W. Y. (1991). Electroluminescence from a pseudomorphic $Si_{0.8}Ge_{0.2}$ alloy. *Appl. Phys. Lett.* **59**, 1350–1352.
Satpathy, S., Martin, R. M., and van de Walle, C. G. (1988). Electronic properties of the (100)(Si)/(Ge) strained-layer superlattices. *Phys. Rev. B* **38**, 13237–13245.
Schäffler, F., Többen, D., Herzog, H.-J., Abstreiter, G., and Holländer, B. (1992). High electron mobility Si/SiGe heterostructures: Influence of the relaxed SiGe buffer layer. *Semicon. Sci. Technol.* **7**, 260–266.
Schittenhelm, P., Gail, M., Brunner, J., Nützel, J. F., and Abstreiter, G. (1995). Photoluminescence study of the crossover from two-dimensional to three-dimensional growth for Ge on Si(100). *Appl. Phys. Lett.* **67**, 1292–1294.
Schmid, U., Lukes, F., Christensen, N. E., Alouani, M., Cardona, M., Kasper, E., Kibbel, H., and Presting, H. (1990). Interband transitions in strain-symmetrized Ge_4Si_6 superlattices. *Phys. Rev. Lett.* **65**, 1933–1936.
Schmid, U., Cristensen, N. E., Alouani, M., and Cardona, M. (1991). Electronic and optical properties of strained Ge/Si superlattices. *Phys. Rev. B* **43**, 14597–14614.
Schmid, U., Humlicek, J, Lukes, F., Cardona, M, Presting, H., Kibbel, H. Kasper, E., Eberl, K., Wegscheider, W., and Abstreiter, G. (1992). Optical transitions in strained Ge/Si superlattices. *Phys. Rev. B* **45**, 6793–6801.
Schorer, R., de Gironcoli, S., Molinari, E., Kibbel, H., Kasper, E., and Abstreiter, G. (1994). In-plane Raman scattering of (001)-Si/Ge superlattices: Theory and experiment. *Phys. Rev. B* **49**, 5406–5414.
Spitzer, J., Thonke, K., Sauer, R., Kibbel, H., Herzog, H.-J., and Kasper, E. (1992). Direct observation of band edge luminescence and alloy luminescence from ultrametastable Si-Ge alloy layers. *Appl. Phys. Lett.* **60**, 1729–1731.
Splett, A., Zinke, T., Petermann, K., Kasper, E., Kibbel, H., and Presting, H. (1994). *IEEE Photon. Technol. Lett.* **6**, 59.
Stenkamp, D. and Jäger, W. (1993). Compositional and structural characterization of Si_xGe_{1-x} alloys and heterostructures by high-resolution transmission electron microscopy. *Ultramicroscopy* **50**, 321–354.
Stimmer, J., Reittinger, A., Nützel, J. F., Abstreiter, G., Holzbrecher, H., and Buchal, Ch. (1996). Electroluminescence of erbium-oxygen-doped Si diodes grown by molecular beam epitaxy. *Appl. Phys. Lett.* **68**, 3290–3292.
Sturm, J. C., Manoharan, H., Lenchyshyn, L. C., Thewalt, M. L. W., Rowell, N. L., Noël, J.-P., and Houghton, D. C. (1991). Well-resolved band edge photoluminescence of excitons confined in strained SiGe quantum wells. *Phys. Rev. Lett.* **66**, 1362–1365.
Sunamura, H., Shiraki, Y., and Fukatsu, S. (1995). Growth mode transition and photoluminescence properties of $Si_{1-x}Ge_x/Si$ quantum well structures with high Ge composition. *Appl. Phys. Lett.* **66**, 953–955.
Terashima, K., Tajima, M., and Tatsumi, T. (1990). Near-bandgap PL of $Si_{1-x}Ge_x$ alloys grown on Si(100) by MBE. *Appl. Phys. Lett.* **57**, 1925–1927.
Tserbak, C., Polatoglou, H. M., and Theodorou, G. (1993). Unified approach to the electronic structure of strained Si/Ge superlattices. *Phys. Rev. B* **47**, 7104.
Turton, R. J., and Jaros, M. (1990). Optimization of growth parameters for direct bandgap Si-Ge superlattices. *Mat. Sci. and Engineering* **B7**, 37–42.

Usami, N., Fukatsu, S., and Shiraki, Y. (1994). Photoluminescence of $Si_{1-x}Ge_x$/Si quantum wells with abrupt interfaces formed by segregant-assisted growth. Jap. J. Appl. Phys. 33, 2304.

Usami, N., Mine, T., Fukatsu, S. and Shiraki, Y. (1994). Fabrication of SiGe/Si quantum wire structures on a V-groove patterned Si substrate by gas-source Si molecular beam epitaxy. Solid State Electronics 37, 539–541.

Usami, N., Issiki, F., Nayak, D. K., Shiraki, Y., and Fukatsu, S. (1995). Enhancement of radiative recombination in Si-based quantum wells with neighboring confinement structure. Appl. Phys. Lett. 67, 524–526.

Usami, N., Shiraki, Y., and Fukatsu, S. (1996). Role of heterointerface on enhancement of no-phonon luminescence in Si-based neighboring confinement structure. Appl. Phys. Lett. 68, 2340–2342.

van de Walle, C. G., and Martin, R. M. (1986). Theoretical calculations of heterojunction discontinuities in the Si/Ge system. Phys. Rev. B 34, 5621–5634.

Vescan, L., Hartmann, A., Schmidt, K., Dieker, Ch., Lüth, H., and Jäger, W. (1992). Optical and structural investigation of SiGe/Si quantum wells. Appl. Phys. Lett. 60, 2183–2185.

Vescan, L. (1994). Selective epitaxial growth of SiGe alloys-influence of growth parameters on film properties. Mater. Sci. Eng. B28, 1–8.

Voigtländer, B., and Zinner, A. (1993). Simultaneous molecular beam epitaxy growth and scanning tunneling microscopy imaging during Ge/Si epitaxy. Appl. Phys. Lett. 63, 3055.

Wachter, M., Thonke, K., Sauer, R., Schäffler, F., Herzog, H.-J., and Kasper, E. (1992). Photoluminescence of confined excitons in MBE-grown $Si_{1-x}Ge_x$/Si(100) single quantum wells. Thin Solid Films 222, 10–14.

Weber, J., and Alonso, M. I. (1989). Near-bandgap photoluminescence of Si-Ge Alloys. Phys. Rev. B 40, 5683–5693.

Wong, K. B., Jaros, M., Morrison, J., and Hagon, J. P. (1988). Electronic structure and optical properties of Si-Ge superlattices. Phys. Rev. Lett. 60, 2221–2224.

Xiao, X., Liu, C. W., Sturm, J. C., Lenchyshyn, L. C., and Thewalt, M. L. W. (1992). Photoluminescence from electron-hole plasmas confined in $Si/Si_{1-x}Ge_x$/Si quantum wells. Appl. Phys. Lett. 60, 1720–1722.

Zachai, R., Friess, E., Abstreiter, G., Kasper, E., and Kibbel, H. (1988). Band structure and optical properties of strain symmetrized short period Si/Ge superlattices on Si(100) substrates, Proc. ICPS, 19th, Warsaw, Poland, pp. 487–490.

Zachai, R., Eberl, K., Abstreiter, G., Kasper, E., and Kibbel, H. (1990). Photoluminescence in short-period Si/Ge strained-layer superlattices. Phys. Rev. Lett. 64, 1055–1058.

CHAPTER 3

Radiative Isoelectronic Impurities in Silicon and Silicon-Germanium Alloys and Superlattices

Thomas G. Brown and Dennis G. Hall

THE INSTITUTE OF OPTICS
UNIVERSITY OF ROCHESTER
ROCHESTER, NEW YORK 14627

I. INTRODUCTORY CONCEPTS	78
1. *Isoelectronic Impurity Atoms and Complexes*	78
2. *Exciton Binding*	79
3. *Historical Perspective: Isoelectronic Impurities in GaP*	82
4. *Isoelectronic Impurities in Si: A Resume*	83
II. ISOELECTRONIC BOUND EXCITON EMISSION FROM c-Si	83
1. *Sample Preparation and Processing*	83
2. *Photoluminescence from Si:In, Si:Al, and Si:Be*	85
3. *Photoluminescence from Chalcogen-related Centers*	91
4. *Electroluminescence at Isoelectronic Centers in c-Si*	93
III. ISOELECTRONIC BOUND EXCITON EMISSION IN BE-DOPED SiGe ALLOYS: A CASE STUDY	94
1. *Photoluminescence from Thick, Be-Doped SiGe Alloys*	95
2. *Photoluminescence from Be-Doped SiGe/Si QWs*	97
3. *Beryllium Doping During Epitaxial Growth*	99
IV. DEVICE CONSIDERATIONS	103
V. CONCLUDING REMARKS	106
References	107

This chapter focuses on recent investigations of the radiative decay of excitons bound to isoelectronic impurities in Si, SiGe alloys, and SiGe/Si superlattices. Because the electron and hole that define the exciton are localized at an impurity site, recombination is not subject to the k-conservation selection rule. The radiative decay of isoelectronic bound excitons (IBEs) can therefore produce intense no-phonon lines at subbandgap photon energies. Relatively few IBE systems have been demonstrated in group IV semiconductors, so the following sections should be interpreted as describing the current status of work in progress.

I. Introductory Concepts

A survey of literature related to the properties of impurities in Si (and related group IV materials) reveals that investigations have been largely divided between shallow, substitutional dopants and deep impurities which incorporate into complexes (*Properties of Silicon*, 1988). While the former are an essential element of electronic materials and device engineering, the latter range from the "nuisance" category to the "destructive", depending on the ensuing nonradiative recombination rate. It was realized rather early in the study of solid state devices, that deep impurities (because of their localized electronic states) could act as traps for charge carriers, decreasing the speed and increasing the noise of electronic devices. The absorption of a photon having energy less than that of the host bandgap could, it was observed, release trapped charges. This observation led to the important development of deep level transient spectroscopy (Pantelides, 1986). It also pointed to a fact which was not well understood until the mid-1960s: The presence of localized impurity states can act to circumvent the momentum conservation selection rule which prevents first-order, band-to-band radiative transitions in indirect-gap semiconductors.

The observation of low-temperature radiative recombination at impurity sites both confirmed this notion and provided an important diagnostic for impurity identification. Further studies showed that an important role was played by the radiative recombination of excitons bound to impurities (Davies, 1989).

A number of groups have, in recent years, engaged in a re-examination of impurity-assisted radiative recombination for the purpose of achieving efficient light emission in a group IV (Si or SiGe) host. Considerable attention has been given to rare earth dopants such as Erbium (Ennen *et al.*, 1983), a topic which is presented elsewhere in this collection. In this chapter we review the concept behind, and the experimental investigation of, radiative recombination at isoelectronic impurities in crystalline Si (c-Si), SiGe alloys, and SiGe/Si superlattices.

1. Isoelectronic Impurity Atoms and Complexes

An isoelectronic impurity is, in its simplest substitutional form, isovalent with the host crystal. Examples of substitutional isoelectronic atoms for a group IV material such as Si would be any other element in the column: C, Ge, and Sn may each substitute for Si without influencing the number of valence electrons. They are electrically inactive in the sense that they do not influence the carrier density in the conduction or valence bands and therefore have a weak influence on the conductivity. Such impurities perturb the crystal in two ways:

1. The core charge distribution is different from that of the host, resulting in a strong, short-range potential difference between the impurity and host.
2. Any size difference between the impurity and the host produces a relaxation of atoms in the surrounding lattice sites.

Both of these effects are short range, and the defect (within the context of the effective mass approximation) has the character of a three dimensional potential well or barrier. A charge carrier (such as an electron in the conduction band) will therefore experience scattering and/or trapping, depending on the magnitude and sign of the potential.

The concept of an isoelectronic impurity as a three dimensional potential well can, despite its simplicity, be instructive. For instance, while a one- or two dimensional well always exhibits a bound state, a three dimensional well does not. In Si, for example, the "point defect" substitutional impurities C, Ge and Sn do not produce bound states (Baldereschi and Hopfield, 1972). However, the substitution of nitrogen for phosphorous in GaP provides a perturbation which is sufficient to trap charge carriers. The latter was the subject of many investigations aimed at the development of efficient, visible semiconductor light emitters (Thomas, Gershenzon, and Hopfield, 1963; Thomas and Hopfield, 1966; Trumbore, Gershenzon, and Thomas, 1966).

An isoelectronic impurity need not however be a simple substitution. A multiple-atom complex, if it is in a configuration with no dangling bonds, may be isoelectronic. In practice, a defect is considered isoelectronic if the threshold for ionization (the energy required to ionize an electron into the conduction band or a hole into the valence band) is comparable to the bandgap energy. Thus, for most temperatures of interest, the isoelectronic impurity adds no free carriers to the host. The total number of free carriers may be reduced, however, through confinement in the short range potential. In this way, an isoelectronic impurity may appear to compensate a lightly doped semiconductor over a limited temperature range, but will not, in general, increase its electrical conductivity.

2. Exciton Binding

An exciton is an electron–hole pair in a state which is hydrogen-like about the center of mass of the pair. An exciton may either exist as a free quasiparticle, in which the pair moves freely about the crystal, or as a localized pair in which one or both of the particles is trapped at a defect. In indirect-gap materials such as Si and Ge, the free exciton obeys optical selection rules similar to that of the free, band-to-band selection rules — the requirement that the electron and hole possess the same crystal momentum (Callaway, 1974). However, if one or both particles are bound to an impurity, the selection rule relaxes with the increased spatial confinement of the particles.

The historical significance of isoelectronic impurities (with regard to the optical properties of indirect-gap semiconductors) lies in the discovery in 1963 of their ability to bind excitons, resulting in efficient radiative recombination of electrons and holes. Most models consider the primary binding mechanism for exciton formation to be the Coulomb attraction between the electron and hole, and for excitons with weak binding energies (these are cases for which the effective mass approximation is valid for one of the particles) this argument appears to be valid. There may be cases (isoelectronic donor-acceptor pairs, for example) in which both electron and hole exist in tightly bound states of close proximity and the exciton binding is not simply described by the electron–hole Coulomb interaction.

Ordinary donor and acceptor impurities may bind excitons (Dean, Haynes, and Flood, 1967). A neutral donor may bind a hole, which subsequently traps an electron. Likewise, a neutral acceptor can trap an electron, which subsequently binds a hole. Such complexes are known to exhibit very rapid nonradiative Auger processes (Lyon *et al.*, 1977; Nelson *et al.*, 1966), in which the energy from the recombination of two particles is imparted to the third carrier in the form of kinetic energy. To avoid the Auger process, the ionization energy of the third particle must exceed the recombination energy. Since the recombination energy is generally a significant percentage of the bandgap, the neutral impurity must be isoelectronic in nature.

A bound exciton may dissociate either by ionization (in which one particle scatters from the bound state to a band state while the other remains trapped), or by escape, in which the exciton leaves the impurity site while maintaining a bound pair. One must take care, therefore, to distinguish between the spectroscopic localization energy (the energy difference between the band edge and the optical transition) and the binding energy (the minimum energy for bound exciton disassociation). Figure 1 illustrates the various energy levels associated with a bound exciton complex.

The binding energy of an exciton can be modified by confinement of one or both particles in a quantum well (QW) (People, 1977). Quantum wells (QWs) are generally distinguished by the character of the conduction and valence band offset between the well and barrier materials. Figure 2 compares type I QWs (which confine both electrons and holes) with type II

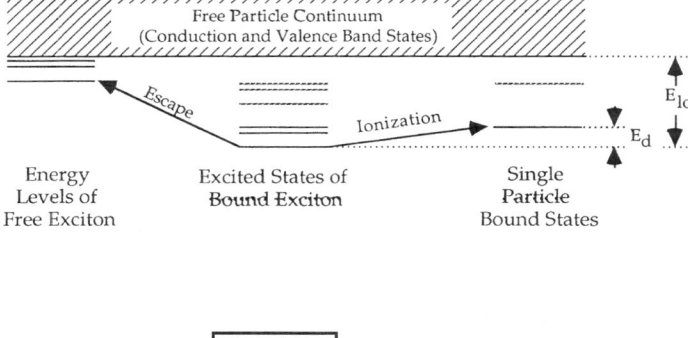

FIG. 1. Energy levels of a bound exciton complex. Each single-particle bound state will, in principle, split into several exciton states. E_{loc} denotes the spectroscopic localization energy while E_d indicates the dissociation energy.

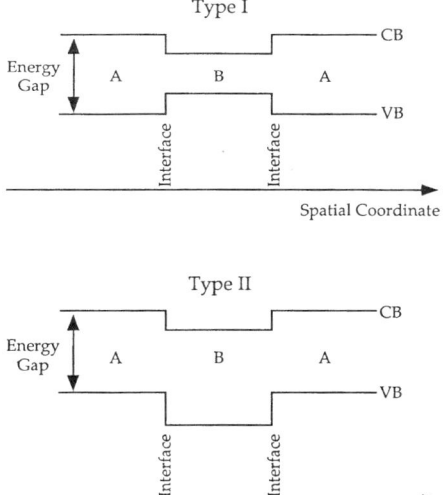

FIG. 2. Comparison of band offsets in type I and type II quantum wells. Two different semiconductors, here labeled A and B, are joined at the interfaces to form a double heterostructure. Energy is implied along the vertical direction; CB identifies the bottom of the conduction band; VB denotes the top of the valence band. In type I alignment, the band edges of B are within those of A, in contrast to type II alignment.

QWs (which confine only one type of carrier). SiGe alloys grown on Si substrates exhibit type I alignments (Bevk et al., 1986; People, 1977), with the largest band offset attributed to the valence band. An IBE inside such a well is therefore expected to exhibit an increase in the binding energy and an accompanying increase in the temperature at which the IBE begins to thermally disassociate.

3. Historical Perspective: Isoelectronic Impurities in GaP

The early success in the fabrication of optoelectronic devices in the III–V material systems coupled naturally to a strong industrial interest in efficient, visible light emitting diodes (LEDs) for display device applications. GaAs exhibits too small a bandgap for visible light emission ($E_g \approx 1.4\,\text{eV}$) but the ternary compound $GaAs_{1-x}P_x$ exhibits a bandgap which monotonically increases with x, yielding a bandgap of 2.3 eV at $x = 1$. The band structure undergoes a transition from direct to indirect at $x \approx 0.4$. For display applications, the bandgap should be as large as possible, with particular interest in pure GaP, which has a bandgap very close to the peak of the human photopic response.

Thomas, Gershenzon, and Hopfield (1963), in the course of investigating donor-acceptor pair recombination in GaP, observed a pair of fluorescence lines together with a series of lower energy transitions which remained unidentified for three years. Explanations for the observed emission included such suggestions as the existence of a low-energy direct bandgap exciton, and the possibility of a three-particle interaction at a neutral acceptor. Thomas and Hopfield (1966) were able to successfully correlate the absorption with intentional nitrogen doping. It was supposed that the nitrogen atoms were occupying isovalent phosphorus sites, and the upper state of the transition was described as that of a bound exciton. Lower energy transitions were explained as originating from nitrogen pairs of varying separation.

The intentional introduction of nitrogen moved GaP from a laboratory curiosity to a serious candidate for the fabrication of visible LEDs, and further investigations yielded evidence for IBE emission at GaP:Bi (Trumbore, Gershenzon, and Thomas, 1966), GaP:Zn-O (Morgan, Weber, and Bhargava, 1968), and GaP:Cd-O (Morgan, Weber, and Bhargava, 1968). The latter were characterized by large exciton binding energies, yielding red emission spectra at room temperature and peak external efficiencies approaching 10%.

Although the efficiency of the recombination was well established in the 1960s, it was not until 1971 that two independent research groups reported the first observation of stimulated emission in indirect semiconductors.

Nahory, Shaklee, and Leheny (1971) reported the observation of gain from bound excitons in GaP:Bi, while almost simultaneously Holnyak et al. (1972) reported stimulated emission in GaP:N and GaP:Zn-O. In addition, the latter group reported multimode laser action in a platelet cavity with an optical pump.

4. Isoelectronic Impurities in Si: A Resume

Isoelectronic bound exciton emission was first observed in c-Si by Vouk and Lightowlers (1977) and identified as such by Mitchard et al. (1979). By their very character, isoelectronic impurities are not easily detected electrically, and are most easily detected in photoluminescence (PL) spectroscopy. This technique had not previously been heavily used in Si, perhaps due to the lack of sensitive detectors in the $\lambda = 1-2\,\mu$m region of the infrared spectrum. Since the substitutional impurities Ge, C, and Sn do not appear to bind excitons, the first isoelectronic impurities observed and identifed in Si were made up of impurity complexes, or clusters, within the Si lattice and were typically present in extremely low concentrations. It is often difficult to make a positive identification of all the constituents in a complex, but the centers may be categorized by the elements that appear to activate the optical centers. A rather complete review of impurity-based transitions, including IBE transitions, can be found in Davies (1989), and a summary of early IBE studies can be found in Sauer and Weber (1983).

II. Isoelectronic Bound Exciton Emission from c-Si

In this section, we provide some experimental details and describe several families of isoelectronic complexes in c-Si. Since our emphasis is on candidates for an impurity-based laser or room-temperature light emitter, attention will be given to: (1) the temperature of peak emission; (2) the radiative lifetime of the bound exciton; (3) the strength of the coupling to local vibrational modes; and (4) the maximum achievable concentration of the constituents. Together, these have a profound influence on the room temperature radiative efficiency, the peak stimulated emission cross section, and the gain per unit length achievable in an optical cavity.

1. Sample Preparation and Processing

The introduction of isoelectronic complexes into c-Si remains an art — after sixteen years of research, the formation kinetics for most centers are still unknown. Certain generic rules of doping apply to isoelectronic complexes:

1. Complexes that form as interstitials or substitutional-interstitials can be easily introduced by diffusion, but the maximum concentration of centers is often severely limited by the solid solubility of one or more constituents.
2. Impurities introduced by ion implantation require a subsequent high temperature (> 900°C) anneal to remove the resulting damage. An exception to this is the Si:Be center, which appears to require some lattice damage to assist in the formation.
3. Impurities introduced during epitaxial growth by an MBE or CVD process cannot be contaminants for nearby electronic components if the light-emitting device is to be very large scale integration (VLSI)-compatible.

It is common for an isoelectronic complex to require a two-stage sample treatment in order to observe IBE. First, the constituents must be introduced into the sample. In most cases, simply introducing the constituents does not activate the center, and a second treatment is required for the center to form. This is somewhat analogous to the formation of a gaseous molecular complex in a reactor—the reactants must be introduced along with a catalyst or heat source which thermodynamically favors the product.

While these general principles can be easily understood, the practice of sample preparation by either diffusion or ion implantation is fraught with uncertainty. A common problem is the lack of complete identification of the constituents of the center—the identification is often left to trial and error sample processing. In the following paragraphs, we list several experimental procedures found to favor the formation of isoelectronic complexes:

a. Heat Treatment/Rapid Quenching

There are several isoelectronic complexes, including both the chalcogen and transition-metal complexes, that require a short heat treatment at a temperature greater than 1000°C followed by a rapid quench. These so-called "quenched-in" defects are generally stable at room temperature, but dissociate under fairly modest annealing temperatures (> 300°C). The quenching must be so rapid that air or liquid nitrogen quenching is insufficient. Water quenching is generally too rapid—even small samples will shatter under the thermal shock. Successful quenching procedures usually involve either immersion in a 50:50 mixture of ethylene glycol and water or an oil bath.

While such a treatment is difficult to employ without risking the introduction of extra contaminants, more controlled quenching treatments such as pulsed laser annealing have failed to achieve equivalent activation of the complexes.

b. *Annealing in the 400–600°C Temperature Range*

There are several complexes whose formation requires a slow anneal at moderate temperatures. Examples include a rather elusive oxygen-related center and the well-studied Si:Be center. Since this is a temperature range that favors vacancy formation as well as the migration of certain interstitials, it is likely that centers of this sort require a high vacancy concentration for formation and involve both substitutional and interstitial impurities.

2. PHOTOLUMINESCENCE FROM Si:In, Si:Al, AND Si:Be

These are centers that exhibit efficient, low-temperature luminescence, extremely sharp spectral features, and comparatively weak coupling to local vibrational modes. As a result, no-phonon emission lines dominate the spectral features.

a. *Si:In*

Figure 3 shows the PL spectrum of a Si:In sample formed by an implantation of In$^+$ at a dose of 10^{13} ions/cm^2 and an implantation energy of 200 keV. Samples were typically annealed at $T = 1200°C$ under various atmospheres in sealed quartz ampoules. The spectrum shows the no-phonon P-line, following the labeling of Mitchard *et al.* (1979) near wavelength $\lambda = 1.11\,\mu m$ attributed to the radiative decay of an exciton bound to an isoelectronic impurity complex. While the IBE emission is detectable under moderate cooling, a short heat treatment and rapid quench is known to increase the luminescence intensity by as much as a factor of 10,000 (Thewalt, Ziemelis, and Parsons, 1981; Weber, Sauer, and Wagner, 1981). Radiative IBEs can also be excited electrically. Figure 3 shows the corresponding electroluminescence (EL) obtained from Si:In. The details of the EL structure are discussed later in this chapter.

Early measurements of the lifetime and temperature dependence of the lines indicated that the major transitions originated from the same excited state, and ended in any of several low-energy states of the defect (Mitchard

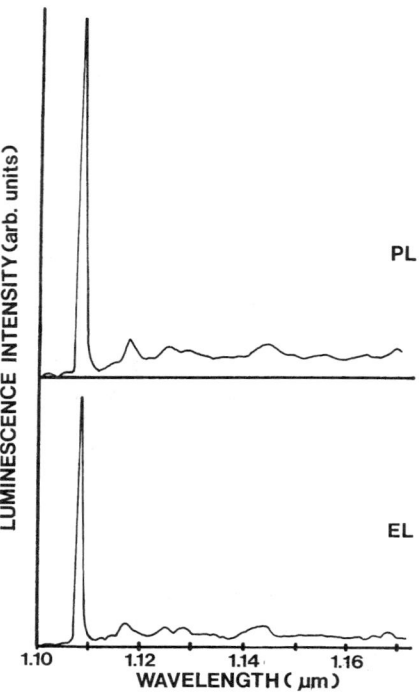

FIG. 3. Photoluminescence and EL spectrum of quenched Si:In at sample temperature $T = 14\,\text{K}$. The strong PL no-phonon line is termed the P line.

et al., 1979; Sauer, Schmid, and Weber, 1978; Thewalt, Ziemelis, and Parsons). These were evidently not states of different total angular momentum, as the lifetimes of all three major lines were very similar ($\approx 200\,\mu s$ at $T = 20\,\text{K}$). Photoluminescence excitation (PLE) spectra taken by Wagner, Weber and Sauer (1981) identified the strongest lines of the system as a no-phonon transition (the "P" line) and its Stokes sideband, the latter transition ending in an excited vibrational mode of the bare impurity ground state. Excited states were observed in the PLE experiments that had not been observed in PL spectra. These states bore a resemblance to the states of a particle bound in a three dimensional square-well potential, a result consistent with the short-range nature of an isoelectronic defect.

The precise identification of the constituents has remained an unsolved problem (Ziemelis, Parsons, and Voos, 1979). Apart from the obvious connection with indium, suggestions have included oxygen (Brown and Smith, 1980), carbon (Ziemelis, Parsons, and Voos, 1979), hydrogen

(Thewalt, Ziemelis, and Parsons, 1981), iron (Weber, Sauer, and Wagner, 1981), and phosphorous. In each of these cases, investigators claimed enhancement of the luminescence signal by intentional doping with that element, but very few cases yielded any corroboration of the reports. Brown and Bradfield (1988) published experimental results that rather convincingly showed the importance of oxygen in the activation of isoelectronic centers in implanted samples, but no such strong correlation has been made for samples having indium incorporated in the growth.

An obvious candidate for isoelectronic pairing would be substitutional nitrogen. Nitrogen shows extremely low solubility at substitutional sites (*Properties of Silicon*, 1988), but could form a nearest-neighbor isoelectronic molecule. Despite some effort, no correlation with nitrogen has yet been reported. However, the following section describes a somewhat similar center involving aluminum that shows strong evidence of pairing with nitrogen.

b. Si:Al-N

Figure 4 shows a PL signature similar to that first reported by Weber *et al.* (1981) who first assigned it to an unknown isoelectronic complex (Weber, Schmid, and Sauer, 1980). The center shows the usual characteristics of IBE emission: a sharp, no-phonon line at low temperatures having a radiative lifetime in the microsecond to millisecond range. High resolution, low temperature spectroscopy showed the no-phonon transition to be a doublet, with a dipole-forbidden transition at the lowest-energy excited state.

Modavis and Hall (1990) carried out a series of experiments that demonstrated the incorporation of nitrogen in the aluminum center and emphasized the difficulty of activating such centers. Samples implanted with Al or N or both Al and N were annealed for 120 h at 450°C following the implantation. A weak IBE signature was obtained with the single-species Al implantation, but a dramatically increased IBE emission was observed in the co-implanted samples, with strong growth continuing for anneals as long as 250 h (Fig. 5).

c. Si:Be

Henry *et al.* (1981) proposed and confirmed that a small double-acceptor such as beryllium could form substitutional-interstitial pairs in such a way as to allow both atoms to occupy the same site and form an isoelectronic molecule. The pair concentrations were sufficient to enable the first absorption measurements at isoelectronic impurity centers

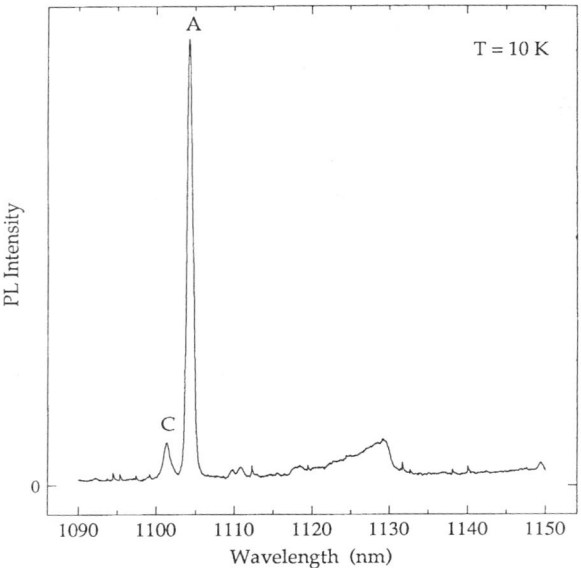

FIG. 4. Photoluminescence from c-Si co-implanted with aluminum and nitrogen, then annealed for 120 h at $T = 450°C$.

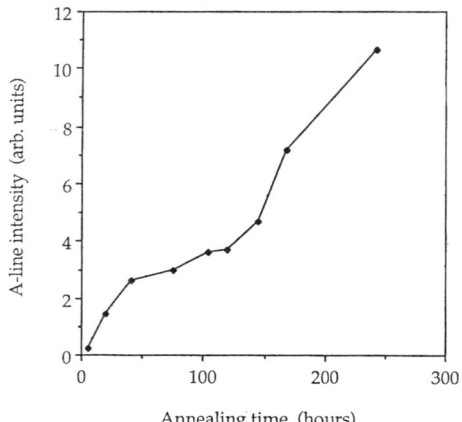

FIG. 5. Growth of IBE emission in Si:Al-N with increasing annealing time at annealing temperature $T = 450°C$.

in Si (Thewalt et al., 1982). Using the absorption data and associated lifetime measurements it was established that the density of optical centers introduced by diffusion exceeded 10^{16} cm^{-3} and that the oscillator strength of the major transition was 1.5×10^{-3}. Zeeman splitting, uniaxial stress measurements, and hydrostatic pressure measurements showed the center to be axially symmetric along the (100) direction and clarified other features of the impurity complex (Henry et al., 1984; Killoran et al., 1982; Kim et al., 1995; Kim et al., 1996).

Figure 6 shows the low-temperature PL spectrum for Si:Be. The "A" line is the lowest energy dipole-allowed no-phonon line. Beryllium can be introduced by diffusion (Henry et al., 1981), ion implantion (Brown et al., 1987; Hall, 1987), or during epitaxial growth (Moore et al., 1994; Moore et al., 1995). Successful incorporation by diffusion requires the formation of a eutectic ($T \approx 1300°C$) on the surface. This can be followed by a "sandwich" diffusion ($T \approx 1100°C$) in which two eutectic surfaces are held in contact during the diffusion.

No further annealing is required for Be–Be pair formation—upon cooling to a few tens of degrees Kelvin, a very strong PL signal is observed near the wavelength 1.149 μm, the external efficiency is found to be approximately 1%, which is an unusually intense emission line for c-Si. Crouch et al. (1972) described resistivity and Hall measurements on samples prepared in this manner. Noticeably less electrical activity was observed in Si:Be than that expected from concentration measurements, and it was suggested that as much as 90% of the available beryllium in such samples existed in electrically inactive (isoelectronic) complexes.

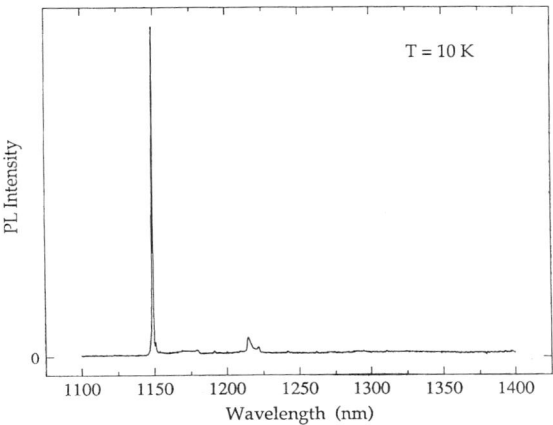

FIG. 6. Photoluminescence spectrum showing the primary ("A") transition in Si:Be.

The first observation of PL from implanted beryllium was made by Gerasimenko et al. (1985), in which the authors reported weak IBE emission similar to that observed in diffused samples. Similar experiments were carried out by the Rochester group (Modavis et al. 1990), in which the annealing behavior of the centers was explored in considerable detail. In contrast to samples prepared by diffusion, implanted samples show no IBE activation by high temperature annealing or quenching. However, the IBE can be activated by annealing for 10–30 min at temperatures in the neighborhood of $T = 500°C$. Figure 7 shows an isochronal annealing curve for Si:Be prepared by an implantation dose of 10^{14} ions/cm² at an energy of 200 keV.

The disappearance of IBE emission at annealing temperatures above $T \approx 600°C$ is irreversible. One cannot carry out a high temperature anneal/diffusion and subsequently recover the PL via a low temperature anneal. It is therefore probable that defects resulting from the implantation damage facilitate the formation of pairs, making the formation kinetics quite different from pair formation in diffusion.

Two additional methods of Si:Be preparation are worthy of mention. One can combine implantation and annealing by performing a single implantation with the wafer held at a temperature of $T = 500°C$ or greater. This yields luminescence intensities comparable to the two-step implant/anneal process. Finally, Be can be incorporated directly in a suitable epitaxial growth, as discussed in Section III.

d. *Influence of Temperature on Efficiency and Lifetime*

Both Si:Al-N and Si:In show an unremarkable decrease in both PL and

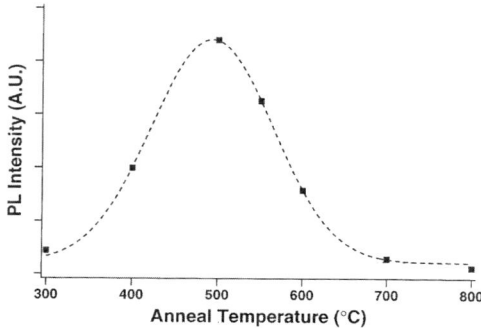

FIG. 7. Photoluminescence intensity under isochronal annealing (30 min) for Si:Be. The dashed line is a fit to a Gaussian, intended as a guide to the eye.

decay time at temperatures above $T = 25$ K. The Si:In center shows a complex low-temperature behavior (which is enhanced in EL) that has been linked to an excitation transfer from nearby acceptors.

The PL from implanted Si:Be peaks near $T = 30$ K, and is nearly undetectable at liquid nitrogen temperatures. (Note: A plot of the PL intensity versus temperature for a Be-doped SiGe alloy appears in Fig. 16. The behavior is very nearly the same for Be-doped Si.) The radiative lifetime associated with the dipole-allowed bound-exciton transition is generally considered to be in the $5-10\,\mu s$ range, and the PL decay time decreases exponentially as the sample temperature increases above $T = 40$ K.

3. PHOTOLUMINESCENCE FROM CHALCOGEN-RELATED CENTERS

Following the success of the heat-treatment/rapid-quench activation of such centers as Si:In (Thewalt, Ziemelis, and Parsons, 1981) and Si:Cu (Weber, Bauch, and Sauer, 1982), Brown and Hall (1986a), and Bradfield, Brown, and Hall (1988) observed that a similar treatment could activate isoelectronic bound excitons in Si:S and Si:Se. While these centers were unquestionably associated with the chalcogens and could be introduced either by diffusion or ion implantation, identification of other constituents has proceded much more slowly.

Figure 8 shows the PL from Si:S. This center is remarkable in several respects. Upon activation of the impurity center, two distinct luminescence systems are evident which peak at different temperatures. The spectrum at temperatures less than 40 K is dominated by transitions near $\lambda = 1.5\,\mu m$, with the strongest (no-phonon) line lying at $\lambda = 1.5071\,\mu m$. This system shows relatively weak coupling to the local vibrational modes and reaches its maximum intensity near $T = 35$ K. In contrast, the higher energy transitions near $\lambda = 1.3\,\mu m$ show strong coupling to the local phonons and a dramatic increase in integrated intensity from $T = 40$ K to 70 K. The low-temperature sharp-line spectrum completely broadens out to a half-width of 120 nm near liquid nitrogen temperatures. Figure 9 illustrates the temperature dependence of the relative PL intensities of the two systems. It was shown by Singh, Lightowlers, and Davies (1989) that the two systems belonged to two distinct configurational arrangements of the same complex, and that the precise arrangement in the temperature range $T = 30-60$ K depended heavily upon the illumination conditions during cooling. Beckett, Nissen, and Thewalt (1989) carried out further PL and excitation spectroscopy studies of Si:S and confirmed that the two systems are indeed part of the same complex.

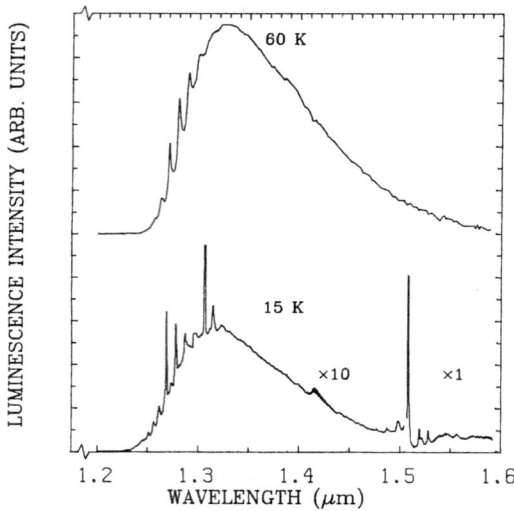

FIG. 8. Photoluminescence spectrum of heated and quenched Si:S at sample temperatures $T = 15\,\text{K}$ and $T = 60\,\text{K}$.

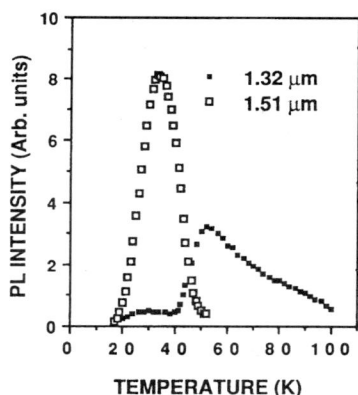

FIG. 9. Temperature dependence of the peak intensity of the two configurations of the isoelectronic center in Si:S.

When Si:Se is prepared under similar conditions, an isoelectronic complex is formed which closely follows that of Si:S. Figure 10 compares the PL spectra obtained from the sulfur and selenium centers in Si (Bradfield, Brown, and Hall, 1989). In practice, selenium is more difficult to introduce. It suffers both from a lower solid solubility and, being a heavier atom, has

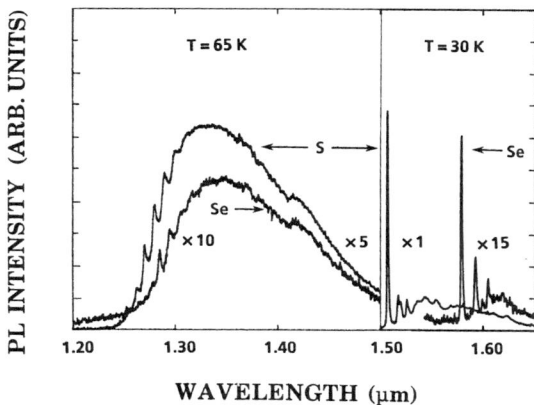

FIG. 10. Comparison of PL spectra from the sulfur- and selenium-related isoelectronic impurities in c-Si.

a very shallow range when introduced by ion implantation. It is therefore necessary to provide a relatively long, high temperature anneal (in this case, $T = 1200°C$ for 70 min) to allow the implanted selenium to diffuse over a volume amenable to PL measurements.

4. Electroluminescence at Isoelectronic Centers in c-Si

Isoelectronic bound exciton emission can be readily excited by electrical excitation. The earliest experiments used simple metal-semiconductor contacts on Si:In samples that had been heated, quenched, and electrically contacted to a copper cold finger using silver paint, which dried under vacuum (Brown and Hall, 1986a). A painted-on top contact was applied, and EL was observed in edge emission. A typical current-voltage curve is shown in Fig. 11. The low temperature current voltage curve demonstrates that carrier freeze-out at low temperature produces rather non-ideal injection conditions. Despite this, a careful comparison of edge-emitting EL with PL (made on the same sample prior to electrode application) showed the EL process having an efficiency as high as 40% that of the PL process.

Junction-based EL experiments were reported by Bradfield, Brown and Hall (1989) on n/n^+ epitaxial Si waveguides with a shallow implanted layer of boron. The junctions, which were also co-implanted with sulfur and oxygen, showed sulfur-related EL similar to the Si:S IBE emission observed

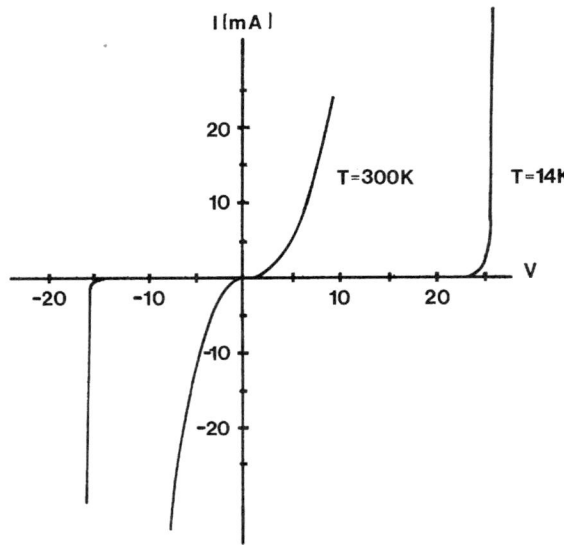

FIG. 11. Typical current voltage curve for Si:In EL sample. The maximum luminescence at wavelength $\lambda = 1.11\,\mu$m is observed just prior to avalanche breakdown. (The EL spectrum for Si:In appears in Fig. 3.)

in PL. The external efficiency from these junctions was shown to be in the 0.2–0.5% range, an order of magnitude below the PL efficiency.

III. Isoelectronic Bound Exciton Emission in Be-doped SiGe Alloys: A Case Study

Because of the importance of heterostructures in optoelectronic devices, it is interesting to inquire whether isoelectronic centers can be formed in group IV alloys and superlattices. The SiGe system is better understood and can be grown on either Si substrates (using Si-rich alloys) or Ge substrates (for Ge-rich alloys). Since (in either case) it represents a strained layer growth, the defect density grows with increasing layer thickness.

Several investigators have explored radiative mechanisms in SiGe alloys and superlattices. Indeed, early investigations focused on the hope that zone-folding in ultrashort-period Si-Ge superlattices would produce an ordered alloy with a direct bandgap. It now seems to be the case that the energy gap in such structures is only quasi-direct, with an oscillator strength that is still too small to be the basis for room-temperature light-emitter technology. Sturm et al. (1991) reported the observation of quantum-confined free excitons in SiGe QWs the features of which attested to the high quality of the epitaxial material. Other investigations examined introducing radiative impurity complexes into SiGe alloys and QWs, as will be discussed in other chapters in this volume. This chapter confines itself to one particular mechanism: the radiative decay of excitons bound to isoelectronic impurity complexes. An extensive study of the influence of the group IV host environment on the Be-related IBE emission was carried out by the Rochester group (Modavis et al. 1990; Modavis et al. 1991; Moore et al. 1994; Moore et al. 1995). Beryllium represents a useful case study, since the make-up of this impurity complex is well understood, and since it can be introduced either by implantation/annealing or incorporated during growth. Further, it exhibits an extremely narrow no-phonon line in a homogeneous host — any spectral broadening can therefore be associated with site inhomogeneities.

1. PHOTOLUMINESCENCE FROM THICK, Be-DOPED SiGe ALLOYS

The samples used to investigate the properties of radiative, isoelectronic Be impurities in thick SiGe alloys consisted of layers, typically 0.3–0.5 μm thick, grown by molecular beam epitaxy (MBE) onto (100) Si substrates. The details of the growth process can be found in Bevk et al. (1986). The fractional Ge content of the Si-rich alloys included pseudomorphic layers with Ge content $x = 0.02$, 0.05, and 0.08 and partially relaxed layers with Ge content $x = 0.13$ and 0.20. The results of Rutherford backscattering measurements suggested the crystal quality of all the samples was excellent.

Each of the alloys, along with a x-Si control sample, was implanted with a dose of 2×10^{13} ions/cm^2 at an energy of 40 keV. Some samples were given a high temperature ($T = 500°$C) implant while others were implanted at ambient temperature and annealed for 10 min in the range $500°$C $< T < 600°$C. Figure 12 shows the resulting PL spectra over the wavelength range 1.1 μm $< \lambda <$ 1.4 μm. The top spectrum is that of the Si:Be no-phonon line ($\lambda = 1.15 \mu$m) and a transverse optic (TO) phonon replica ($\lambda = 1.215 \mu$m), included for reference. The no-phonon emission peak is clearly shifted to longer wavelengths with increasing Ge content in the alloy. This wavelength shift is nearly linear over the composition range, with the energy-shift varying at the approximate rate 6 meV/atm%). The emission also broadens considerably with increasing Ge fraction. For $x = 0.20$, Be-related PL was observed only for samples implanted at $T = 500°$C; no

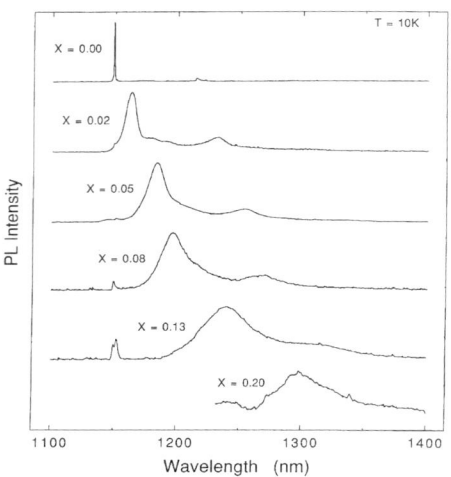

FIG. 12. Measured PL spectra for several thick, Be-implanted, Si-rich $Si_{1-x}Ge_x$ alloys grown by molecular beam epitaxy onto (100) Si substrates. The top spectrum is that for pure Si ($x = 0$). The alloy thicknesses ranged from 0.3–0.5 µm.

Be-related PL was observed for samples implanted at ambient temperature, although those samples did exhibit previously reported dislocation lines near the energies 0.82 and 0.88 eV (Tajima and Matsushita, 1983).

Modavis et al. (1990) observed that the spectroscopic localization energy (as measured between the transition energy and the bandgap of a strained alloy) decreased somewhat with increasing Ge fraction. It is not clear whether this effect is due to a higher-than-expected bandgap of the host layer (due, perhaps, to relaxation of the lattice under implantation and annealing) or a reduced binding energy. The thermal decay was shown to behave similarly for all alloys implanted at room temperature, indicating no significant reduction in thermal binding energy. They also observed a correlation between the luminescence efficiency, the temperature-dependence of the intensity of the measured PL, and the defect density in Be-doped alloys (Modavis et al., 1990). While the configuration of the center is rather simple, the formation kinetics are not. Isoelectronic centers in implanted Si:Be evidently require some defect damage for formation, since high temperature annealing irreversibly removes the luminescent centers. The same appears to be true for Be-doped SiGe alloys, with the added complication that a strained alloy exhibits a defect density that increases with layer thickness and lattice mismatch.

2. PHOTOLUMINESCENCE FROM Be-DOPED SiGe/Si QWs

In Si and related group IV materials, the beryllium pair behaves as an isoelectronic acceptor, in which the exciton can be thought of as the combination of a hole, weakly bound in the Coulomb field of the more tightly bound electron. It is interesting to ask to what degree the presence of a QW can modify the properties of such an exciton via quantum confinement. The conduction band offset in a Si/SiGe/Si QW is very small for Si-rich alloys (People, 1977). The full band offset resides almost entirely in the valence band, suggesting these QWs will confine holes more effectively than electrons for all excitons, either free or bound. One advantage of the isoelectronic bound exciton is that with the impurity complex introduced into the well, the electron will be localized (trapped) by the impurity and the hole will be localized by the quantum-well structure, a type of cooperative quantum confinement.

A simple, single-particle, single-impurity model rather successfully describes the physics of the combined Coulomb/well trapping (Modavis et al., 1991). Figure 13 shows the binding energy versus QW width for a family of barrier heights, with the impurity located at the center of the well. The

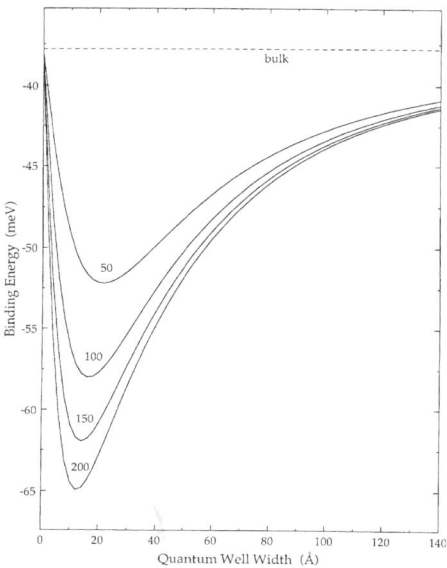

FIG. 13. Binding energy vs. quantum-well width for four different barrier heights, labeled in meV near each curve.

horizontal dashed line represents the binding energy for a bare impurity in a bulk host material. The largest enhancements are predicted for rather narrow well widths and large barrier heights (this is the limit in which the hole becomes highly localized).

If the impurities are introduced into a Si/SiGe multi-quantum-well system by diffusion or ion implantation, they will not be confined to the center of or even within the wells (alloy layers), but will be distributed relatively uniformly throughout both SiGe wells and Si barriers. Because the exciton binding energy varies considerably with impurity location, the broad spatial distribution of the impurities produces an inhomogeneously broadened and shifted bound-exciton emission spectrum. This effect is illustrated in Fig. 14, which compares the measured bound-exciton emission from two samples implanted with Be: a 0.5 μm-thick layer of $Si_{0.92}Ge_{0.08}$ alloy (dashed line) and a $Si_{0.92}Ge_{0.08}$/Si superlattice (solid line) with 50 Å-thick alloy wells and 100 Å-thick. Both the shift and the broadening are quite evident. The effect of the spatially distributed Be impurities on the IBE emission can be modeled rather well using a Coulomb/well model that allows the impurity location to vary randomly throughout the well and barrier regions in a superlattice. Figure 15 compares the measured PL lineshapes obtained from a pair of 8% SiGe superlattices with the theoretical prediction. (The low-energy deviation from the theory is due to a TO phonon replica not included in the model.) Both $Si_{0.92}Ge_{0.8}$/Si superlattices have 100 Å-thick Si barrier layers, but one has 50 Å-thick alloy wells and the other has 100 Å-thick wells. The theory describes the inhomogeneous broadening of

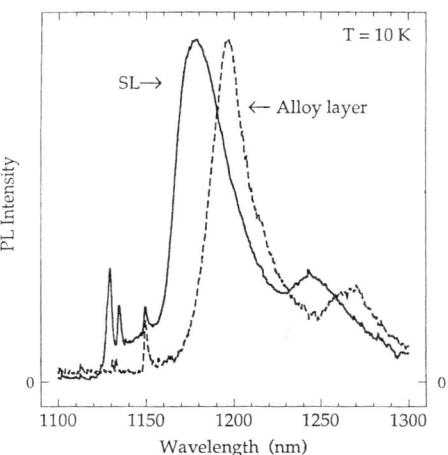

FIG. 14. Measured low-temperature PL spectrum for a thick, Be-implanted $Si_{0.92}Ge_{0.08}$ alloy and a Be-implanted $Si_{0.92}Ge_{0.08}$/Si superlattice (solid curve) with 50 Å-thick SiGe quantum wells and 100 Å-thick Si barriers.

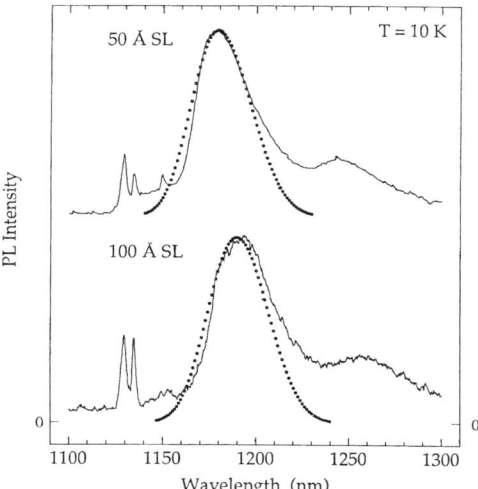

Fig. 15. Comparison of measured (solid lines) and calculated (dashed lines) PL lineshapes for $Si_{0.92}Ge_{0.08}$/Si superlattices with 100 Å-thick Si barriers and two $Si_{0.92}Ge_{0.08}$ quantum-well thicknesses: 50 Å (upper plots) and 100 Å (lower plots).

the no-phonon line rather well, and suggests that by properly controlling the impurity placement during growth (e.g., δ-doping), one can tailor the spectral width and location of the IBE transition.

A measurement of the luminescence intensity versus temperature shows the impact of the increased effective binding energy. Figure 16 compares the temperature behavior of the peak PL intensity for several superlattices and a thick 8% SiGe alloy. For the superlattices, the measurements were made for the emission wavelengths corresponding to impurity positions that produce the maximum predicted confinement. There is a clear increase in the temperature to which the luminescence persists, indicating that careful optimization of both the growth and doping could yield even more significant gains.

3. BERYLLIUM DOPING DURING EPITAXIAL GROWTH

Controlling the placement of Be impurities within a superlattice is more easily accomplished during sample growth, when one has the chance to minimize the spread of the impurities by regulating the rate at which Be atoms are delivered to the sample. Some degree of impurity migration is in all likelihood inevitable, of course, for the range of ambient temperatures used for epitaxial growth. True delta doping will be difficult to achieve.

Moore *et al.* (1994) reported the results of investigations of the behavior of Be-related bound-exciton emission from MBE-grown Si/SiGe super-

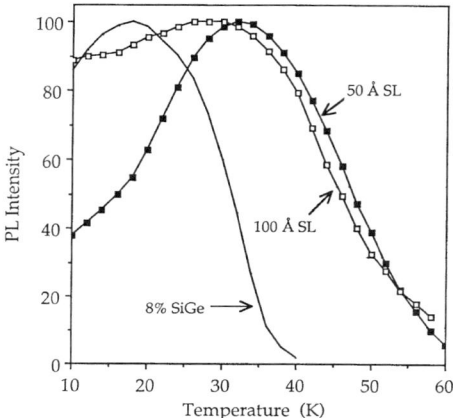

FIG. 16. Temperature dependence of the Be-related isoelectronic bound exciton emission from a thick $Si_{0.92}Ge_{0.08}$ alloy layer and from two $Si_{0.92}Ge_{0.08}/Si$ superlattices with 100 Å-thick Si barriers and either 50 Å- or 100 Å-thick $Si_{0.92}Ge_{0.08}$ quantum-wells. All three samples were grown by molecular beam epitaxy.

lattices. Be was introduced during MBE growth by using a shuttered boron nitride effusion cell containing approximately one gram of solid 99.999%-pure Be. Several Be-cell temperatures and Be-outgassing procedures were attempted before finding a set of Be-deposition parameters that produced acceptably reproducible results. Figure 17 demonstrates the success of the method by showing, for sample temperature $T = 11$ K, the PL spectrum for a 50-nm-thick layer of *in situ* doped Si grown on an undoped (100) Si substrate. The Be cell temperature during growth was $T_{Be} = 775°C$ and the growth temperature was $T \sim 500°C$. The spectrum showing the familiar no-phonon line near wavelength $\lambda = 1.149\,\mu m$ and the accompanying TO-phonon replica, is virtually identical to that obtained from Si wafers ion implanted with Be. No post-growth annealing was required to obtain this spectrum.

The lower solid curve in Fig. 18 shows the measured PL spectrum at sample temperature $T = 11$ K for a ten-period $Si_{0.92}Ge_{0.08}/Si$ superlattice, with each period consisting of 50 Å of alloy and 100 Å of Si. The alloy layers were grown at a temperature of $T \sim 450°C$, and the shutter in front of the Be source was opened only during growth of the central 17 Å of each alloy layer. The first 20 Å of each Si layer were grown at temperature $T \sim 300°C$, after which the growth temperature was raised to $T \sim 550°C$ to complete the Si layer growth. Secondary ion mass spectrometry (SIMS) measurements show that the peaks in the Be concentration line up with the peaks

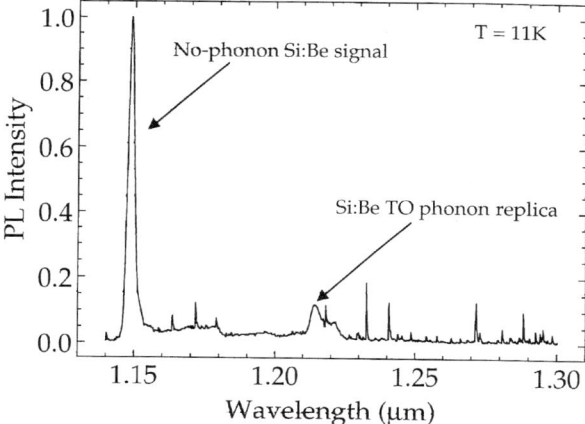

FIG. 17. Low-temperature PL spectrum from an MBE-grown Si layer into which Be was introduced during growth.

in the Ge concentration, supporting the notion the Be atoms are concentrated in the alloy layers, as desired. The dashed curve is the corresponding spectrum for a Be-implanted thick alloy, and the uppermost curve (solid line) gives the PL spectrum for an ion-implanted superlattice. As discussed in the preceding section, the broad distribution of Be impurities characteristic of ion-implanted samples broadens and shifts the spectrum because of

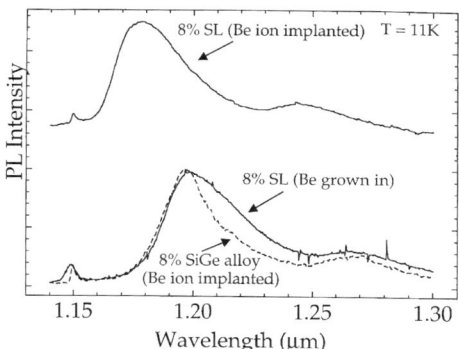

FIG. 18. Low-temperature PL spectra from three MBE-grown samples doped with Be. The solid curves are those for $Si_{0.92}Ge_{0.08}$/Si superlattices (50 Å/100 Å), the upper curve for implanted Be and the lower curve for Be introduced during growth. The dashed curve is that for a thick $Si_{0.92}Ge_{0.08}$ alloy layer.

the position-dependent exciton binding energy. When the Be is introduced during growth to attempt to restrict the Be to the central portion of the alloy layer, the peak PL wavelength is found to occur at essentially the same value as that for an implanted alloy of the same composition. This new result confirms the predictions of the model discussed in the preceding section of this chapter, and suggests the combination of an 8%-alloy and a 50 Å well thickness is insufficient to produce spectral shifts attributable to quantum confinement. The origin of the additional broadening exhibited by the emission line from the superlattice with grown-in Be, over that from the thick, Be-implanted alloy, is uncertain at this time.

That thin SiGe QWs will indeed produce a true spectral shift of the IBE emission line toward shorter wavelengths is apparently verified by the spectra shown in Fig. 19 which compares the PL measurements for the 50 Å/100 Å $Si_{0.92}Ge_{0.08}$/Si superlattice to that for a 20-period 20 Å/100 Å superlattice (Moore et al., 1995). In both cases, Be was introduced during MBE growth in a manner similar to that described earlier, except that the shutter in front of the Be-cell was opened only during deposition of the central 11 Å of each 20 Å-thick QW. For the thinner QWs, both the no-phonon line and the phonon replica have shifted toward the blue end of the spectrum.

These initial experiments demonstrate that controlling the placement of isoelectronic impurity complexes can be accomplished in SiGe/Si superlattices using existing technology and growth techniques. The Be isoelectronic impurity complex is admittedly a low-temperature center, radiating strong

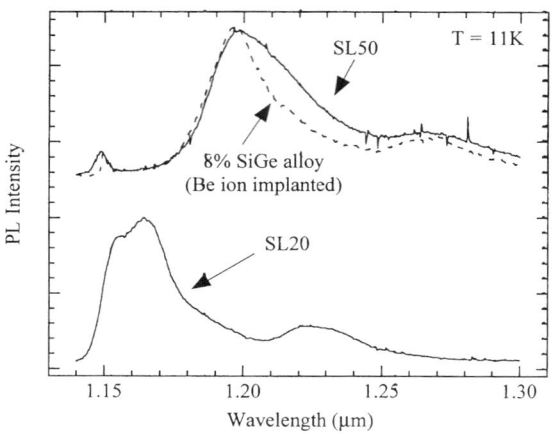

FIG. 19. Lower curve: low-temperature PL spectrum from a $Si_{0.92}Ge_{0.08}$/Si superlattice (20 Å/100 Å). The upper curves are identical to the two lower curves in Fig. 18.

bound-exciton emission only for temperatures below that of liquid nitrogen. It is, however, interesting in its own right, and makes a useful case study for future technologies in the event another isoelectronic complex capable of emitting bound-exciton radiation efficiently at room temperature is found.

IV. Device Considerations

Investigations of optical emission from Si-based materials derive their motivation from the idea that a breakthrough might stimulate the development of a Si light-source technology. In the ideal case, a Si-based emitter technology would be compatible with Si VLSI technology, a combination presumed to be powerful enough to open up many new opportunities. A Si-based light source need not outperform the best III–V semiconductor laser to be considered a success. Rather, its value would be judged more by the capability it provides when combined with Si microelectronics than by its value as a stand-alone component.

In broad terms, a semiconductor-emitter technology requires two main features to achieve good "wall-plug" efficiency, a light-emission mechanism and a means for making device structures. In III–V emitters, band-to-band radiative recombination provides the emission mechanism and lattice-matched heterostructures provide the low-resistance carrier injection and the carrier and optical confinement that lead to practical, high-performance devices. In Si, both aspects present their own challenges. The previous sections of this chapter focus on investigations of one interesting emission mechanism, the radiative decay of excitons bound to isoelectronic impurities, but the structural aspects merit attention, too. The search for a heterostructure technology is made more difficult by the absence of a group IV material that is lattice-matched to Si. The Si-Ge family of alloys is an attractive option, but strain produced by the lattice mismatch limits the thickness of the SiGe layers that can be grown relatively defect-free.

It is instructive to review briefly the types of optical waveguides that have been suggested or demonstrated to be compatible with Si substrates. Because a waveguide is likely to be essential for both optical confinement and compatibility with other optical and optoelectronic devices, examining the available waveguide structures is one way of surveying the possibilities for Si-based device geometries. Figure 20 shows five possibilities. Four of the five structures share the familiar requirement that the light-guiding layer has a larger refractive index than the adjacent layers, as required for total internal reflection.

Silicon-Based Optical Waveguides

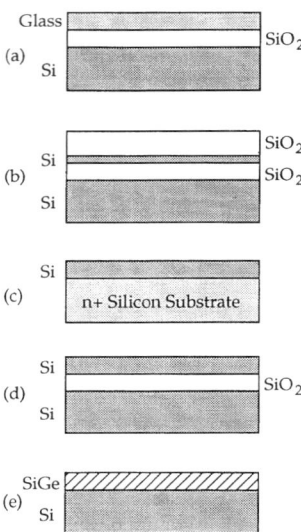

FIG. 20. Examples of five different types of Si-based optical waveguides: (a) glass on SiO_2, formed on a Si substrate; (b) anti-resonant reflecting optical waveguide (ARROW); (c) epitaxial Si waveguide; (d) Si-on-insulator (SOI) optical waveguide; and (e) SiGe alloy grown epitaxially onto an Si substrate.

Figure 20(a) illustrates the use of a layer of SiO_2 to isolate, in this case, a glass optical waveguide from the Si substrate. The idea that one may deposit a waveguide layer directly onto Si is thwarted by Si's large refractive index ($n \sim 3.5$). It is difficult to find a layer with good optical quality with a refractive index higher than 3.5, as necessary for total internal reflection within the overlayer. Glasses, for example, have refractive indices in the range $1.5 \leqslant n \leqslant 2.0$ in most cases. A sufficiently thick spacer-layer of SiO_2 ($n \sim 1.46$), easily grown or deposited onto a Si substrate, provides the low refractive index needed to permit the glass layer in Fig. 20(a) to function as an optical waveguide. The thickness of the oxide layer needs to be sufficient, however, to keep the exponential tail from the waveguide mode from spilling over significantly into the Si substrate. Such a thick oxide effectively rules out the possibility of radiative impurities in the Si substrate exciting waveguide modes in the glass. The very mechanism that makes waveguiding possible, in this instance, cuts off optical communication between the Si substrate and the waveguide layer. The structure in Fig. 20(a) is likely to find greatest interest for applications in which Si plays only the role of a convenient, inexpensive substrate.

Figure 20(b) illustrates a novel approach to forming an optical waveguide that does not rely exclusively on total internal reflection. Dubbed an ARROW waveguide (Anti-Resonant Reflecting Optical Waveguide), this structure traps light in the uppermost SiO_2 layer by the combination of total internal reflection at the air/SiO_2 interface and high reflectivity at the Si/SiO_2 interface (Duguay et al., 1986). The thin Si layer shown in Fig. 20(b) can actually be any high-index layer; it's function is to combine with the layers below it to form a multilayer reflector with thicknesses chosen to provide a high compound reflectivity. Reflectivities in excess of 99% have been observed, resulting in acceptably small attenuation for wavelengths in the visible and near-infrared regions of the spectrum. The schemes in Figs. 20(a, b) are interesting, but somewhat less so than the remaining three for the purposes of this chapter because the optical wave does not propagate in the same medium where the impurities are to be placed. The next three possibilities offer this feature.

Figure 20(c) shows a Si optical waveguide formed by depositing an epitaxial Si layer onto a heavily doped Si substrate (Soref and Lorenzo, 1986). The presence of a surplus of free carriers in the substrate reduces the refractive index below that of the undoped or less-heavily doped overlayer, satisfying the above-mentioned criterion for total internal reflection. In the original demonstration, epilayers ranging in thickness from several microns to a few tens of microns, exhibited waveguide losses as small as 5–10 dB/cm at wavelength $\lambda = 1.3\,\mu m$. Most of that loss arises from free-carrier absorption in the substrate, which is unavoidable in this case because the very mechanism that permits waveguiding in the first place produces the loss. And because the refractive-index difference between substrate and epilayer is very small (~ 0.01), the evanescent tails of the waveguide modes overlap those carriers significantly, producing the observed absorption. Despite the high residual absorption, the epitaxial Si waveguide has been used to demonstrate edge-emission (PL) from excitons bound to Be-related impurities introduced into the waveguiding layer by ion implantation (Brown et al., 1987) and as the basis for a $T = 77\,K$ LED employing sulfur-related bound excitons (Bradfield, Brown, and Hall, 1989).

Figure 20(d) illustrates a Si-on-insulator (SOI) structure, in which a single-crystal layer of Si is formed on a layer of SiO_2. The two most prominent methods for preparing SOI wafers are the SIMOX (Separation by the IMplantation of OXygen) and the bond-and-etchback processes (Colinge et al., 1986; Maszara, 1988). SOI material is under development for use in microelectronics applications, but precisely because of the refractive-index arrangement, it was recognized some time ago as an ideal optical waveguide structure for sub-bandgap wavelengths for which Si is nominally transparent (Kurdi and Hall, 1988). Since that time, a number of investigators have demonstrated propagation losses of a few dB/cm or less for light of wavelength $\lambda = 1.3\,\mu$m. Typical layer thicknesses are a few tenths of a micron for the Si layer and up to a micron or more for the oxide layer, depending on the application. The high-quality Si layer in an SOI wafer is an attractive platform for a Si-based emitter technology because it supports optical waveguide modes, is suitable for forming p-n junctions and other electronic devices, and, returning to the subject of this chapter, is an acceptable host for radiative isoelectronic impurities, something we have demonstrated in our own laboratories.

Figure 20(*e*) represents the final option, an epitaxial layer of a Si-Ge alloy grown onto a Si substrate. The alloy has the higher index of refraction, but the lattice mismatch limits the layer thickness to a value determined by the Ge-fraction in the alloy. Nevertheless, the SiGe/Si waveguide has been demonstrated to perform reasonably well in the transparent regime (Schmidtchen *et al.*, 1992) and possess features that make it an attractive option to consider.

No matter which Si-based optical waveguide geometry proves optimal, good carrier injection will be essential, which requires the fabrication of a high quality p-n junction, something rather easily accomplished in c-Si. In addition, the injected carriers must not merely flow "through" the junction, but must be confined in a recombination region long enough for radiative recombination to take place. Isoelectronic impurities can, in principle, provide the necessary carrier confinement in a p-i-n junction geometry, with the isoelectronic impurities placed in the intrinsic region. (In most such diodes, the intrinsic region is actually a lightly-doped p-type or n-type layer, and has a low space-charge field under moderate bias voltage.) The isoelectronic impurities can then trap the injected carriers to form bound excitons. Carrier confinement can be further improved by using a Si/SiGe heterojunction, as was described in the previous section. The principles of operation of such a junction are similar to those for III–V heterojunctions, although one must take care to understand the role of band offsets for junctions between dissimilar group IV materials.

V. Concluding Remarks

Isoelectronic impurity complexes that bind excitons can be formed in Si, in SiGe alloys, and in SiGe/Si superlattices (multi-quantum-wells) by a number of methods. The accompanying bound-exciton emission is intense at low temperatures, with external efficiencies of several percent in some cases, but it has proved difficult to form centers that emit strongly at room temperature, something that will be essential if IBE emission is to become a technologically important radiative mechanism. What is not clear at this time is whether a fundamental limitation precludes strong IBE emission at room temperature in Si or a SiGe alloy. Because IBE emission does occur at room temperature in GaP, another indirect semiconductor, any fundamental limitation must involve an argument specific to Si, not to indirect semiconductors in general. In the future, it will be important to develop a clear, quantitative understanding of group IV semiconductors as environments for impurity-related radiative processes.

References

Baldereschi, A., and Hopfield, J. J. (1972). "Binding to isoelectronic impurities in silicon," *Phys. Rev. Lett.* **28**, 171.

Beckett, D. J. S., Nissen, M. K., and Thewalt, M. L. W. (1989). "Optical properties of the sulfur-related isoelectronic bound excitons in Si," *Phys. Rev. B* **40**, 9618.

Bevk, J., Mannaerts, J. P., Ourmazd, A., Feldman, L. C., and Davidson, B. A. (1986). "Ge-Si layered structures: artificial crystals and complex cell ordered superlattices," *Appl. Phys. Lett.* **49**, 286.

Bradfield, P. L., Brown, T. G., and Hall, D. G. (1988). "Radiative decay of excitons bound to chalcogen-related isoelectronic impurity complexes in silicon," *Phys. Rev. B* **38**, 3533.

Bradfield, P. L., Brown, T. G., and Hall, D. G. (1989). "Electroluminescence from sulfur impurities in a p-n junction formed in epitaxial silicon," *Appl. Phys. Lett.* **55**, 100.

Brown, D. H., and Smith, S. R. (1980). "Evidence for the existence of a complex isoelectronic center in Si:In," *J. Lumin.* **21**, 329.

Brown, T. G., and Hall, D. G. (1986). "Observation of electroluminescence from excitons bound to isoelectronic impurities in crystalline silicon," *J. Appl. Phys.* **59**, 1399.

Brown, T. G., and Hall, D. G. (1986). "Optical emission at 1.32 μm from sulfur-doped crystalline silicon," *Appl. Phys. Lett.* **49**, 245.

Brown, T. G., Bradfield, P. L., Hall, D. G., and Soref, R. A. (1987). "Optical emission from impurities within the epitaxial-silicon optical waveguide," *Optics Lett.* **12**, 753.

Brown, T. G., and Bradfield, P. L. (1988). "Influence of oxygen in the formation of isoelectronic complexes in implanted Si:In," *Phys. Rev. B* **37**, 2699.

Callaway, J. (1974). *Quantum Theory of the Solid State.* New York: Academic Press.

Canham, L. T., Davies, G., Lightowlers, E. C., and Blackmore, G. W. (1983). "Complex isotope splitting of the no-phonon lines associated with exciton decay at a four-lithium atom isoelectronic centre in silicon," *Physica B+C* **117/118**, 119.

Colinge, J-P., Hashimoto, K., Kamins, T., Chiang, S-Y., Liu, E-D., Peng, S., and Rissman, P. (1986). "High-Speed, Low-Power, Implanted-Buried-Oxide CMOS Circuits," IEEE Electron Device Lett. **EDL-7**, 279.

Crouch, R. K., Robertson, J. B., and Gilmer, J. T. E. (1972). "Study of beryllium and beryllium-lithium complexes in single-crystal silicon," *Phys. Rev. B* **5**, 3111.

Davies, G. (1989). "The optical properties of luminescence centres in silicon," *Physics Reports* **176**, 83.
Dean, P. J., Haynes, J. R., and Flood, W. F. (1967). "New radiative recombination processes involving neutral donors and acceptors in silicon and germanium," *Phys. Rev.* **161**, 711.
Duguay, M. A., Kokubun, Y., Koch, T. L., and Pfeiffer, L. (1986). "Antiresonant refcting optical waveguides in SiO_2/Si multilayer structures," *Appl. Phys. Lett.* **49**, 13.
Ennen, H., Schneider, J., Pomrenke, G., and Axman, A. (1983). "1.54 μm luminescence from erbium-implanted III-V semiconductors and silicon," *Appl. Phys. Lett.* **43**, 943.
Gerasimenko, N. N. (1985). "Radiative recombination centers in silicon irradiated with beryllium ions," *Sov. Phys. Semiconductors* **19**, 762.
Hall, D. G. (1987). "The role of silicon in optoelectronics," *Proceedings of the Materials Research Society: Silicon Based Optoelectronic Materials*, San Francisco, p. 367 and references therein.
Henry, M. O., Lightowlers, E. C., Killoran, N., Dunstan, D. J., and Cavenett, B. C. (1981). "Bound exciton recombination in beryllium-doped silicon," *J. Phys. C: Solid State Phys.* **14**, L255.
Henry, M. O., Moloney, K. A., Treacy, J., Julligan, F. J., and Lightowlers, E. C. (1984). "Uniaxial stress studies of the Be pair bound exciton absorption spectrum in silicon," *J. Phys. C: Solid State Phys.* **17**, 6245.
Holonyak, N., and co-workers (1972). "Stimulated emission and laser operation (cw, 77 K) associated with deep isoelectronic traps in indirect semiconductors," *Phys. Rev. Lett.* **1972**, 230.
Killoran, N., Dunstan, D. J., Henry, M. O., Lightowlers, E. C., and Cavenett, B. C. (1982). "The isoelectronic centre in beryllium-doped silicon: I. Zeeman study," *J. Phys. C.: Solid State Phys.* **15**, 6067.
Kim, S., Herman, I. P., Moore, K. L., Hall, D. G., and Bevk, J. (1995). "Use of hydrostatic pressure to resolve phonon replicalike features in the photoluminescence spectrum of beryllium-doped silicon," *Phys. Rev. B* **52**, 16309.
Kim, S., Herman, I. P., Moore, K. L., Hall, D. G., and Bevk, J. (1996). "Hydrostatic pressure dependence of isoelectronic bound excitons in beryllium-doped silicon," *Phys. Rev. B* **53**, 4434.
Kurdi, B. N., and Hall, D. G. (1988). "Optical waveguides in oxygen-implanted buried-oxide silicon-on-insulator structures," *Optics Lett.* **13**, 175.
Lyon, S. A., Osbourn, G. C., Smith, D. L., and McGill, T. C. (1977). "Bound exciton lifetimes for acceptors in Si," *Solid State Commun.* **23**, 425.
Maszara, W. P., Goetz, G., Caviglia, A., and McKitterick, J. B. (1988). "Bonding of silicon wafers for silicon-on-insulator," *J. Appl. Phys.* **64**, 4943.
Mitchard, G. S., Lyon, S. A., Elliott, K. R., and McGill, T. C. (1979). "Observation of long lifetime lines in photoluminescence from Si:In," *Solid State Commun.* **29**, 425.
Modavis, R. A., and Hall, D. G. (1990). "Aluminum-nitrogen isoelectronic trap in crystalline silicon," *J. Appl. Phys.* **67**, 545.
Modavis, R. A., and co-workers (1990). "Isoelectronic bound exciton emission from Si-rich SiGe alloys," *Appl. Phys. Lett.* **57**, 954.
Modavis, R. A., Hall, D. G., Bevk, J., and Freer, B. S. (1991). "Cooperative quantum confinement of excitons bound to isoelectronic impurity complexes in $Si_{1-x}Ge_x$/Si superlattices," *Appl. Phys. Lett.* **59**, 1230.
Moore, K. L., King, O., Hall, D. G., Bevk, J., and Furtsch, M. (1994). "Radiative isoelectronic complexes introduced during the growth of Si and $Si_{1-x}Ge_x$/Si superlattices by molecular beam epitaxy," *Appl. Phys. Lett.* **65**, 2705.

Moore, K. L., King, O., Hall, D. G., Bevk, J., and Furtsch, M. (1995). "Introduction of radiative isoelectronic complexes during molecular beam epitaxial growth of Si and $Si_{1-x}Ge_x$ superlattices," *Proc. Mat. Res. Soc.* **378**, 867.

Morgan, T. N., Weber, B., and Bhargava, R. N. (1968). "Optical properties of Cd-O and Zn-O complexes in GaP," *Phys. Rev.* **166**, 751.

Nahory, R. E., Shaklee, K. L., and Leheny, R. F. (1971). "Indirect bandgap superradiant laser in GaP containing isoelectronic traps," *Phys. Rev. Lett.* **27**, 1647.

Nelson, D. F., Cuthbert, J. D., Dean, P. J., and Thomas, D. G. (1966). "Auger recombination of excitons bound to neutral donors in gallium phosphide and silicon," *Phys. Rev. Lett.* **17**, 1262.

Pantelides, S. T. (Ed.) (1986). *Deep Centers in Semiconductors*. Gordon and Breach, New York.

People, R. (1977). "Physics and application of Ge_xSi_{1-x}/Si strained-layer heterostructures," *IEEE J. of Quant. Elec.* **QE-22**, 1696.

Properties of Silicon (4 ed.). London: Institution of Electrical Engineers, 1988.

Sauer, R., Schmid, W., and Weber, J. (1978). "Indium bound exciton luminescence in silicon," *Solid State Commun.* **27**, 705.

Sauer, R., and Weber, J. (1983). "Photoluminescence of defect complexes in silicon" In (ed. J. Giber), *Defect Complexes in Semiconductor Structures*, Springer Verlag, New York.

Schlesinger, T. E., and McGill, T. C. (1983). "Isotope-shift experiments on luminescence attributed to (Fe,B) pairs in silicon," *Phys. Rev. B* **28**, 3643.

Schmidtchen, J., Schuppert, B., Splett, A., and Petermann, K. (1992). "Germanium-diffused waveguides in silicon for $\lambda = 1.3\,\mu m$ and $\lambda = 1.55\,\mu m$ with losses below 0.5 dB/cm," *IEEE Photon. Technol. Lett.* **4**, 875.

Singh, M., Lightowlers, E. C., and Davies, G. (1989) In *Science and Technology of Defects in Silicon*, (eds. A. C. Chantre, C. A. J. Ammerlaan, and P. Wagner), North Holland, Paris, p. 303.

Soref, R. A., and Lorenzo, J. P. (1986). "All-silicon active and passive guided-wave components for $\lambda = 1.3$ and $1.6\,\mu m$," *IEEE J. Quant. Elec.* **QE-22**, 873.

Sturm, J. C., and co-workers (1991). "Well-resolved band-edge photoluminescence of excitons confined in strained $Si_{1-x}Ge_x$ quantum wells," *Phys. Rev. Lett.* **66**, 1362.

Tajima, M., and Matsushita, Y. (1983). "Photoluminescence related to dislocations in annealed Czochralski-grown Si crystals," *Jpn. J. Appl. Phys.* **22**, L589.

Thewalt, M. L. W., Ziemelis, U. O., and Parsons, R. R. (1981). "Enhancement of long lifetimes lines in photoluminescence from Si:In," *Solid State Commun.* **39**, 27.

Thewalt, M. L. W., Watkins, S. P., Ziemelis, U. O., Lightowlers, E. C., and Henry, M. O. (1982). "Photoluminescence lifetime, aborption and excitation spectroscopy measurements on isoelectronic bound excitons in beryllium-doped silicon," *Solid State Commun.* **44**, 573.

Thewalt, M. L. W., Labrie, D., and Timusk, T. (1985). "The far-infrared absorption spectra of bound excitons in silicon," *Solid State Commun.* **53**, 1049.

Thomas, D. G., Gershenzon, M., and Hopfield, J. J. (1963). "Bound excitons in GaP," *Phys. Rev.* **131**, 2397.

Thomas, D. G., and Hopfield, J. J. (1966). "Isoelectronic traps due to nitrogen in gallium phosphide," *Phys. Rev.* **150**, 680.

Trumbore, F. A., Gershenzon, M., and Thomas, D. G. (1966). "Luminescence due to the substitution of bismuth for phosphorus in gallium phosphide," *Appl. Phys. Lett.* **9**, 4.

Vouk, M. A., and Lightowlers, E. C. (1977). "A high-resolution investigation of the recombination radiation from Si containing the acceptors B, Al, Ga, In, and Tl," *J. Lumin.* **15**, 357.

Wagner, J., Weber, J., and Sauer, R. (1981). "Excitation spectroscopy on the P, Q, and R isoelectronic lines in indium-doped silicon," *Solid State Commun.* **39**, 1273.

Watkins, S. P., and Thewalt, M. L. W. (1986). "Excitation spectroscopy of the In-related isoelectronic bound excitons under uniaxial stress and magnetic-field perturbations," *Phys. Rev. B* **34**, 2598.

Weber, J., Schmid, W., and Sauer, R. (1980). "Localized exciton bound to an isoelectronic trap in silicon," *Phys. Rev. B* **21**, 2401.

Weber, J., Sauer, R., and Wagner, P. (1981). "Photoluminescence from a thermally induced indium complex in silicon," *J. Lumin.* **24/25**, 155.

Weber, J., Bauch, H., and Sauer, R. (1982). "Optical properties of copper in silicon: excitons bound to isoelectronic copper pairs," *Phys. Rev. B* **25**, 7688.

Weber, J., and Sauer, R. (1983). "Identification of electronic and vibronic states of the deep indium-related defect in quenched Si:In by excitation spectroscopy," *Phys. Rev. B* **27**, 6568.

Ziemelis, U. O., Parsons, R. R., and Voos, M. (1979). "The puzzle of double-doped Si(B,In): sharp line series in near band edge photoluminescence," *Solid State Commun.* **32**, 445.

CHAPTER 4

Erbium in Silicon

J. Michel, L. V. C. Assali, M. T. Morse, and L. C. Kimerling

DEPARTMENT OF MATERIALS SCIENCE AND ENGINEERING
MASSACHUSSETTS INSTITUTE OF TECHNOLOGY
CAMBRIDGE, MA 02139

I.	INTRODUCTION	111
II.	Er DOPING OF Si	113
	1. *Ion Implantation*	113
	2. *Solid Phase Epitaxy (SPE)*	117
	3. *Molecular Beam Epitaxy (MBE)*	118
	4. *Chemical Vapor Deposition (CVD)*	118
	5. *Ion-beam Epitaxy (IBE)*	120
III.	DIFFUSIVITY AND SOLUBILITY	121
IV.	LIGHT EMISSION	127
	1. *Physics of Light Emission*	127
	2. *Ligands*	133
	3. *Electrical Properties*	136
	4. *Activation and Deactivation Processes*	139
V.	ELECTRONIC STRUCTURE	142
	1. *Electronic Structure of Er-related Impurities in Si*	144
	2. *Isolated Er Impurity in Si*	145
	3. *Er-related Complexes in Si*	148
VI.	LIGHT EMITTING DIODE DESIGN	150
VII.	SUMMARY	153
	Acknowledgments	153
	References	153

I. Introduction

Silicon is the dominating material in semiconductor manufacturing. Ninety-five percent of all semiconductor devices fabricated are based on Si. Although Si is technologically superb for electronic applications in terms of integration, control and cost, it is nonexistent in the emerging optoelectronic market. This shortcoming is due to its indirect bandgap which leads to a relatively long carrier lifetime, compared to direct bandgap III–V semiconductors. Recombination of carriers, therefore, occurs mainly nonradiatively that is, without light emission. The attempts to bypass these shortcomings are mainly driven by the vision of a low cost, high integration optoelectronic semiconductor material. Silicon is the material of choice because of its unmatched development infrastructure that has outpaced every potential competitor. Optoelectronic capabilities would open the door to even faster data transfer and higher integration densities at very low cost.

Several approaches were taken in the past to increase Si emission efficiency in order to reach output power levels that meet commercial applications. Among the first attempts was the use of isoelectronic centers, a concept that led to the development of commercial green light emitting devices (LEDs) in GaP, also an indirect bandgap material. Though initially successful at low temperatures with quantum efficiencies up to 5%, an isoelectronic center in Si that emits light efficiently at room temperature has not been found. Another intriguing approach was the development of Si-Ge superlattices in order to make direct bandgap material by band folding. The efficiency of these devices, however, is very small because of a limited carrier density at the band edge. The integration of silicon and III–V semiconductors is a more practical approach where the strength of both materials contributes to the light emitting device. But the lattice mismatch between these materials results in a large dislocation density that degrades device performance. It is now possible to grow SiGe buffer layers on Si with low threading dislocation densities of 10^6 cm^{-2}. Depending on the SiGe composition, different III–V materials can therefore be grown lattice matched on SiGe. Strong light emission in the visible has been demonstrated for devices grown on SiGe buffer layers.

Erbium in Si has several advantages over other Si based light emitters. Erbium, like other rare-earth metals, emits light at the same wavelength independently of the host. This emission is due to the excited states of the 4f manifold of the trivalent Er. The transition from the lowest excited state yields light at 1.54 µm, compatible with low loss optical fibers. The emission energy is independent of the temperature and the linewidth is only about 100 Å at room temperature compared to several thousand Ångstroms in III–V materials. Room temperature photoluminescence (PL) and electroluminescence (EL) has been reported from Er doped Si by several research groups.

This chapter will review the properties of Er in Si and its possible applications in Si technology. Section II deals with the formation and growth of Er doped Si layers. Implantation of Er with additional co-implantation of oxygen will be reviewed. Progress in epitaxial growth of Er doped layers using chemical vapor deposition (CVD) and molecular beam epitaxy (MBE) will be discussed. Section III discusses the diffusion of Er in

Si as an important parameter for processing technology. The physics of light emission from Er, the influence of ligands on the light emission and activation and deactivation processes of Er in the excited state are discussed in Section IV. *Ab initio* calculations to understand the microscopic structure of Er complexes are presented in Section V. Section VI, finally, will review the progress made in constructing LEDs.

II. Er Doping of Si

Since the first report of Er-doped Si luminescence in 1983 (Ennen *et al.*, 1983), several constraints on the luminescence of Er in Si have been discovered. Erbium's equilibrium solid solubility in Si (see Section III) of $\leqslant 1 \times 10^{18}$ cm^{-3} (at 900°C) is one to two orders of magnitude below the concentration estimated to be needed for LEDs (Xie, Fitzgerald, and Mie, 1991). Above this concentration, optically inactive precipitates of $ErSi_{2-x}$ form when Er is implanted into Si (Eaglesham *et al.*, 1991). High temperature processing can also be limited by the need to avoid the dissociation of the optically active Er-ligand complexes (see Section IV.2). The other critical issue in the growth of Er-doped Si is to minimize damage induced defects caused by the incorporation of Er. These defects can act as nonradiative recombination centers which decrease the pumping efficiency of Er. Any growth technique useful for production of commercial devices must address these issues.

Several methods have been used to incorporate Er into Si including ion implantation (Ennen *et al.*, 1983; Favennec *et al.*, 1990; Eaglesham *et al.*, 1991; Tang *et al.*, 1989a; Ren *et al.*, 1993), SPE (Polman *et al.*, 1993; Custer, Polman, and van Pixteren, 1994; Liu *et al.*, 1995), MBE (Efeoglu *et al.*, 1993; Serna *et al.*, 1995; Miyashita *et al.*, 1995), CVD (Beach *et al.*, 1992; Rogers *et al.*, 1995a; Morse *et al.*, 1996), ion-beam epitaxy (Matsuoka and Tohno, 1995a), diffusion (Sobolev *et al.*, 1993), pulsed laser ablation (Thilderkvist *et al.*, 1996), laser irradiation (Nakashima, 1993) and sputtering (Kim *et al.*, 1993). We will focus on the first five since they are processes most easily compatible with integrated circuits and have the highest crystalline quality.

1. ION IMPLANTATION

Most of the research done on Er-doped Si has used ion implantation to incorporate the Er. The implants were typically done with ^{166}Er at ion energies ranging from 200 keV to 5.25 MeV. Implantation doses ranged from 10^{12} cm^{-2} to 10^{16} cm^{-2} using the appropriate substrate temperatures to avoid amorphization. Annealing is always done after the implant to improve the crystalline quality and activate the Er. Studies have shown that luminescence is affected by the temperature and environment during annealing through changes in the defect concentration (Michel et al., 1991; Sobolev et al., 1994). For samples implanted at high energies (MeV), annealing at 900°C maximized the intensity at 1.54 μm (Michel et al., 1991). This has been attributed to the competition of two processes. At lower annealing temperatures the crystallinity has not yet recovered from the implantation and nonradiative recombination pathways hinder Er luminescence. At temperatures above 900°C, the Er-ligand complexes dissociate, reducing the optically active Er concentration. This behavior is seen in lower energy implants (400 keV) as well, although the optimum annealing temperature shifts downward to about 800°C (Fig. 1). The lowering of the optimal annealing temperature with lower implantation energy is believed to be associated with different defect profiles after annealing.

The microstructure of Si implanted with Er at various energies and concentrations has been extensively studied (Eaglesham et al., 1991; Duan

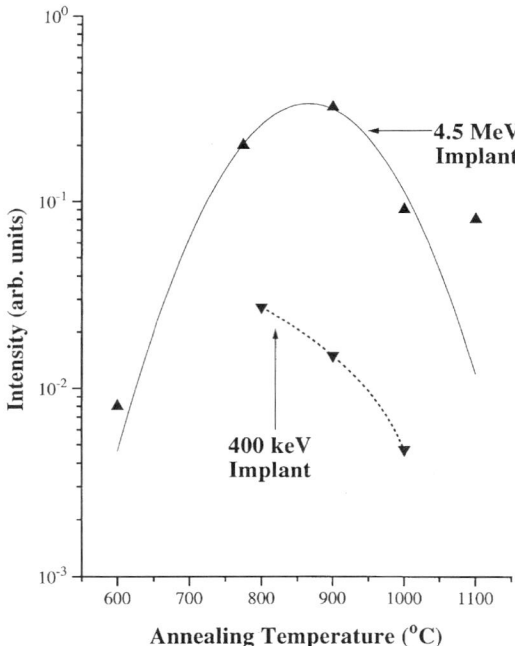

FIG. 1. Effect of temperature on the Er PL intensity at 1.54 μm for different implant energies. All anneals were done for 30 min in an inert atmosphere.

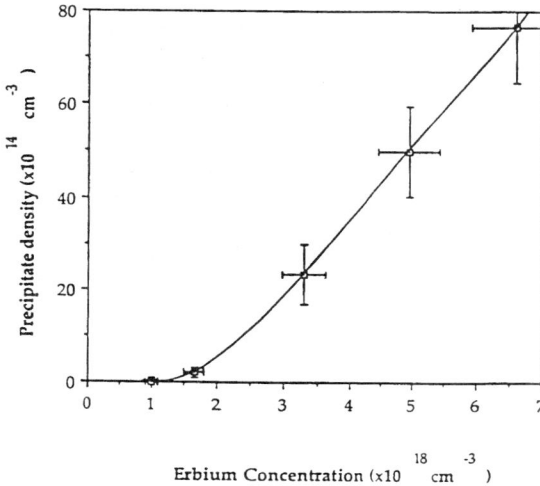

FIG. 2. Precipitate density as a function of peak erbium concentration as determined by plane view TEM. The implantation energy was 500 keV with an anneal at 900°C for 30 min. (From Eaglesham et al., 1991.)

et al., 1997). Eaglesham et al. (1991) found a threshold Er concentration for precipitation of 1.3×10^{18} cm^{-3} in 500 keV implanted Si annealed at 900°C for 30 min. Above this concentration, Er-rich platelets (tentatively identified as ErSi$_2$), formed on [111] planes near the peak concentration of Er during annealing. The precipitate density increases rapidly as the Er concentration rises above the threshold value (Fig. 2). Implantation energy can affect the onset of precipitation however. Silicon implanted with Er at 400 keV to achieve a peak concentration of 5×10^{17} cm^{-3}, and oxygen to a peak concentration of 3×10^{18} cm^{-3} at the same depth, did not have precipitates after annealing (Duan et al., 1997). When the Er implantation was done at 4.5 MeV for the same concentrations, precipitates did form. Cross-sectional transmission electron microscopy (TEM) was used to record the evolution of the microstructure for this sample as a function of annealing temperature as shown in Fig. 3. Small dislocation loops at the end of range have formed after an 800°C anneal for 30 min, which continue to grow in size as the annealing temperature is increased. Much longer anneals on this material show a second band of dislocation loops forming deeper into the substrate (Fig. 4). A strong interaction between oxygen and both bands of defects can be seen by secondary ion mass spectroscopy (SIMS), suggesting that the oxygen segregates to the dislocation loops.

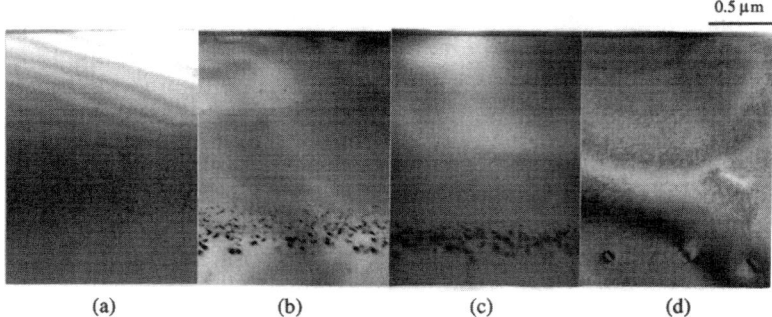

FIG. 3. Evolution of defect microstructure after a 4.5 MeV Er implant with increasing temperature. Damage clusters, evident after an anneal at 600°C for 30 min, give way to the formation of dislocation loops at higher temperatures. This loop grows in size at higher temperatures while narrowing in depth distribution. Anneals were done at (a) 600°C, (b) 800°C, (c) 900°C, and (d) 1000°C for 30 min. (From Duan et al., 1997.)

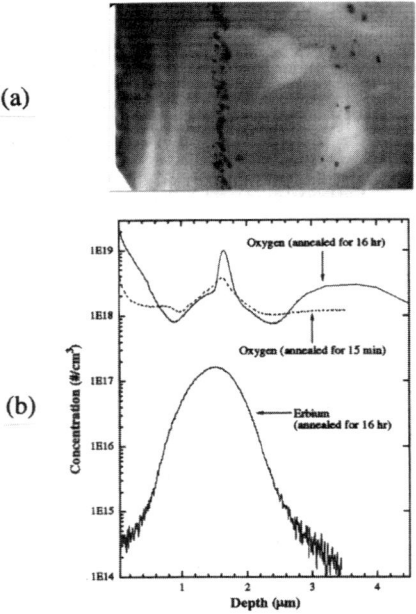

FIG. 4. Interaction of oxygen with defect microstructure. (a) Cross-sectional TEM shows two defect bands in a sample implanted at 4.5 MeV and annealed at 900°C for 16 h. (b) SIMS plot of erbium and oxygen concentrations in the same sample at an identical depth scale. The near surface defect band, consisting of $ErSi_2$ precipitates, forms near peak concentration of erbium. The deeper band is made of dislocation loops. Oxygen segregates to both defect bands during annealing.

2. SOLID PHASE EPITAXY (SPE)

In an effort to improve the crystal quality of high-dose Er implants and to increase the amount of Er incorporated into the Si, some research has been focused on SPE. In the earliest report Er was implanted with high doses at room temperature to amorphize the surface (Eaglesham et al., 1991). The same annealing that was typically used for implantation, 900°C for 30 min, was used to regrow the crystalline Si (c-Si) from the single crystal/amorphous interface. Erbium segregated with the moving interface, reaching a maximum concentration of 8.5×10^{19} cm^{-3} before precipitation occurred and regrowth broke down. Above this point the Er concentration fell to pre-regrowth values and the subsequent growth was highly defective. TEM revealed stacking faults and twinning extending to the surface. Later work on SPE modified the annealing procedure to optimize it for regrowth. In these studies Er was implanted at 250 keV into c-Si or pre-amorphized Si (a-Si) at 77 K (Polman et al., 1993). Post-implantation anneals at 600–650°C for 3 h were done to regrow single crystal Si through the Er-doped amorphous Si. The Er again segregated in front of the a-Si/c-Si interface until it reached high concentration, where it became trapped in c-Si. Erbium concentrations as high as 2×10^{20} cm^{-3} have been achieved in very thin regions (~ 20 nm). The minimum channeling yield (χ_{min}), which is the ratio between the backscattered intensity of channeled and random sample orientation, was $<5\%$ indicating "high quality" material. This is a qualitative measure of film perfection, however, and is not sensitive to differences in structure when χ_{min} is near 5%. Cross-sectional TEM of this sample did show a thin disordered region near the surface. A final rapid thermal anneal at temperatures $\geqslant 900$°C for 15 sec. was done in an attempt to optically activate the Er. No evidence of optical or electrical activation was reported however.

More recent work has used multiple Er implants at energies from 0.5 to 5 MeV with corresponding oxygen co-implants for spatial overlap to increase the thickness of the highly doped material (Coffa et al., 1993). Regrowth in samples without O only preceded for 800 nm before breaking down leaving large amounts of twins. The presence of O ($\sim 10^{20}$ cm^{-3}) during the anneal stabilized the regrowth over the entire amorphous thickness leaving a 2 μm region with an Er concentration of 10^{19} cm^{-3}. End of range defects and threading dislocations were still evident in TEM. Spreading resistance measurements showed a donor concentration of 8×10^{18} cm^{-3}, similar to values reported by Priolo et al. (1995) by ion implantation. The PL intensity at 3 K of the SPE sample was approximately 4 times that of the ion implanted sample with similar concentrations of Er, O and donors.

3. MOLECULAR BEAM EPITAXY (MBE)

The previous two methods have been able to incorporate large amounts of Er while simultaneously degrading the crystalline quality. High temperature anneals can remove the implantation damage at the cost of precipitation of metastable Er concentrations. To avoid the problem of damage, research into the nonequilibrium growth process of MBE is being done. Initial work used a metallic Er source in a standard MBE system (Efeoglu et al., 1993) at growth temperatures of 500 or 700°C. Cross-sectional TEM showed precipitates at an Er concentration of 2×10^{18} cm^{-3} even at these low processing temperatures. After growth, the samples were implanted with O to achieve a constant concentration (10^{19} cm^{-3}) throughout the Er profile, then annealed at 800°C to remove the implantation damage. The PL intensity of this material was comparable to implanted Er standards at very low temperatures, but it was quenched near 150 K.

A more promising method which incorporates both Er and O during the growth of Si has been reported. (Serna et al., 1995; Serna et al., 1996; Stimmer et al., 1996a,b) When O is not present during growth on (100) Si, Er segregates to the surface, possibly forming precipitates above areal densities of 2×10^{14} cm^{-2}. With the introduction of O through a leak valve during growth, Er segregation was suppressed due to a reduction in Er mobility or the formation of Er/O complexes. Four times the amount of Er was incorporated in the presence of O by this method. An O partial pressure of 4×10^{-10} mbar results in a $(4 \pm 1) \times 10^{19}$ cm^{-3} O concentration in the film. A maximum Er concentration of 2×10^{19} cm^{-3} before precipitation was reported based on Rutherford backscattering spectroscopy. Light emitting diodes have been made with similar material (Stimmer et al., 1996b). Maximum PL intensity occurred in samples grown at 500°C with an Er concentration of 4×10^{19} cm^{-3}. Oxygen incorporation was limited to pressures below 4×10^{-9} mbar due to the emergence of 3-D growth.

4. CHEMICAL VAPOR DEPOSITION (CVD)

Like MBE, CVD of Er-doped Si offers metastable Er concentrations due to low processing temperatures without lattice damage. Electron cyclotron resonance plasma enhanced CVD (ECR-PECVD) and ultra-high vacuum CVD (UHV-CVD) have grown Er-doped Si at temperatures from 400 to 650°C (Beach et al., 1992; Rogers et al., 1995a; Andry et al., 1996; Morse et al., 1996). Both types of systems used organometallic compounds which contained both the Er and the ligands needed for optical activation. An inherent problem with these sources is the incorporation of large concentrations of carbon coming from the precursor which can degrade the crystal quality of the film. Control of the decomposition of the precursor through processing parameters is crucial to suppress this problem. While the difficulties associated with precursor cracking are common to both systems, the method of film growth is different.

ECR-PECVD uses a high-density plasma to excite the gas molecules to a reactive state which are then accelerated toward the Si surface with energies $\geqslant 10\,\text{eV}$, hence thermal energy is not required for decomposition and growth. Since energetic particles impinge on the surface, however, damage can occur above the energy necessary to displace a Si atom from a lattice site, estimated to be 13 eV from Monte Carlo simulations (Smith, 1994). Threading dislocations were observed by cross-sectional TEM in undoped Si films grown at 430°C with an ionic impingement energy of approximately 11 eV, but not in films grown at 500°C, where only a slightly disordered interface was seen (Rogers et al., 1995b). The Si device community has not yet accepted this technique for production of single crystal Si due to this problem, but further refinement might produce integrated circuit device quality material. Maximum PL was seen in films grown at 430°C containing an Er concentration of $6 \times 10^{18}\,\text{cm}^{-3}$ and a χ_{min} of 6%, indicating fairly high crystal quality (Andry et al., 1996). Nitrogen was the Er ligand in the precursor and had concentrations of approximately $2 \times 10^{19}\,\text{cm}^{-3}$ in the film. High levels of carbon contamination were reported (15 times that of the Er), but it is possible that the carbon is also complexing with Er, forming optically active centers. Rocking curve analysis of the Si (044) shows a narrow FWHM (full width at half maximum) of 71 arcsec for the film. The PL intensity decreased by a factor of 40 from 4 K to room temperature.

UHV-CVD is a low-temperature (500–650°C) epitaxial growth method which relies on low partial pressures of oxidizing species and hydrogen passivation to maintain an oxide free Si surface before the initiation of growth. The low growth pressures, typically between 10^{-3} and 10^{-2} torr, and system geometry result in molecular flow of the gas at the surface of the substrates, and hence highly uniform films. Temperatures as low as 500°C supply enough thermal energy to decompose the gases at the substrate for film growth, although temperatures of 550 to 600°C are normally used for faster growth rates. Undoped Si growth rates still tend to be smaller than growth rates for ECR-PECVD (10 Å/min vs. 40 Å/min), but over 20 wafers can be deposited on at a time in UHV-CVD as compared to one wafer in ECR-PECVD. It is still unclear, however, if the large batch sizes quoted for Si growth in UHV-CVD can be extrapolated to Er-doped films, since the precursor could be depleted at the end of the batch. Erbium concentrations have been greater than $10^{20}\,\text{cm}^{-3}$ in UHV-CVD grown films, but contamination from other species in the precursor can be large, exceeding $10^{21}\,\text{cm}^{-3}$

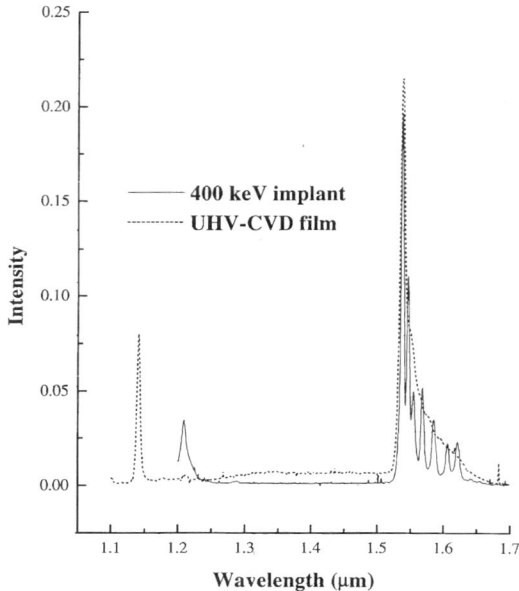

FIG. 5. Comparison of PL taken at 4 K for implanted and CVD-grown material. The implanted sample had peak concentrations of erbium and oxygen of 5×10^{17} cm^{-3} and 2×10^{18} cm^{-3} respectively, and was annealed to 800°C for 30 min. The CVD-grown film is 50 nm and was grown at 560°C.

in some cases. Beach *et al.* (1992) reported high concentrations of threading defects in the Er-doped films from TEM, most likely due to the high concentration of carbon (10^{20} cm^{-3}). Figure 5 shows PL spectra taken under identical conditions of a standard 400 keV Er/O implanted sample and a CVD grown film. The intensity of the 1.54 μm peak is larger for the CVD film even though the film thickness is 50 nm, less than the straggle of the implanted Er profile. Luminescence from these UHV-CVD films is reduced in intensity by two orders of magnitude from 4 K to room temperature.

5. Ion-beam Epitaxy (IBE)

IBE is an ultra-high vacuum sputtering process with a plasma beam in which single crystal Si can be grown at temperatures as low as 320°C (Matsuoka and Tohno, 1995a). Matsuoka and Tohno (1995b) have grown

Er-doped Si films with low energy ions (~ 10 eV) at 500°C with Er concentrations as high as 6×10^{20} cm^{-3}. Oxygen was incorporated into the film by flowing argon with trace amounts of O during growth. The incorporation of Er was strongly dependent on the amount of O in the argon gas, suggesting that the O is increasing the sticking coefficient of Er and selectively oxidizing it. As with MBE, the presence of O suppressed segregation of Er during growth.

The crystal quality of IBE films has been characterized by TEM, scanning electron microscopy (SEM), reflective high energy electron diffraction (RHEED), SIMS and PL. No dislocations or extended defects were seen by TEM. This was also confirmed by the more sensitive technique of etch pitting with examination in the SEM. RHEED patterns of the surface show a typical Si 2×1 reconstruction for films with an Er concentration below 3×10^{19} cm^{-3}, after which a diffuse pattern is observed, indicating poor crystallinity. Metal contamination, a common problem for sputtering, was below the detection limits of SIMS for growth temperatures above 400°C. Maximum intensity occurred for samples with an Er concentration of 2×10^{18} cm^{-3} at 4 K. Weak room temperature PL has also been seen in samples with 5×10^{18} Er/cm^3 and 3×10^{18} O/cm^3.

III. Diffusivity and Solubility

In order to use Er as a dopant in Si, its diffusion behavior has to be well understood. Especially in view of possible cross-contamination in wafer processing, a low diffusivity of Er in Si is desirable.

Direct measurement of the Er diffusivity is experimentally difficult. Several methods have been reported by using either ion implantation or diffusion from a surface source. Diffusion experiments from the surface have to deal with the high reactivity of Er with O, N, and Si. Experiments showed that Er forms very stable compounds, mainly silicides and, in the presence of O, erbium roxide or ternary Er-Si-O phases (Ren et al., 1993). Therefore the conditions for an unlimited surface source is difficult to meet. Furthermore the Er solubility limits the detection of Er especially for SIMS, where the detection limit of Er is at about 10^{16} cm^{-3}.

Different techniques were used to overcome these obstacles. Indirect measurements were used by Ageev et al. (1977), Nazyrov, Kulikov, and Malkovich (1991), and Sobolov (1995b). Nazyrov, Kulikov, and Malkovich (1991) used the β-activity of radioactive ^{169}Er to measure the diffusion profile of indiffused Er. Er was deposited on the Si surface and a diffusion annealing was then carried out at temperatures between 1100 and 1280°C in air. Sobolev (1995b) used a surface film based on tetraethoxysilane containing Er_2O_3 or ErCl for the diffusion step. The diffusion was carried out in different ambients (Ar, O_2; dry and wet) at 1100 to 1250°C. The Er profile was measured by using differential conductivity and the assumption that Er forms acceptors. The diffusivity was then derived from the Er concentration profiles. Roberts and Parker (1995) and Ren et al. (1993) used SIMS to measure Er profiles. Only Ren et al. (1993) used ion implantation to introduce Er into Si and subsequently measured the change of the implantation profile. All other experiments were done by diffusing Er from the Si surface.

Figure 6 shows the diffusivity as a function of 1/T compiled from the different measurement techniques. Obviously there is major disagreement in the measured values of the Er diffusivity, as shown in Table 1, and in the derived activation energies. Activation energies have been reported from 2.9 eV (Sobolev et al., 1993) to about 5 eV (Ageev et al., 1977; Ren 1994b). One of the major reasons for these large variations can be found in the formation of precipitates. The precipitation of Er was shown to have a significant impact on the extraction of Er diffusivity from experimental data (Ren et al., 1993).

Precipitation as a possible reason for poor reproducibility of surface diffused Er SIMS profiles was also suggested by Roberts and Parker (1995). Erbium was diffused from a Si cavity in a controlled atmosphere. Very shallow concentration profiles were found with diffusion constants around

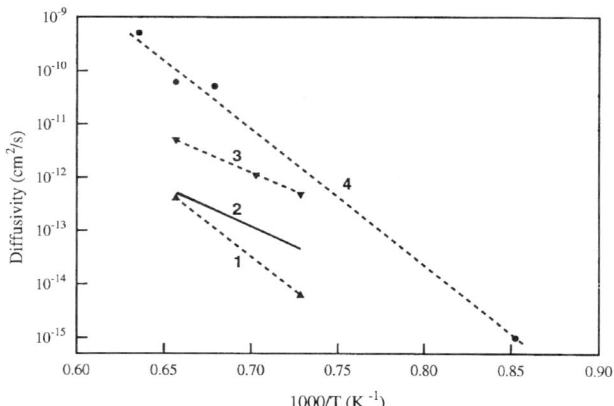

FIG. 6. Temperature dependence of the Er diffusion coefficient in Si (from Sobolev, 1995) 1: data from Ageev et al. (1977); 2: data from Nazyrov, Kulikov, and Malkovich (1991); 3: data from Sobolev et al. (1993); 4: Ren (1994b).

TABLE I

DIFFUSIVITY OF Er IN Si AT 1250°C, ACCORDING
TO DIFFERENT AUTHORS

Diffusivity (cm^2/s)	Reference
4×10^{-13}	Ageev et al., 1977
5×10^{-13}	Nazyrov, Kulikov, and Malkovich, 1991
5×10^{-12}	Sobolev et al., 1993
6×10^{-11}	Ren et al., 1993

2×10^{-16} cm^2/s at 1315°C. The diffusion depth of Er varied between 200 and 1500 nm where the smaller depth showed single exponential decay while deep indiffused Er profiles showed step behavior, that is, rapid changes in Er concentration.

The influence of precipitation on the Er diffusivity was discussed by Ren et al. (1993). Figure 7 shows a typical SIMS profile of Er implanted Si before and after diffusion annealing. Erbium was implanted at an energy of 5.25 MeV. After heat treatment at 850°C for 30 min no change from the as-implanted profile was observed. A subsequent heat treatment of 1300°C, 1 h, results in contraction of the profile rather than the broadening normally observed for Group III and V dopants in Si (dashed line). The profile shows

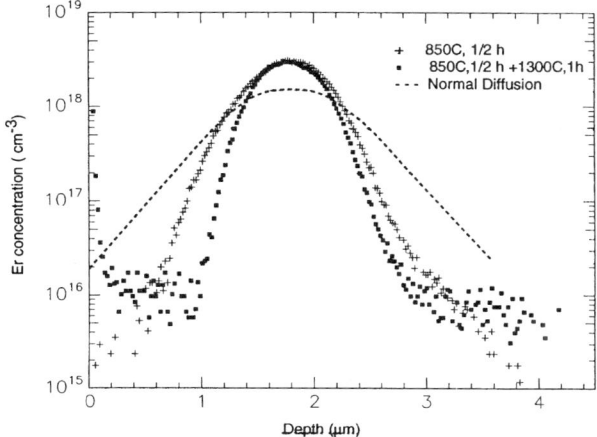

FIG. 7. SIMS depth profiles for Er in Si before and after diffusion annealing at 1300°C for 1 h. The dashed line represents an expected normal diffusion profile. (From Ren et al., 1993.)

furthermore a constant Er concentration [Er] of 10^{16} cm^{-3} away from the implanted distribution.

The narrowing of the distribution unambiguously connotes precipitation at the peak of the distribution. Thus, the precipitates, as well as the surfaces, act as sinks for the Er to reach its equilibrium solubility. By incorporating the dissolution of the precipitates the diffusivity of Er in Si can be extracted. Assuming that complete point defect equilibrium is reached by 1300°C, 1 h, we define the flat portion of the distribution as the solid solubility ($[Er]_{sol}$) of Er in Si.

In the diffusion model, Er diffusion is assumed to be the rate limiting step for the narrowing of the Er profile. In other words, the dissolution process of subcritical Er-Si precipitates at the maximum of the Er profile is fast enough to maintain local equilibrium with solute Er, which out-diffuses to the surface. This assumption is critical. Otherwise, the observed narrowing of the Er profile would be determined by the rate of precipitates dissolution instead of Er out-diffusion. In this respect, the extracted diffusivity from the model may not represent the true diffusivity, but instead indicate only a lower bound of Er diffusivity at these temperatures. Furthermore the Er-Si precipitation is assumed to take place instantaneously for concentrations above its equilibrium solid solubility at such high temperatures (1150–1300°C).

Erbium in the form of the precipitates is immobile and assumed to remain in a gaussian distribution. The peak concentration is unchanged after short time anneals. This assumption is consistent with our SIMS observations after a 1 h anneal. However, the assumption would break down at much longer anneal times when significant depletion of the precipitates reduce the Er peak concentrations. To maintain mass conservation, the Er out-diffusion rate J is equal to the rate of Er reduction in the precipitation region. Er diffusivity in Si is then obtained by comparing simulated profiles with SIMS profiles. The results are shown by the full circles in Fig. 6.

The observed precipitation results from Eaglesham *et al.* (1991) (see Section II.1) can be further analyzed by diffusion limited precipitation theory to derive Er diffusivity at 900°C. In the homogeneous nucleation precipitation theory, the density of precipitates N_p depends on annealing time and temperature, the Er diffusivity, and the nucleation barrier (F) for Er-Si precipitates (Ren *et al.*, 1993).

Figure 8 shows a fit of the data from Eaglesham *et al.* (1991) to a diffusion limited precipitation model. The data exhibit two key features: an "incubation" stage of [Er] at 2×10^{18} cm^{-3} for a 900°C, 30 min anneal, and a linear increase in precipitation density with [Er] beyond that concentration. The linear increase of precipitates with [Er] is consistent with a nucleation limited precipitate density. The incubation stage represents a minimum time

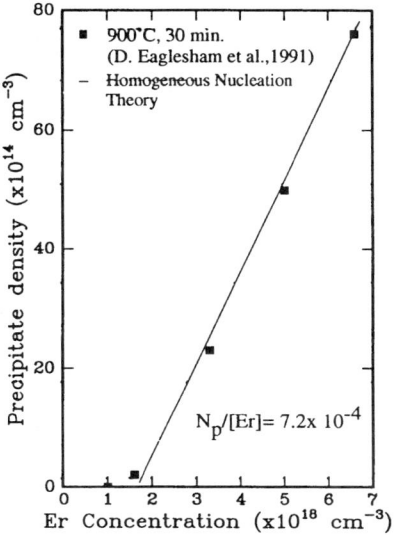

FIG. 8. Precipitation density of Er in Si for a 900°C, 30 min anneal vs. Er peak concentration. The line shows a fit using homogeneous nucleation theory.

required for Er atoms to associate at the temperature and is also dependent on the Er concentration. The observed value of 10^{18} cm^{-3} sets a lower limit of the diffusivity $D_{Er} > 6 \times 10^{-16}$ cm^2/s at 900°C. The slope of the data $N_p/[Er] = 7.2 \times 10^{-4}$ is very sensitive to D_{Er} and F in the analysis. A two parameter fit yields $D_{Er} = 10^{-15}$ cm^2/s and $F = 1.57$ eV. The nucleation barrier is similar to values discussed for O precipitation in Si (Freeland et al., 1977).

A migration enthalpy for Er diffusion, derived from the diffusion experiments, of 5.1 ± 0.3 eV is obtained. The migration enthalpy of Er is similar to that of Ge diffusion in Si (Sze, 1981). Since Er has a large atomic radius $r_a = 1.78$ Å, it is not surprising that Er is a slow diffuser in Si. The migration enthalpy for Er diffusion in Si can be estimated using an elastic energy model (Utzig, 1989). In the model, a diffuser migrates interstitially and moves from tetrahedral interstitial sites (T site) through the hexagonal interstitial sites (H site) in Si. The migration enthalpy for interstitial diffusion is derived mainly from the elastic energy difference at the saddle point, H site, and at the equilibrium point, T site. The same approach has successfully explained the diffusion of dissolved 3d transition metal atoms in Si (Utzig, 1989). For Er diffusion in Si the model predicts a migration energy of $E_m = 5.7$ eV. The agreement between model and experimentally obtained migration is good. This agreement suggests that Er diffuses interstitially in Si.

The flat plateau in the Er profiles in Fig. 7 represent the equilibrium solubility of Er in Si. Figure 9 shows the Er equilibrium solubilities in Si at different annealing temperatures. They are in the range of 10^{16} cm^{-3}, similar to S in Si and retrograded at $\sim 1200°$C (Sze, 1981). Yang (1991) reported solid solubility of Er in Si at 900°C. The measured doses correspond to concentrations of about 3×10^{16} cm^{-3}. This equilibrium solid solubility $[\text{Er}]_{\text{sol}}$ is significantly less than the value of 1×10^{18} cm^{-3} reported by Eaglesham *et al.* (1991) as an onset for Er–Si precipitation at 900°C. Because of the slow Er diffusivity, its equilibrium solid solubility can be significantly different from the threshold concentration for the onset of precipitation. The metastable threshold concentration may be controlled by either kinetics or equilibrium factors.

The metastable threshold concentration can be further altered by modifying either kinetics or equilibrium factors in Si:Er. Recently, Polman *et al.* (1993) reported (see Section II.2) an incorporation of much higher concentrations ($\sim 10^{20}$ cm^{-3}) of Er in c-Si by solid phase epitaxy of Er-implanted amorphized Si. The limit for Er incorporation was temperature dependent, decreasing from 1.2×10^{20} cm^{-3} after 15 min anneal at 600°C to 6×10^{19} cm^{-3} after 15 s anneal at 900°C. A much higher migrational barrier for Er precipitation must be in place than in Eaglesham's observation. Conversely, the precipitation kinetics can also be enhanced and therefore reduce the metastable Er concentration. In a Si:Er sample with a flat Er concentration of 5×10^{17} cm^{-3}, created by multiple implantations, less than 20% of ErSi$_2$ precipitates were reported (Polman *et al.*, 1993). The precipitation enhancement was likely due to high defect densities created during the

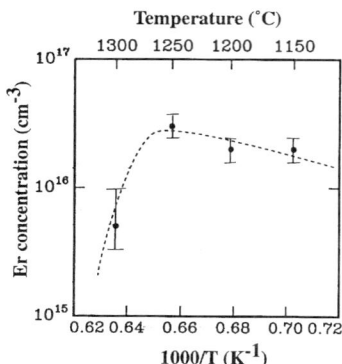

FIG. 9. Equilibrium solubility limit of Er in Si at different annealing temperatures. The dotted line serves to guide the eye. (From Ren, 1994b.)

multiple Er implants. Impurity doping can therefore alter the equilibrium conditions in Si:Er.

IV. Light Emission

1. PHYSICS OF LIGHT EMISSION

Rare earth (RE) atoms exhibit a wide variety of interesting properties. They are unusual in that they have an unfilled core shell of electrons (the 4f shell) embedded in the interior of the atom. Because of the shielding effect, it was expected, and indeed observed, that they are relatively unaffected by the ion's external environment. Most of the free rare earth atoms have a $4f^{n+1}6s^2$ configuration. In metals, the three free-atom valence electrons form the conduction band, while in salts, the valence electrons are transferred to other ions; in both cases, it often appears to be a reasonable approximation to consider the constituent RE atoms as behaving like trivalent ions and having the ionic configuration $4f^n$ ($n = 1$ to 13 from Ce^{+3} to Yb^{+3}). Sharp absorption and fluorescence spectra in the visible or infrared regions have been observed for RE atoms in ionic materials where the RE ions are mostly in the 3^+ charge state (Crosswhite and Moos, 1967). It is known that, for free ions, intra-4f electric dipole transitions are strictly forbidden by the parity rule. Therefore the observation of luminescence spectra is due to the influence of the crystal field on energy levels and wavefunctions which reduces the degeneracy of the levels and modifies the electric dipole rule. Because RE spectra show sharp lines, it is reasonable to interpret them as transitions between intra-4f orbitals slightly perturbed by the crystalline environment. When an RE atom is placed in the crystal it is subject to a number of forces, which are absent for the free ion, due to interaction with neighboring ions through molecular orbital interactions.

The identification of free ion levels from ionic material spectra was highly successful and the lower levels for almost all rare earth ions have been identified by interpreting the lines as transitions between the ionic excited and ground spin–orbit states weakly split by the crystal field (Crosswhite and Moos, 1967; Dieke, 1968).

The 4f electrons in RE ions exhibit a type of spin–orbit coupling that follows the Russell-Saunders (RS) approximation fairly closely. The energy level scheme for any ion thus consists of a number of multiplets, each of which can be described by two quantum numbers L and S corresponding to the total orbital and total spin angular momentum, respectively. The individual levels that make up a multiplet are characterized in addition by different values of J, corresponding to the total angular momentum. For spin–orbit coupling of the RS type, the spacings between the various levels that comprise a multiplet are given by the Landé interval rule, which states that the energy separation between two neighboring levels is proportional to the higher J value of the pair.

The positions of many of the ground multiplet levels are known from experiments with rare earth salts. The observed multiplet structures may be compared with those calculated on the basis of RS coupling, by arranging the calculated results to have the same overall splittings as the experimental results.

Rare earth energy levels in the weak field approximation, in which J is a good quantum number, are generally used. This has been the natural approach because the crystal field is normally small compared to the spin–orbit coupling energy. However, in principle it is possible to consider the other limiting case as well. An example of the energy levels of the $4f^{13}$ configuration, in T_d symmetry, is schematically shown in Fig. 10 as a function of the relative strength of the crystal field and spin–orbit interactions. As can be seen, the 4f wavefunctions, in a T_d crystal field in the absence of spin–orbit coupling (the extreme right of Fig. 10), transform as the single group representations (a_1, t_1, and t_2). When the spin functions are connected to the crystal field states, double group representations are required (Γ_6, Γ_7, and Γ_8). Naturally, the ordering in Fig. 10 will change depending on the magnitude of the crystal field, but symmetry classifications of the states will remain since they are appropriated to the T_d group.

Following the RS coupling scheme, the Er^{3+} ion ($4f^{11}$ configuration) is

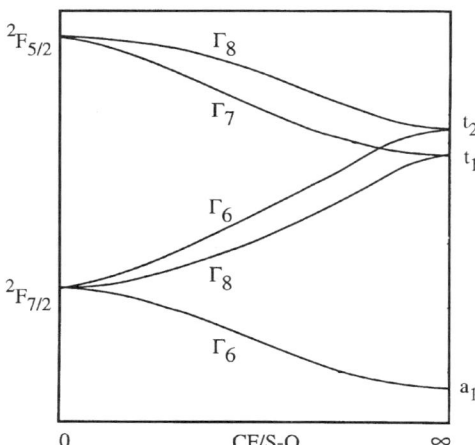

FIG. 10. Schematic of energy levels of the $4f^{13}$ configuration, in T_d symmetry, as a function of the relative strength of the crystal field and spin–orbit interactions.

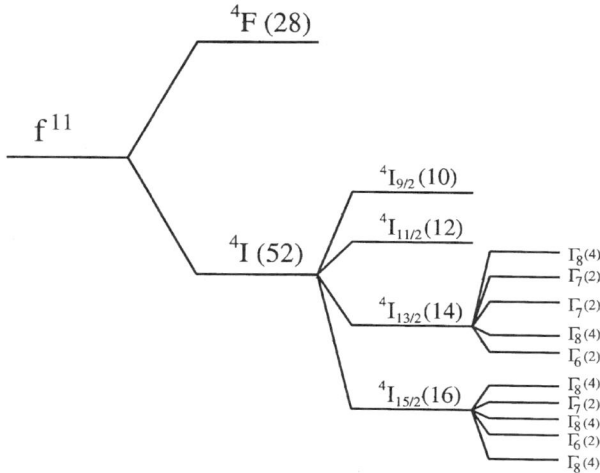

FIG. 11. Stark level splitting of the 4f^{11} configuration for T$_d$ symmetry.

represented by the ground state multiplet ^4I and the first excited multiplet ^4F. The energy difference between them is of the order of 2.0 eV. As the ground multiplet configuration has $L = 6$ and $S = 3/2$, the possible J values are 15/2, 13/2, 11/2, and 9/2, such that 15/2 is the lowest spin–orbit state. Using the Landé interval rule one could find the difference energy between the two lowest spin–orbit states. In a weak crystal field the spin–orbit manifolds split into a number of Stark levels which are represented by two-fold (Γ_6 and Γ_7) or four-fold (Γ_8) irreducible representations of the double T$_d$ group. The ordering of these Stark levels would change depending on the magnitude of the crystal field. This picture is schematically shown in Fig. 11. If the crystal field symmetry is lower than T$_d$, the number of Stark levels from $^4I_{15/2}$ is eight and from $^4I_{13/2}$ is seven. Transitions between these levels are expected to be observed in absorption or emission.

The first Er related PL in Si was observed by Ennen et al. (1983) and is shown in Fig. 12. Er was implanted at low energies during Si MBE growth. The luminescence spectrum exhibits several sharp peaks around 1.54 μm, corresponding to the $^4I_{13/2}$–$^4I_{15/2}$ transition. The large number of lines observed indicate that several Er centers with different symmetries are present. As described before, a single Er center with low symmetry will show only up to eight transitions at low temperatures due to the maximum splitting of the spin–orbit $^4I_{15/2}$ ground state. These transitions involve the first excited $^4I_{13/2}$ and the $^4I_{15/2}$ ground states, weakly split by the crystal field with an energy of about 0.1 eV. At low temperatures, the linewidth of

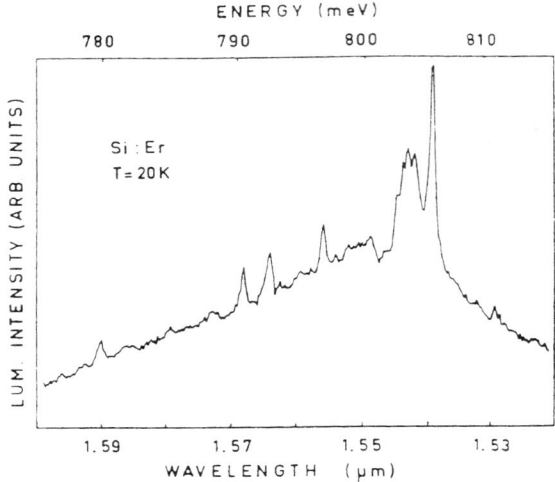

FIG. 12. The luminescence spectrum at 20 K of Er-implanted and annealed Si. The half-width of the intense line at 1.53904 μm (805.6 meV) is 3.4 cm^{-1}. (From Ennen et al., 1983.)

these PL lines is generally below 1 Å, as was demonstrated for Er in GaAs (Thonke, Hermann, and Schneider, 1988), due to the long lifetime (~1 ms) of the partially forbidden transition. At higher temperatures, transitions from higher $^4I_{13/2}$ split states can be observed (Ren et al., 1994a). Furthermore a broadening of the lines is reported. At 300 K the Er PL linewidth is 140 Å (Michel et al., 1991), significantly smaller than, for example, band edge transitions from III–V semiconductors. Both PL and, by fabricating a p-n diode, EL was observed. The observation of EL demonstrates that carriers are used to excite Er contrary to direct optical excitation, as used to excite rare earth metals in ionic materials.

Tang et al. (1989a) interpreted a dominant five-line spectrum (Fig. 13), measured after implantation of Er into bulk Si, as Er in T_d symmetry due to the Stark splitting. This five-line spectrum was generally observed in Czochralski grown (Cz) Si with an O concentration of about 10^{18} cm^{-3} (Michel et al., 1992). High resolution spectra as depicted in Fig. 14 revealed beside the five dominant lines a number of weaker lines, again pointing to additional low symmetry centers (Michel et al., 1991). The five-line spectrum was not observed in float zone grown (FZ) Si (O-concentration ~10^{16} cm^{-3}). Figure 15 shows PL spectra of Er co-implanted with N in FZ and Cz Si. As reference, a spectrum of Er in Cz Si is shown. Four lines of the five T_d related lines are visible and marked by arrows. In the FZ sample, the Er PL consists of one broad line around 1.54 μm, possibly inhomo-

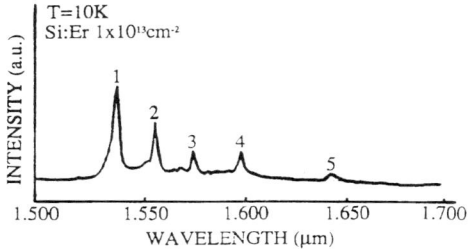

FIG. 13. PL of a sample annealed at 900°C for 30 min. (From Tang *et al.*, 1989.).

geneously broadened. C and F co-implanted FZ material shows the same broad PL line while the same elements, co-implanted in Cz material, yield as dominant PL features the five-line spectrum. It was therefore concluded that the five-line spectrum is due to Er surrounded by O as ligands. Recent reports gave an opposite view (Pryzbylinska *et al.*, 1995). A PL study of Er implanted Si showed that the five-line spectrum was also observed in FZ material (Fig. 16). It is not clear if this spectrum is due to Er surrounded by residual O or if it is indeed substitutional Er in T_d symmetry. Furthermore, for Er in Cz material a seven-line spectrum is reported, contrary to the generally observed five-line spectrum. It appears that the implantation and

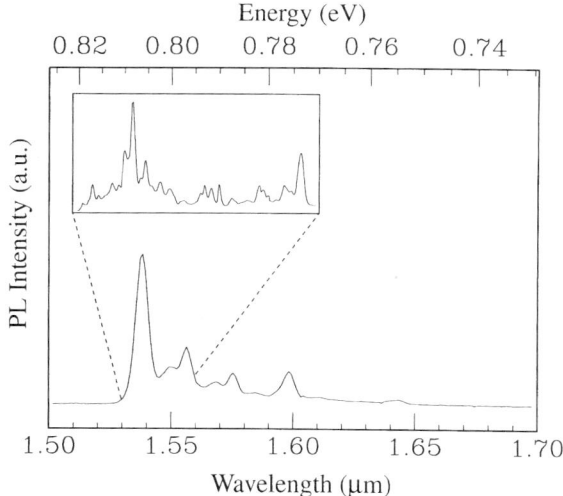

FIG. 14. High resolution PL spectrum shown in the insert with a spectrometer resolution of 2 Å. $T = 4.2$ K.

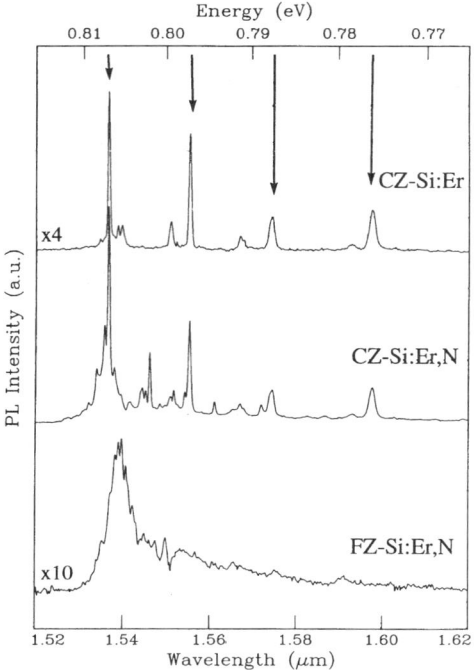

FIG. 15. PL spectra of Er co-implanted with N. The peak concentration for Er is 1×10^{18} cm^{-3} and for N is 5×10^{17} cm^{-3}. The oxygen related lines are marked. $T = 4.2$ K.

annealing conditions are controlling the different Er centers that are formed. It is still impossible to draw a conclusive picture of the crystal field splitting and local, microscopic environment. This situation can change with the controlled growth of Er in Si by MBE or CVD, as described in Sections II.3 and II.4.

In FZ and Cz Si, extended X-ray absorption fine structure (EXAFS) measurements were performed to resolve the microscopic structure of the Er centers. It was concluded that there are two possible structures for the centers in Si:Er samples (Adler et al., 1992). Comparing the structure around the Er impurity for FZ and Cz Si with ErSi$_2$ and Er$_2$O$_3$, respectively, it was proposed that in Cz material, Er is surrounded by six O atoms as first neighbors in quasi-octahedral symmetry, while in FZ material Er is surrounded by 12 Si atoms. The Er–O configuration was proposed to be the optically active center. So far, these experimental data have not been directly correlated with any other measurement method.

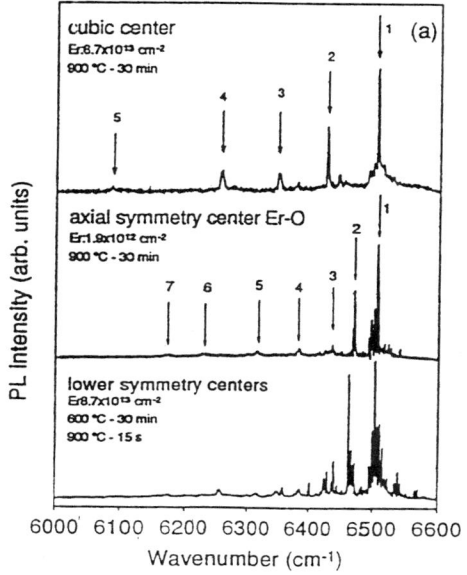

FIG. 16. Spectra of Er implanted into Si at 2 MeV (top and bottom) and 320 keV (middle). The Er dose and the annealing conditions are indicated in the figure. The middle trace spectrum is amplified by a factor of 50. (From Jantsch et al., 1996.)

2. LIGANDS

Rare earth metals are known to be very reactive. They show especially a high affinity to O. Therefore it is not surprising that Er also acts as an O getterer in Si. Favennec et al. (1990) were the first to show that the O concentration in Si influences the luminescence intensity of Er in Si. Co-implantation of Er with O yielded higher light emission than a simple Er implant. Annealing experiments of Er implanted Si support this observation. Figure 17 shows the PL intensity of Er in Cz and FZ Si at different annealing temperatures. While Er in FZ material shows very little luminescence after a 900°C anneal, co-implantation with O improves the luminescence by two orders of magnitude to reach Cz luminescence levels (Michel et al., 1991). Different processes dominate at different temperatures. Below 800°C the light emission is mainly due to Er-defect complexes, generated after implantation and insufficient annealing. At about 900°C the dominating process is the formation of Er–O complexes. The implantation defects are removed due to the anneal. The decrease in luminescence above 900°C

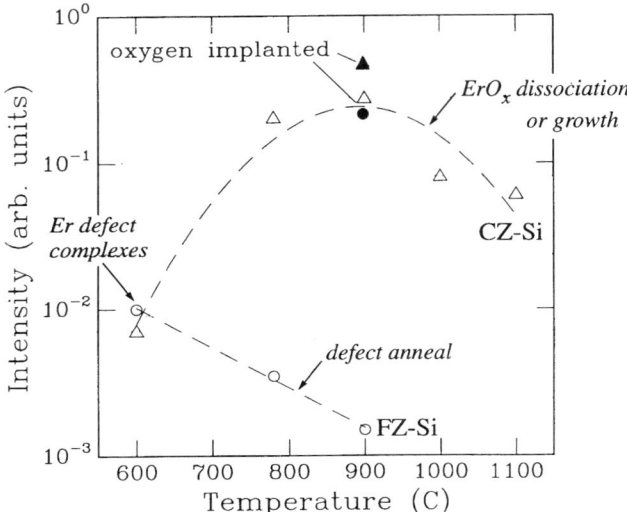

FIG. 17. Annealing temperature dependence of the Er PL intensity at 4.2 K for FZ and Cz silicon. The filled data points are for FZ samples with additional oxygen implantation. (From Michel et al., 1991.)

occurs because of the dissociation of the Er–O complexes. A detailed study was done for Er–F complexes, which show a very similar annealing behavior (Ren et al., 1994a).

Further experiments widened the number of elements that enhance the Er light emission. Experiments have shown that the presence of ligands with large electronegativity greatly enhance the PL intensity, suggesting that ligands increase the concentration of optically active Er centers. Nitrogen, carbon, and fluorine showed a luminescence enhancement that surpassed O. Figure 18 shows the PL intensity of the Er 1.54 μm line in dependence of the ligands. This comparison was taken at 4.2 K. An influence of the ligand on the PL spectrum was observed (Michel et al., 1992). Some PL lines could be correlated to boron and nitrogen in the immediate neighborhood of Er, implying a change of the crystal field due to the different co-implants. While all samples containing O always show the previously mentioned five-line spectrum, FZ samples show a significant smaller crystal field splitting of the spectrum. All lines are located in a very small energy range, giving the appearance of a single broadened line (see Fig. 15).

It is not clear yet what the role of the ligands is. Four different mechanisms are possible:

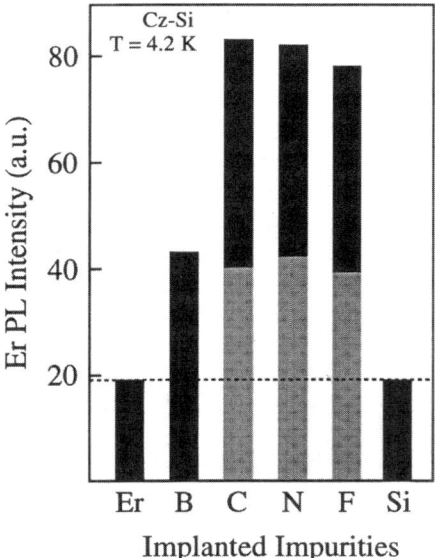

FIG. 18. The effect of co-implantation of Er with other elements in Cz and FZ silicon on the Er PL intensity. The peak concentration of all elements was 1×10^{18} cm^{-3}. The grey bars represent the PL intensity in FZ.

1. The Er-ligand complex provides dipole coupling to channel e-h recombination energy to the 4f manifold.
2. Local phonon modes of the ligands dissipate energy to promote resonant coupling of a portion of the recombination energy to the 4f manifold.
3. The ligand field adds to the Si lattice contribution in breaking the inversion symmetry of the ion site. The symmetry breaking allows admixing of states of opposite parity to allow the optical transition that is dipole forbidden for the isolated ion.
4. The ligand ions enhance the solubility of Er in Si by the formation of localized complexes.

Recently, it was proposed that the role of co-dopants in the luminescence from Er in Si is not related to varying the character of the wavefunctions and consequently altering the transition probabilities, but shielding the Er ion from the rest of the lattice, increases the availability of the radiative paths of relaxation for the f shell (Anderson, 1996).

3. Electrical Properties

When dealing with defects in Si, important information about the defects always includes the energy levels in the bandgap, correlated with the defects. If the defects are thought to be used as dopants, the knowledge of their electrical activity is crucial for device performance. The experimental findings for the electrical activity of Er in Si have been controversial and the optical centers have been labeled donor, acceptor, and electrically neutral. In the following part we will critically review the experimental evidence of the electrical activity of Er in Si.

In an effort to make p-n diodes in Er doped Si, Ennen et al. (1985) noted that the Er doped epitaxial Si layer turned p-type, indicating the formation of acceptors, while undoped Si showed n-type behavior. The source of the p-type doping was not determined but Er related defects were suggested as cause for the electrical behavior.

Widdershoven and Naus (1989) characterized Er implanted Si with deep level transient spectroscopy (DLTS) and capacitance-voltage (CV) profiling and correlated the results with SIMS profiles. After Er implantation and annealing, n-type samples showed an increase in the donor concentration while in p-type samples the acceptor concentration decreased. CV profiles showed a strong correlation between the change in carrier concentration and the Er concentration profile taken by SIMS. They concluded that the excess donor concentration after implantation is most likely due to Er related centers where Er occupies several different lattice positions.

An extensive study on the electrical activity of Er in Si by Benton et al. (1991) confirmed that Er implantation leads to donor states. Spreading resistance, Hall effect, and CV profile measurements showed an increase in the mobile carrier concentration with increasing Er concentration up to 10^{18} Er/cm^3 as depicted in Fig. 19. The donor concentration was approximately equal to the Er concentration until near the solubility limit, where it saturated at a maximum value of 4×10^{16} cm^{-3}. The experiments were performed on ion implanted material. Several deep states were identified by DLTS and correlated with Er related defects.

Priolo et al. (1995) increased the donor concentration by co-implanting with O. With an Er dose of 6×10^{15} cm^{-2} and an O concentration of 2.5×10^{21} cm^{-3}, the donor concentration reached a maximum value of 2×10^{19} cm^{-3}. However, the Er dose was above the solubility limit and precipitates were evident in TEM. The PL for this sample was weaker than for samples in the previous study, reportedly due to the defects in the material. Similar studies were also undertaken with carbon to maximize the donor concentration (Priolo et al., 1995). The highest donor concentration of 2×10^{18} cm^{-3} occurred with Er and carbon concentrations of

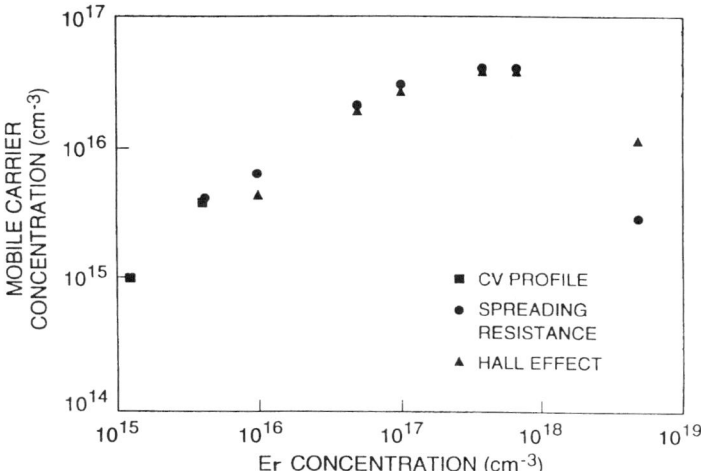

FIG. 19. Free-carrier concentration measured as a function of implanted Er concentration. Data marked by the circles are spreading resistance results, data marked with a triangle are the results of Hall effect measurements. All samples are CZ p-type silicon. (From Benton et al., 1991.)

7×10^{18} cm^{-3} and 1×10^{18} cm^{-3}, respectively. The film quality was much better than for Er/O samples, possibly due to carbon's ability to reduce secondary defects during annealing (Tamura and Suzuki, 1989). The integrated photoluminescence intensity was three times higher than Er-implanted Cz Si.

Libertino et al. (1995) studied the influence of O on Er related deep states with DLTS and CV measurements. They again observed donor behavior of Er related defects. Figure 20 depicts three major traps identified as Er related. These traps are localized at $E_c - 0.51$ eV, $E_c - 0.26$ eV, and $E_c - 0.20$ eV. The peak at $E_c - 0.34$ eV was attributed to residual damage from the implantation process. An increase in O concentration modified the DLTS spectra and promoted most of the deepest levels to shallower ones. An additional peak at $E_c - 0.15$ eV was observed in O-rich material. This activation energy is identical with the activation energy for the Er–PL decay and was therefore attributed to Er–O complexes.

Although most authors report donor behavior for Er related centers in Si, there are serious challenges to this view. Sobolev et al. (1993) reported that Er indiffused from the surface always generates acceptor states. He related the acceptor concentration to the Er concentration. Ren (1994b) observed p-type conversion of an n-type Si wafer after high temperature surface

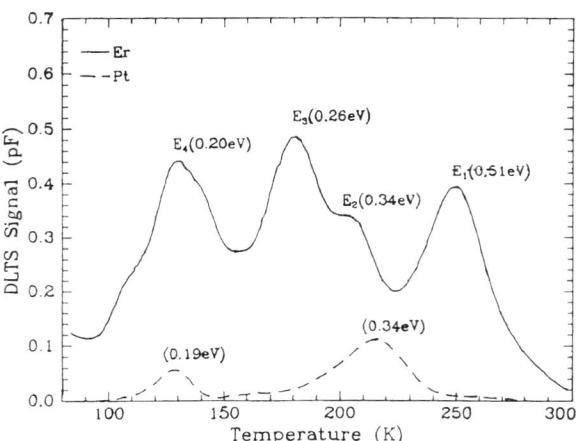

FIG. 20. DLTS spectrum for Er in epitaxial c-Si (solid line). Erbium was implanted at 5 MeV to a fluence of 6×10^{11} ions/cm^2 and annealing was performed at 900°C for 30 sec. The DLTS spectrum for a sample in which Pt was introduced using identical implant and annealing conditions is also reported (dashed line). (From Libertino et al., 1995.)

indiffusion. An attempt to correlate the acceptor concentration with an Er SIMS profile failed because of the very low Er concentration. As mentioned in Section III, Er indiffusion may lead to Er clusters and precipitates that make interpretation of the measurements very difficult.

The most serious challenge comes from experimental evidence reported by Benton et al. (1993). Annealing experiments on Er implanted Si showed a significant change in donor concentration while the Er PL intensity was unchanged. It was also possible to observe a strong PL signal after the layer showed compensation due to the annealing process. These experiments led to the conclusion that the optically active Er centers are not correlated with any electrical activity.

Recent experiments on 400 keV Er implanted Si show a linear dependence between donor activity and PL intensity for different annealing temperatures as depicted in Fig. 21 (Morse et al., 1996). The donor profile was measured by spreading resistance and follows precisely the Er SIMS profile. Cross-sectional TEM performed after the anneal does not show any end-of-range secondary defect structure prevalent after high energy implantation.

Apparently the sample preparation plays a significant role in the electrical characteristic of the Er containing layer. The majority of the experimental evidence points to the correlation between optically active Er and donors, observed by different measurement techniques. Nevertheless, experimental

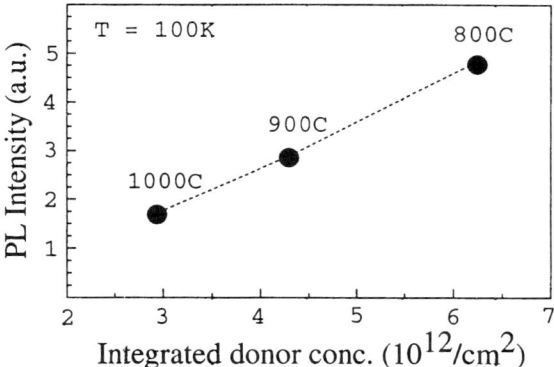

FIG. 21. PL intensity dependence on the added donor concentration, measured by spreading resistance. The donor concentration was dependent on the annealing temperature.

evidence that shows different electrical characteristic for Er, needs to be explained.

4. ACTIVATION AND DEACTIVATION PROCESSES

One of the major obstacles for Er to be used for the fabrication of Si LEDs is the strong luminescence quenching towards higher temperatures. It was noted by Ennen *et al.* (1983) that the luminescence disappears at temperatures above 200 K. More careful investigations showed a decay process that is controlled by three activation energies as shown in Fig. 22 (Efeoglu *et al.*, 1993). These activation energies were correlated with possible Er related defect levels in the bandgap and the luminescence decay interpreted as a result of the thermalization of these levels.

Favennec *et al.* (1989) compared the PL decay of Er in several different semiconductor materials. The data implied that the onset of the decay depends on the bandgap of the semiconductor material. Semiconductors with large bandgaps showed a later onset than small bandgap semiconductors. Later experiments showed that the onset of the PL decay is not simply controlled by the bandgap. Room temperature luminescence in Si was observed by improving annealing conditions and thereby surpassing the performance of Er in GaAs, a semiconductor with a larger bandgap than Si. The activation energy of the decay has been shown to vary depending on the annealing and implantation conditions (Michel *et al.*, 1992). Variation of the decay with processing conditions made obvious that de-excitation of Er

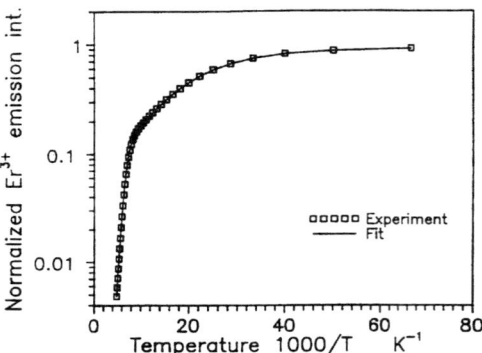

FIG. 22. Temperature dependence of the integrated intensity of the 1.54 μm line of the Er^{3+} transition. The sample was implanted with oxygen at the $10^{19} cm^{-3}$ level and annealed at 950°C for 30 min. A fit of the data shows three quenching mechanisms with distinct activation energies of 4 meV, 12 meV, and 140 meV, and coupling coefficients 0.7, 17 and 5×10^5, respectively. (From Efeoglu et al., 1993.)

is not due to a single process. It is rather, a competition of several serial and parallel processes.

Figure 23 depicts the different processes involved in the excitation and de-excitation of Er in Si at low temperatures (Palm et al., 1995). Processes listed in regimes I–III compete with the Er excitation. Carriers can be easily trapped by deep defect centers and are therefore lost for the formation of excitons. Excitons then have different paths to recombine. The most significant de-excitation, besides the excitation of Er, occurs through electrons in an Auger recombination because of the high carrier densities during Er excitation. These processes determine the pumping efficiency. After Er is excited, two processes can occur to relax the Er ion. Radiative recombination leads to light emission at 1.54 μm while an Auger-impurity effect can relax the Er nonradiatively. For serial processes the slowest process will dominate, while for parallel processes, the fastest mechanism will control the energy flux.

Palm et al. (1996) showed that Er is excited through excitons at low temperatures. The excitation is competing with an electronic Auger recombination where electrons loose their energy through interaction with other electrons. The significance of this effect was shown by using different bias conditions in an Er LED. Figure 24 shows in a concise form the excitation and recombination processes in two temperature regimes. Regime I refers to a temperature below 100 K while regime II comprises temperatures above 100 K. While in regime I the excess energy after e-h recombination and Er

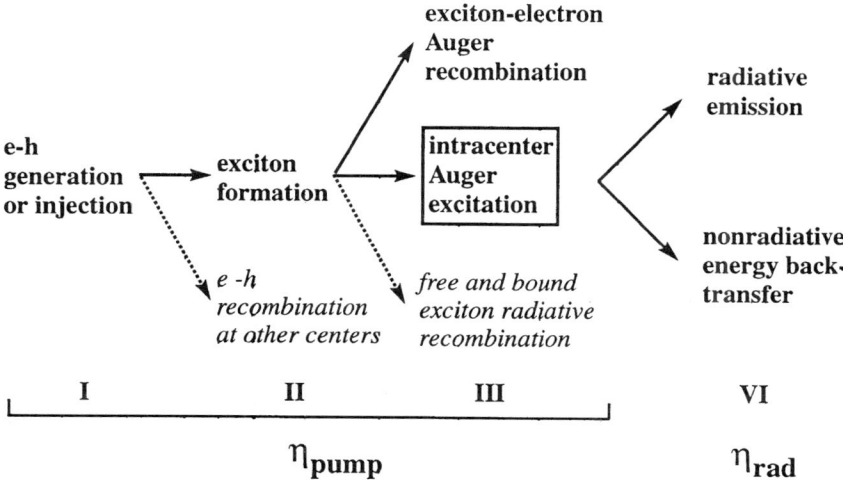

FIG. 23. Excitation and de-excitation processes of Er in Si. All processes in bold letters are comprised in a model by Palm et al. (1996) to describe this low-temperature Er luminescence phenomena.

excitation is used to ionize electrons bound to donor levels, the excess energy in regime II is released by the emission of phonons. A level at ~150 meV was correlated with the optically active Er and an energy backtransfer was observed by junction photocurrent spectroscopy (JPCS) (Palm et al., 1997; Michel et al., 1996; De Dood et al., 1996). The photocurrent of an Er LED was observed depending on the excitation

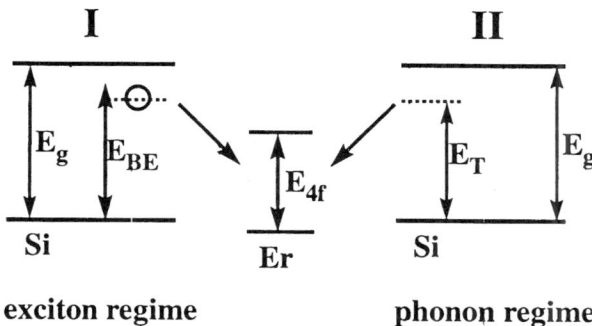

FIG. 24. Excitation of Er in Si different temperature regimes. Regime I shows excitonic excitation for $T < 100$ K. Regime II shows phonon assisted excitation for $T > 100$ K.

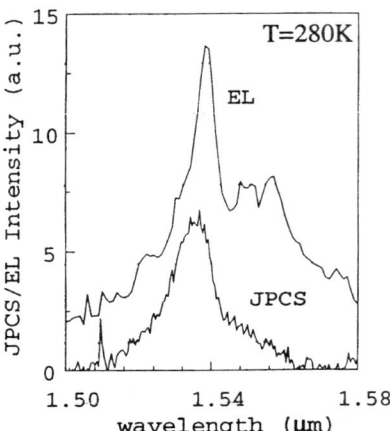

FIG. 25. Junction photocurrent spectroscopy (JPCS) on a Si:Er LED. The JPCS spectrum is compared to an EL spectrum from the same LED. (From Michel et al., 1996.)

wavelength. At 1.54 μm the photocurrent shows a peak due to the energy backtransfer from the excited Er as shown in Figure 25. The temperature dependence of the JPCS signal shows a thermally activated process with an activation energy of 170 meV. This energy is interpreted as the donor related binding energy of the Er related defect and represents the bottleneck of the energy flux.

Several recent publication support the approach of Palm et al. (1996) by studying different aspects of the excitation and de-excitation processes (Gregorkiewicz et al., 1996; Hartung et al., 1996).

V. Electronic Structure

A vast amount of experimental work has been done in the Si:Er system to understand the structure, energy transfer processes in the excitation, de-excitation, and backtransfer mechanisms for characteristic 4f shell luminescence, device process, high quality film growth, ligand enhancement of excitation mechanism, and electrical properties. In spite of these efforts the microscopic structure of the optically active Er-ligand complexes is still unknown. It is desirable to understand the lattice site of the Er impurity after it is incorporated in Si because this information could help to reduce the defect density and grow high quality Er-doped optoelectronic Si.

The theoretical treatment of the interaction of a RE ion with its covalent environment is more complicated than that for a d-shell ion. The chemical nature of Er impurities in Si remains controversial. Although the luminescence spectra have been identified as intra-4f transitions, the 3^+ oxidation state is misunderstood. It is important to know why, when Er is incorporated in the material, it *prefers* to have a $4f^{11}$ ionic configuration and why the luminescence linewidth is dependent on the nature of the host. It seems that the covalent environment has an effective role in determining the optical properties of the centers. The understanding of RE impurities in semiconductors is still a challenge from both theoretical and experimental points of view. The tendency of RE ions to behave like free ions has greatly simplified the interpretation, but it has been observed that experiments can *see* the effects of the environment on the RE transitions. This suggests that more accurate calculations are needed to interpret the optical spectra and crystal field splittings, since they depend on the precise form of the 4f orbital behavior.

In spite of active experimental research about Er related impurities in Si, there have been few fundamental theoretical studies of these centers. This is in part because of the complexity of the problem to treat the Er 4f orbitals, as was pointed out before. Another reason is the lack of direct experimental observations related to the microscopic structure of the center. Electron paramagnetic resonance (EPR) studies, usually the most effective way to establish the structure and identity of an impurity, have not been reported for optically active Er related centers in Si although optically inactive centers were identified (Pryzbylinska et al., 1994; Jantsch et al., 1996).

The physical properties of impurities in covalent semiconductors are determined by the nature of the interaction between the impurity and the environment. A theoretical study of impurity and environment is too complex to be treated exactly. The method of calculating the structure and energy of an impurity by solving the many-body electronic Schröedinger equation is to transform it into coupled *one-electron* equations using the Hartree-Fock (HF) or density-functional theories. The HF theory is a variational wavefunction procedure and the density-functional theory is a variational charge density procedure. Local and nonlocal potentials have been proposed to simplify the computational problems associated with the calculations. The development of pseudopotentials, which remove core electrons from the problem and therefore leaving only the valence electrons to be considered, has been essential in allowing impurities in solids to be simulated. The results of the electronic structure calculations related to impurities and defects in crystals, using one of the various theoretical models existing in literature (DeLeo and Fowler, 1991), can give energies and wavefunctions or total energies with respect to atomic positions. The results of one-particle wavefunctions and energies can be used to estimate energy level positions. These results can also be compared with EPR and optical excitation measurements. From total energy surfaces one can obtain stable and metastable ground state configurations and then, for example, estimate migration barriers, formation energies, and vibrational frequencies. However, in the evaluation of any calculation it is important to keep in mind the origin of the approximations used to solve the electronic problem and what can be extracted from the results, knowing their limitations, advantages, and disadvantages. From computational simulations of the electronic structure of impurities in semiconductors, one can obtain information that serves to complement or guide the corresponding experimental study to better characterize or understand the system's main properties. This kind of complement between theoretical studies and experimental observations has been demonstrated throughout the years and has been essential to characterize deep light and heavy impurities in semiconductors with special emphasis on Si.

1. Electronic Structure of Er-related Impurities in Si

As was pointed out, electron–electron interaction and spin–orbit coupling are important to describe the Er 4f electrons. Even though free RE atoms have a 4f unfilled core shell of electrons embedded in the interior of the atom, they are affected by the ion's crystalline environment. Therefore, they can be thought of as a special class of impurities showing an atomic-like character and at the same time effects from the local environment. One of the approaches to describe the electronic structure of Er in Si is to assume that the 4f orbitals can be treated as a frozen core ignoring the overlap between them and the s and p states of the neighboring atoms. In this way, the atomic character of the 4f electrons is preserved. As a frozen core description, the 4f multielectron interaction and spin–orbit coupling can be left out from the crystal-impurity interaction description and one can consider only the 6s and 5d atomic states to be responsible for the impurity coupling. This approach cannot explain the crystal field splittings and the observed luminescence spectra, since they are due to the influence of the crystal field on Er 4f energy levels and wavefunctions.

Another approach is to solve the Schröedinger equation to describe interactions between Er 4f and 6s states and the Si crystal. This methodology can be classified as a strong crystal field approach and is convenient to discuss covalent bonding and to describe the directionality of the bonds, and thus the ability of the wavefunctions to overlap with the neighboring atom wavefunctions. The evaluation of this interaction is required to achieve the crystal field splitting energy. The general indication is that the covalent bonding effects can be expected to be of very great importance in understanding RE crystal field splittings (Crosswhite and Moos, 1967).

2. ISOLATED Er IMPURITY IN Si

The first study of the isolated substitutional Er impurity in Si was done by using a self-consistent tight-binding Green's function formulation (Delerue and Lannoo, 1991). These investigations studied the chemical trends of substitutional rare earth impurities in semiconductors. It was assumed that the 4f shell could be kept as a frozen core. It was further assumed that the 6s states were so extended that they would couple strongly and would be spread equally over the valence band bonding states and the conduction band antibonding states. Using this convention, the electron population of the 6s orbitals for the atom in the crystal was assumed to be $6s^1$. Therefore, it was considered that the rare earth impurity could be described by the atomic configuration $4f^{n+3-m}5d^{m-1}6s^1$, where m is the number of valence electrons ($m = 2, 3, 4$ corresponding to oxidation states). It was argued that such an atomic-like point of view was developed successfully for RE metals and this approximation could be justified by the luminescence experiments. Based on the frozen-4f-shell approach, it was considered that only the 5d atomic states were responsible for the impurity coupling. Therefore, using this picture, the essential difference between 3d transition metal and RE impurities is that the electron population, in the case of RE elements, is determined by the 4f core. Discarding the 4f and 6s orbitals from the tight binding matrix, the problem that had to be solved was the same as for 3d transition metal impurities. The results for substitutional Er in Si showed no 5d derived gap states.

It was also found that the $+3/+4$ 4f core energy level transition was located in the conduction band while the $+2/+3$ transition was found resonant in the valence band. It was concluded that substitutional Er in Si should be in the 3^+ state as observed experimentally, even though isolated Er impurities could be on interstitial sites (Ennen, Pomrenke, and Axmann, 1988; Prybylinska et al., 1995; Jantsch et al., 1995).

Another theoretical simulation, using a 32-atom supercell of Si, was performed by calculating the total energy of isolated Er at several sites in the Si lattice to predict the lowest energy configuration (Needles, Schlüter, and Lannoo, 1993). Three high symmetry geometries for the Er atom and two oxidation states were investigated. The total energy of the system was evaluated for the impurity in the tetrahedral and the hexagonal interstitial sites and the substitutional site. The simulations were performed within the framework of the density-functional theory and norm-conserving nonlocal pseudopotentials. The f electrons were treated as core electrons and the scalar relativistic Er pseudopotential was generated from a relativistic atomic calculation. A quantum molecular dynamics technique was utilized to minimize the total energy of the supercell with respect to atomic and electronic configurations. Since the calculations were performed within the pseudopotential approach where the f orbitals were included in the core of the Er atom, the oxidation states investigated were $+2$ ($4f^{12}6s^2$ isolated ground state; Er^{2+}) and $+3$ ($4f^{11}6s^2 5d^1$; Er^{3+}). It is important to note here that even though these calculations assumed a frozen f core, the starting atomic configurations are different from those assumed in the first referred theoretical simulation (Delerue and Lannoo, 1994).

It was found that Er in the tetrahedral interstitial site presents the lowest energy for both oxidation states. This conclusion was reached by fitting the electron promotion from the lowest energy configurations $4f^{12}6s^2$ and $4f^{11}6s^2 5d^1$ with the experimental transition energy (0.89 eV) between the multiplets $^3H_6$ and $(15/2, 3/2)^0$ with $J = 6$ (Martin, Zalubas, and Hagan, 1971).

Using this approximation, the cohesive energy of an isolated Er impurity at various positions was calculated. For Er^{3+}, the minimum energy configuration showed a relaxation in the cage of ten Si atoms surrounding the Er atom. It was found that the four tetrahedrally coordinated Si atoms move 0.16 Å from their ideal crystalline positions and that the Er–Si distance was 2.48 Å. The six octahedrally coordinated Si atoms move 0.04 Å and the Er–Si distance was 2.73 Å. By using the valence charge densities, the Er related states and the defect energy levels related to the Si bands were identified. A t_2 level related to Er 5d orbitals at $E_c - 0.2$ eV was found. It was argued that this same behavior was found in the tight-binding calculations for the isolated interstitial Er impurity (Lannoo and Delerue, 1993). The main conclusion is that Er as an isolated impurity in Si is most stable in the interstitial tetrahedral site in a charge state $+3$ and introduces an energy level in the gap related to Er 5d states. It is worth mentioning that even though this theoretical simulation shows that the isolated tetrahedral interstitial Er impurity in the $+3$ charge state is the most stable one, it cannot explain the crystal field splittings observed in the luminescence spectra.

The third theoretical investigation related to the electronic structure of isolated Er impurities in Si was based on a cluster model without assuming that the 4f electrons should be treated as a frozen core (Gan, Assali, and Kimerling, 1995). Using a multiple-scattering formalism, the electronic structure of the system was obtained. The cluster surface dangling-bonds were saturated by hydrogen atoms and a nonempirical muffin-tin overlapping-spheres choice was used in connection with quasirelativistic theory.

The isolated tetrahedral interstitial Er impurity (Er_i) was simulated by a 39-atom cluster ($Er_iSi_{14}H_{24}$) which comprises 14 Si atoms (three shells of neighbors), 24 H terminators, and the Er atom which is placed at the center of the cluster. The cluster used to simulate the substitutional Er impurity (Er_s) is a 71-atom cluster ($Er_sSi_{34}H_{36}$) with a central Er atom surrounded by 34 Si atoms (four shells of neighbors) and 36 H terminators. As Er doped Si PL measurements show an atomic-like characteristic and consequently a very small 4f crystal field energy splitting, this information was used to analyze the results of the electronic structure simulations.

The first interesting results that could be extracted from these simulations were that, besides 4f and 6s states, the 5d and 6p atomic excited states played an important role in describing the interactions between the Er impurity and the Si crystal. Hybridization among Er 4f, 6s, 5d, and 6p atomic orbitals was observed and is thought to be responsible for the weak interaction between the impurity and the Si crystal.

It was found that the Er_s center is electronically unstable, since the results showed occupied energy levels in the conduction band. Therefore Er_s could not be the microscopic configuration to describe experimental evidence related to optically active Er centers in Si, contrary to the tight-binding simulation results (Delerue and Lannoo, 1991).

Among the results obtained for the electronic structure of the Er_i center, the 3^+ cluster charge state displayed an electronic configuration close to $4f^{11}$ in the gap of the material and a crystal field energy splitting of 0.18 eV. Comparing the results obtained by interstitial isolated Er with PL spectra it was suggested that this center could be related to one observed in FZ Si which is optically active, in agreement with the interpretation of recent experimental results (Przbylinska *et al.*, 1995; Jantsch *et al.*, 1996). It was also concluded that the Er_i PL intensity is charge transfer controlled and that ligands, such as O, N, F, and C, were not required for parity breaking, since the Er_i electronic structure results showed a weak interaction between Er and Si host atoms, which was enough to split the 4f states.

It was concluded that among the configurations studied for isolated Er in Si, the tetrahedral interstitial Er^{3+} was the most stable in agreement with the supercell calculations (Needles, Schlüter, and Lannoo, 1993). However, the gap states were different since they are related to Er 5d orbitals for the supercell calculations compared to Er 4f orbitals.

The analysis of the results also led to the conclusion that when 4f orbitals were treated as valence states, hybridization among impurity 4f, 6s, 5d, and

6p states drives the 4f related energy levels towards the Si bandgap for the Er_i center and towards the Si conduction band for the Er_s center.

3. Er-related Complexes in Si

To understand the role of ligands in the electronic structure of the centers, using the multiple-scattering formalism as described above, different simple models were proposed to verify changes caused by the introduction of light atoms into the cluster. (Assali, Gan, and Kimerling, 1997).

Oxygen and fluorine atoms surrounding the Er center were built in to analyze the modifications in the electronic spectra caused by these ligands.

The first cluster studied was composed of an interstitial tetrahedral Er atom surrounded by four interstitial tetrahedral O atoms (first neighbors), 14 Si atoms, and 24 hydrogen atoms, used for saturation, designated by $Er_i(4O_i)Si_{14}H_{24}$. To study the modifications in the substitutional Er impurity when in the presence of interstitial O atoms, two different clusters were constructed in T_d symmetry: four and six O atoms were put into interstitial tetrahedral positions, designed by the clusters $Er_s(4O_i)Si_{34}H_{36}$ and $Er_s(6O_i)Si_{34}H_{36}$, respectively.

The results of the electronic structure of the $Er_i(4O_i)Si_{14}H_{24}$ cluster indicated a large delocalization of Er 4f electrons and covalent interactions among Er, Si, and O. The simulations showed that the four interstitial O atoms as Er_i first neighbors were not able to enhance the supposed optically active Er_i center. This ineffectiveness is due to covalence interaction of the crystal and Er states. The O atoms were not able either to take away electrons from the Er impurity or to decrease the crystal field splitting of the Er_i 4f derived gap orbitals. Therefore, it was concluded that this center could not be the Er center observed in the experiments that gives sharp PL lines in the Si:Er:O samples. This statement was based on the delocalization of the Er 4f electrons and the large value of the crystal field splitting obtained. It was argued that even if such a complex structure could exist in the samples, the PL intensity from this center could be very weak, broadened, and have a different energy luminescence peak since the f electrons do not have a $4f^{11}$ configuration and are very delocalized.

For the $Er_s(4O_i)Si_{34}H_{36}$ cluster, the presence of O atoms around the Er_s center reduced the interactions between Er and Si, but a large 4f derived crystal field splitting and delocalization was observed that stems from the covalent interactions among O atoms and Er_s. The majority of the Er 4f related energy levels were scattered in the valence band, interacting mostly with O atoms and some contributions from Si atoms, resulting in a large crystal field energy splitting. Despite the fact that the $Er_s(4O_i)$ center is stable and is a closed shell system, it was concluded that it could not be associated with the PL spectra, since the electronic structure results displayed a large delocalization of the Er 4f electrons and a large value of the crystal field splitting.

The electronic structure of the $Er_s(6O_i)Si_{34}H_{36}$ cluster, with six interstitial O atoms in octahedral symmetry, showed that the majority of Er 4f derived levels were in the gap and with a more localized character, than for the $Er_s(4O_i)Si_{34}H_{36}$ cluster. When a $+1$ charge state cluster was simulated, the role of the six O atoms in octahedral symmetry around Er_s was to stabilize the substitutional center, locating the 4f states close to the top of the Si valence band and introducing a deep level in the bandgap with some 4f character which could be a donor level since it was simulated in the $+1$ charge state. When O atoms were present, it was suggested by Adler et al. (1992) that a possible structural configuration is a substitutional Er surrounded by six interstitial O atoms arranged in an octahedron configuration. It was also proposed that a possible mechanism to explain the role of the O atoms is that when O atoms are present in the sample, they stabilize Er impurities in substitutional sites through covalent bond interaction and localized f electrons. It was also proposed that this center could be related to the PL lines in Si:Er:O samples with a donor character, and the intensity of the PL lines could be explained by an ionization mechanism that is Fermi level controlled.

To understand the role of F ligands in the luminescence of Er in Si, clusters in T_d symmetry were built and the self-consistent field (SCF) electronic spectra showed that these clusters were not stable centers because there are always some occupied levels in the Si conduction band, even though charged clusters were simulated. This study suggested that interstitial fluorine atoms tetrahedrally placed around interstitial Er are not a stable configuration and cannot be associated with any optically active center in Si:Er:F samples.

It was demonstrated that the formation of optically active Er centers in Si samples co-doped with fluorine, requires an association of one Er atom with three F ligands (Ren et al., 1994). A center comprised of an interstitial Er surrounded by three interstitial F atoms in trigonal symmetry, labeled $Er_i(3F)_iSi_{14}H_{24}$, was simulated. The 4f derived energy levels were found to be located in the Si conduction band leading to the conclusion that this center was not stable. It was pointed out that although Er 4f electrons in C_{3v} symmetry preserved much of their atomic character with three fluorine atoms around it, the reason why fluorine co-implanted Si:Er samples show strong PL was unclear. It was argued that the structural configuration is possibly different from the one proposed or the role of F atoms could be other than forming complexes with Er.

VI. Light Emitting Diode Design

The potential that Er-doped Si could serve as a material to fabricate LEDs, amplifiers, and eventually lasers, has been recognized by many researchers. Because of the materials challenges described in earlier sections, only recently has a more serious approach design and fabrication of LEDs been taken. To date, work reported on electrical devices in Er doped Si has been performed on LEDs. This section will discuss issues more specific to device design.

The first LEDs were made through the implantation of Er at 20 keV during MBE growth (Ennen *et al.*, 1985). The films were then implanted with boron to form a junction with the n-type substrate, and mesa structures were etched to define the device. EL of Er was detected at 77 K with a forward bias resulting in 1 mA of current. An estimate for the quantum efficiency of these devices was 5×10^{-4} at 77 K.

LEDs operating at room temperature have been made from both high energy (4.5 MeV) and low energy (400 keV) Er implanted Si (Zheng *et al.*, 1994; Michel *et al.*, 1996). Oxygen was co-implanted in these samples to overlap the Er profile at a concentration of 2×10^{18} cm^{-3}. Extensive care was taken to keep the device processing compatible with very large scale integration (VLSI) standards. Because of the dependency of the optically active Er on the annealing conditions, the control of the thermal budget was critical. Low temperature oxide was deposited in order to comply with the stringent conditions. It was furthermore shown that cross contamination in device processing due to Er was negligible. The fabrication of Er LED and MOS FET on the same chip proved the concept of device/LED integration. Figure 26 shows an edge emitting diode, coupling into a waveguide as a possible scheme for future interconnects on Si. Operational LEDs on Si-on-insulator (SOI) wafers show light emission up to 200 K (Zheng, 1996)

Luminescence was seen only in forward bias. For low drive current densities, the internal quantum efficiency of the high energy implanted LED was 4×10^{-4} at 100 K. The total light output power at 100 K was estimated to 3.4 μW. The low energy implanted samples produced 2.5 times more light per Er atom than the high energy implants. Cross-sectional TEM revealed end of range defects in the high energy implants which were not present in the 400 keV implanted diodes. This difference in crystal quality, as well as the lower annealing temperatures, which enhance the stability of the Er complexes, could explain the improvement for the low energy implants.

Junction placement with respect to the Er distribution is critical in determining light output. Maximum efficiency occurred when the Er distribution peak was just below a p$^+$-n junction as depicted in Fig. 27. In this

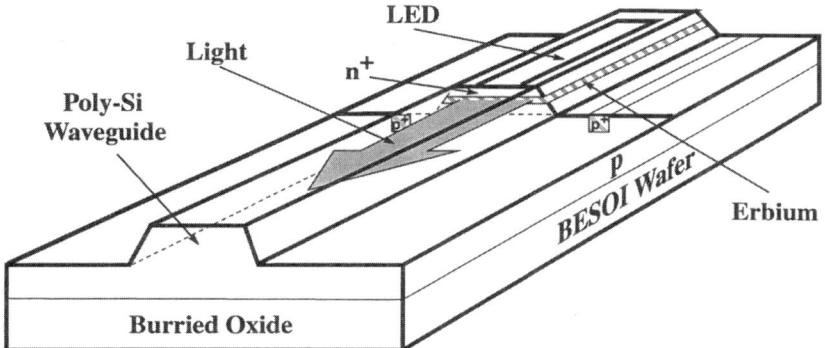

FIG. 26. Integrated LED-waveguide design for edge emitting LED and multimode waveguide.

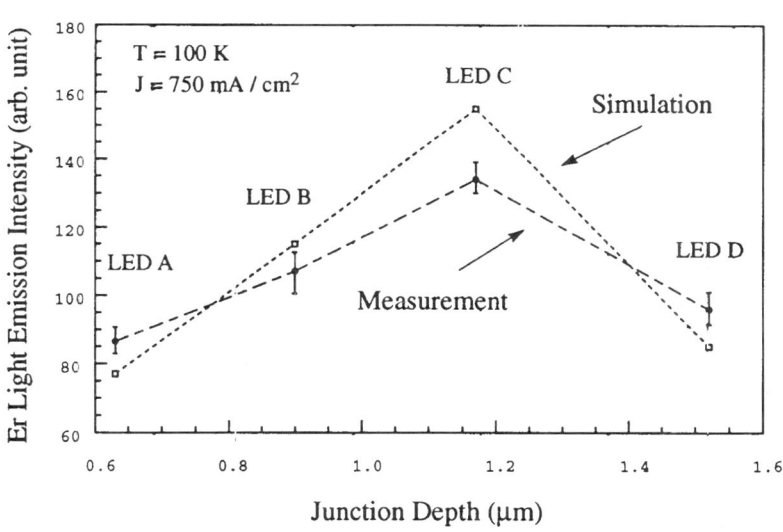

FIG. 27. Si:Er LED light emission intensity vs. junction depth. The center line with error bars is the measurement and the dashed line is a simulation. (From Zheng, 1996.)

position, the largest minority carrier concentration exists that can excite the Er. If the Er is outside the depletion region, luminescence is effectively quenched. EL from devices with varying junction depths was successfully modeled with these assumptions (Zheng, 1996).

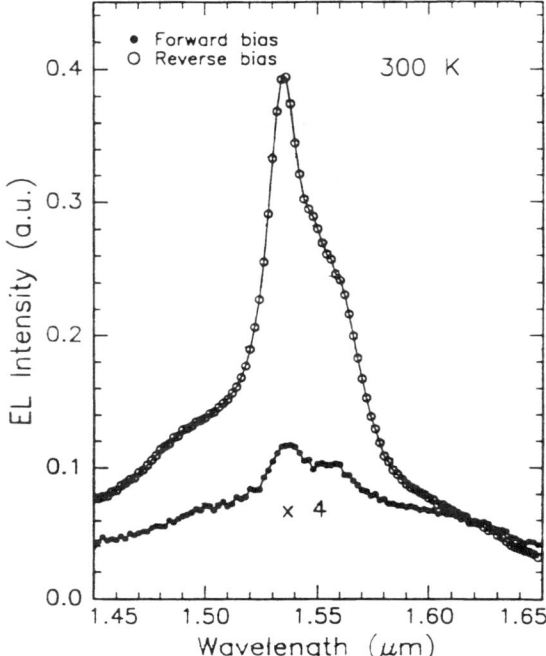

FIG. 28. Room-temperature EL spectra under forward bias (●) and reverse bias conditions (○). The spectra are taken at a constant current density of 2.5 A/cm^2 (corresponding to a total current of 600 mA). The spectrum under forward bias conditions is multiplied by a factor of 4. (From Franzo et al., 1994.)

In contrast to the devices described above, some p^+-n^+ devices have shown improved luminescence when operated in reverse bias (Franzo et al., 1994). Figure 28 shows the experimental results. Er was implanted at energies ranging from 0.5 to 5 MeV to achieve a constant concentration of approximately 1×10^{19} cm^{-3} over a range of 1.5 µm. Oxygen was co-implanted at 1×10^{20} cm^{-3} to overlap the Er distribution. Boron and phosphorus were implanted to concentrations of 1×10^{20} cm^{-3} to form the junction. Current-voltage (I-V) curves show the diode breakdown under a reverse bias of greater than 5 V, and room-temperature EL taken above this bias was 16 times higher than forward bias EL. Temperature quenching under reverse bias conditions was less sensitive than in forward bias indicating that two different pumping mechanisms were responsible for the luminescence. Since hot carriers could have energies of 0.9 eV in these samples, impact ionization is the likely excitation method for diodes in reverse bias.

Both forward and reverse bias EL were observed for MBE grown layers, although the reverse bias intensity was over 30 times larger than that of forward bias (Stimmer *et al.*, 1996a). Reverse biased EL exhibited no temperature quenching to 293 K and the forward bias luminescence only decreased by a factor of five. The estimated power output was several microwatts with a quantum efficiency greater than 10^{-5}.

VII. Summary

Over the last ten years, research on Er in Si has developed from gaining basic understanding and knowledge to the fabrication of simple device structures. Early on it was recognized that this materials system has potential applications in Si optoelectronics. This notion was the driving force especially for the most recent research. It was recognized that without a basic understanding of the physical processes involving Er in Si, little progress could be made in device design and fabrication.

Significant progress has been made in the understanding of excitation and de-excitation processes of the optically active Er ion. These insights, combined with new growth techniques like MBE and UHV-CVD for high quality Er doped Si layers should provide the basis to overcome the most serious obstacle, the strong luminescence quenching above 200 K. Reducing this luminescence decay would bring this defect system much closer to device applications. Thin film growth might furthermore change some of the conclusions, drawn from implantation experiments, where residual implantation damage screens the properties of Er in Si. It is expected that in the near future Er in Si will be investigated for light amplification as a first step towards developing a Si based laser.

Acknowledgments

The authors would like to thank A. Thilderkvist and T. Cheng for their help in preparing the manuscript.

References

Adler, D. L., Jacobson, D. C., Eaglesham, D. J., Marcus, M. A., Benton, J. L., Poate, J. M., and Citrin, P. H. (1992). *Appl Phys. Lett.* **61**, 2118.
Ageev, V. V., Aksenova, N. S., Kokovina, V. N., and Troshina, E. P. (1977). *ozv. Leningr. Elektrotekh. Inst.* **221**, 80.

Alexandrov, O. V., Emtsev, V. V., Poloskin, D. S., Sobolev, N. A., and Shek, E. I. *Semicond.* **28**, 1126.
Anderson, F. G. (1996). *Appl. Phys. Lett.* **68**, 2421.
Andry, P. S., Varhue, W. J., Ladipo, F., Ahmed, K., Adams, E., Lavoie, M., Klein, P. B., Hengehold, R., and Hunter, J. (1996). *J. Appl. Phys.* **80**, 551.
Beach, D. B., Collins, R. T., Legoues, F. K., and Chu, J. O. (1992). *Mat. Res. Soc. Symp. Proc.* **282**, 397.
Benton, J. L., Michel, J., Kimerling, L. C., Jacobson, D. C., Xie, Y. H., Eaglesham, D. J., Fitzgerald, E. A., and Poate, J. M. (1991). *J. Appl. Phys.* **70**, 2667.
Benton, J. L., Eaglesham, D. J., Almonte, M., Citrin, P. H., Marcus, M. A., Adler, D. L., Jacobson, D. C., and Poate, J. M. (1993). *Mat. Res. Soc. Symp. Proc.* **301**, 119.
Coffa, S., Priolo, F., Franzo, G., Bellani, V., Carnera, A., and Spinella, C. (1993). *Phys. Rev. B* **48**, 11782.
Coffa, S., Franzo, G., Priolo, F., Polman, A., and Serna, R. (1994). *Phys. Rev. B* **49**, 16313.
Crosswhite, H. M., and Moos, H. W. (1967) (eds.). *Optical Properties of Ions in Crystals.* Wiley, New York.
Custer, J. S., Polman, A., and van Pinxteren, H. M. (1994). *J. Appl. Phys.* **75**, 2809.
De Dood, M. J. A., Kik, P. G., Shin, J. H., and Polman, A. (1996). *Mat. Res. Soc. Symp. Proc.* **422**, 219.
DeLeo, G. G., and Fowler, W. B. (1991). *Semicond. and Semimet.* **34**, 511.
Delerue, C., and Lannoo, M. (1991). *Phys. Rev. Lett.* **67**, 3006.
Delerue, C., and Lannoo, M. (1994). *Mater. Sci. Forum* **143–147**, 699.
Dieke, G. H. (1968). *Spectra and Energy Levels of Rare Earth Ions in Crystals.* Wiley, New York.
Duan, X., Palm, J., Zheng, B., Morse, M. T., Michel, J., and L. C. Kimerling (1997). *Mat. Res. Soc. Symp. Proc.* **442**, 249.
Eaglesham, D. J., Michel, J., Fitzgerald, E. A., Jacobson, D. C., Poate, J. M., Benton, J. L., Polman, A., Xie, Y.-H., and Kimerling, L. C. (1991). *Appl. Phys. Lett.* **58**, 2797.
Efeoglu, H., Evans, J. H., Jackman, T. E., Hamilton, B., Houghton, D. C., Langer, J. M., Peaker, A. R., Perovic, D., Poole, I., Ravel, N., Hamment, P., and Chan, C. W. (1993). *Semicond. Sci. Technol.* **8**, 236.
Emtsev, V. V., Alexandrov, O. V., Poloskin, D. S., Shek, E. I., and Sobolev, N. A. (1995). *Mater. Sci. Forum* **196–201**, 615.
Ennen, H., Schneider, J., Pomrenke, G., and Axmann, A. (1983). *Appl. Phys. Lett.* **43**, 943.
Ennen, H., Pomrenke, G., Axmann, A., Eisele, K., Hayde, W., and Schneider, J. (1985). *Appl. Phys. Lett.* **46**, 381.
Ennen, H., Pomrenke, G., and Axmann, A. (1988). *J. Appl. Phys.* **57**, 2182.
Favennec, P. N., L'Haridon, H., Salvi, M., Moutonnet, D., and Le Guillou (1989). *Electronic Lett.* **25**, 718.
Favennec, P. N., L'Haridon, H., Moutonnet, D., Salvi, M., and Gauneau, M. (1990). *Jpn. J. Appl. Phys.* **29**, L524.
Franzo, G., Priolo, F., Coffa, S., Polman, A., Carnera, A. (1994). *Appl. Phys. Lett.* **64**, 2235.
Freeland, P. E., Jackson, K. A., Lowe, C. W., and Patel, J. R. (1977). *Appl. Phys. Lett.* **30**, 31.
Gan, F., Assali, L. V. C., and Kimerling, L. C. (1995). *Mater. Sci. Forum* **196–201**, 579.
Assali, L. V. C., Gan, F., and Kimerling, L. C. (1992). To be published.
Gregorkiewicz, T., Tsimperidis, I., Ammerlaan, C. A. J., Widderhoven, F. P., and Sobolev, N. A. (1996). *Mat. Res. Soc. Symp. Proc.* **422**, 207.
Hartung, J., Evans, J. H., Dawson, P., Scholes, A. P., Taskin, T., Huda, M. Q., Jeynes, C., and Peaker, A. R. (1996). *Mat. Res. Soc. Symp. Proc.* **422**, 119.
Jantsch, W., Przybylinska, H., Suprun-Belevich, Yu, Stepikhova, M., Hendorfer, G., and Palmetshofer, L. (1995). *Mater. Sci. Forum* **196–201**, 609.

Jantsch, W., Przybylinska, H., Skierbiszewski, C., Lanzerstorfer, S., and Palmetshofer, L. (1996). *Mater. Res. Soc. Symp. Proc.* **422**, 101.
Kim, H. K., Li, C. C., Fang, X. M., Solomon, J., Nykolak, G., and Becker, P. C. (1993). *Mat. Res. Soc. Symp. Proc.* **301**, 55.
Lannoo, M., and Delerue, C. (1993). *Mat. Res. Soc. Sym. Proc.* **301**, 385.
Libertino, S., Coffa, S., Franzo, G., and Priolo, F. (1995). *J. Appl. Phys.* **78**, 3867.
Liu, P., Zhang, J. P., Wilson, R. J., Curello, G., Rao, S. S., and Hemment, P. L. F. (1995). *Appl. Phys. Lett.* **66**, 3158.
Maat-Gersdorf, I., Gregorkiewicz, T., Ammerlaan, C. A. J., and Sobolev, N. A. (1995). *Semicond. Sci. Technol.* **10**, 666.
Martin, W. C., Zalubas, R., and Hagan, L. (1971). In *Atomic Energy Levels* (ed. C. E. Moore). U.S. Government Printing Office, Washington, DC.
Matsuoka, M., and Tohno, S.-i. (1995a). *J. Vac. Sci. Technol A* **13**, 305.
Matsuoka, M., and Tohno, S.-i. (1995b). *J. Appl. Phys.* **78**, 2751.
Michel, J., Benton, J. L., Ferrante, R. F., Jacobson, D. C., Eaglesham, D. J., Fitzgerald, E. A., Xie, Y. H., Poate, J. M., and Kimerling, L. C. (1991). *J. Appl. Phys.* **70**, 2672.
Michel, J., Kimerling, L. C., Benton, J. L., Eaglesham, D. J., Fitzgerald, E. A., Jacobson, D. C., Poate, J. M., Xie, Y. H., and Ferrante, R. F. (1992). *Mater. Sci. Forum* **83–87**, 653.
Michel, J., Zheng, B., Palm, J., Ouellette, E., Gan, F., and Kimerling, L. C. (1996). *Mat. Res. Soc. Symp. Proc.* **422**, 317.
Miyashita, K., Shiraki, Y., Houghton, D. C., and Fukatsu, S. (1995). *Appl. Phys. Lett.* **67**, 235.
Morse, M. T., Zheng, B., Palm, J., Duan, X., and Kimerling, L. C. (1996). *Mat. Res. Soc. Symp. Proc.* **422**, 41.
Nainov, V. O., Sobolev, N. A., Alexandrov, O. B., Bresler, M. S., Gusev, O. V., Gusinskii, G. M., Shek, E. I., Makaviichuk, M. I., and Parshin, E. O. (1995). *Nucl. Instr. and Meth. in Phys. Res. B* **99**, 587.
Nakashima, K. (1993). *Mat. Res. Soc. Symp. Proc.* **301**, 61.
Nazyrov, D. E., Kulikov, G. S., and Malkovich, R. Sh. (1991). *Sov. Phys. Semicond.* **25**, 997.
Needles, M., Schluter, M., and Lannoo, M. (1993). *Phys. Rev. B* **47**, 155.
Palm, J., Gan, F., and Kimerling, L. C. (1995). *SPIE Proc.* **2706**, 31.
Palm, J., Gan, F., Michel, J., and Kimerling, L. C. (1996). *Phys. Rev. B* **54**, 17603.
Palm, J., Gan, F., Michel, J., and Kimerling, L. C. (1997). To be published.
Polman, A., Custer, J. S., Snoeke, E., and van den Hoven, G. N. (1993). *Appl. Phys. Lett.* **62**, 507.
Priolo, F., Franzo, G., Coffa, S., Polman, A., Libertino, S., Barklie, R., and Carey, D. (1995). *J. Appl. Phys.* **78**, 3874.
Przybylinska, H., Enzenhofer, J., Hendorfer, G., Schisswohl, M., Palmetshofer, L., and Jantsch, W. (1994). *Mater. Sci. Forum* **143–147**, 715.
Przybylinska, H., Hendorfer, G., Bruckner, M., Palmetshofer, L., and Jantsch, W. (1995). *Appl. Phys. Lett.* **66**, 490.
Ren, F. Y. G., Michel, J., Sun-Paduano, Q., Zheng, B., Kitagawa, H., Jacobson, D. C., Poate, J. M., and Kimerling, L. C. (1993). *Mater. Res. Soc. Symp. Proc.* **301**, 87.
Ren, F. Y. G., Michel, J., Jacobson, D. C., Poate, J. M., and Kimerling, L. C. (1994a). *Mat. Res. Soc. Symp. Proc.* **316**, 493.
Ren, F. Y. G. (1994b). Ph.D. thesis, Massachusetts Institute of Technology.
Roberts, S., and Parker, G. (1995). *Materials Lett.* **24**, 307.
Rogers, J. L., Andry, P. S., Varhue, W. J., Adams, E., Lavoie, M., and Klein, P. B. (1995a). *J. Appl. Phys.* **78**, 6241.
Rogers, J. L., Andry, P. S., Varhue, W. J., McGaughnea, P., Adams, E., and Kontra, R. (1995b). *Appl. Phys. Lett.* **67**, 971.

Schmitt-Rink, S., Varma, C. M., and Levi, A. F. J. (1991). *Phys. Rev. Lett.* **66**, 2782.
Serna, R., Lohmeier, M. P., Zagwijn, M., Vlieg, E., and Polman, A. (1995). *Appl. Phys. Lett.* **66**, 1385.
Serna, R., Shin, J. H., Lohmeier, M., Vlieg, E., Polman, A., and Alkemade, P. F. A. (1996). *J. Appl. Phys.* **79**, 2658.
Smith, D. L. (1994). *Thin-Film Deposition.* McGraw-Hill, New York.
Sobolev, N. A., Alexandrov, O. V., Gresserov, B. N., Gusinskii, G. M., Naidenov, V. O., Shick, E. I., Stepanov, V. I., Vyzhigin, Yu, V., Chepik, L. F., and Troshina, E. P. (1993). *Solid State Phenom.* **32–33**, 83.
Sobolev, N. A., Bresler, M. S., Gusev, O. B., Shek, E. I., Makoviichuk, M. I., and Parshin, E. O. (1994). *Semicond.* **28**, 1100.
Sobolev, N. A. (1995). *Semicond.* **29**, 595.
Sobolev, N. A., Alexandrov, O. V., Bresler, M. S., Gusev, O. B., Shek, E. I., Makoviichuck, M. I., and Parskin, E. O. (1995). *Mater. Sci. Forum* **196–201**, 597.
Stimmer, J., Reittinger, A., Absteiter, G., Holzbrecher, H., and Buchal, C. (1996a). *Mat. Res. Soc. Symp. Proc.* **422**, 15.
Stimmer, J., Reittinger, A., Nutzel, J. F., Absteiter, G. Holzbrecher, H., and Buchal, C. (1996b). *Appl. Phys. Lett.* **68**, 3280.

Sze, S. M. (1981). *Physics of Semiconductor Device*, 2nd edn. p. 69. Wiley, New York.
Tamura, M., and Suzuki, T. (1989). *Nucl. Instrum. Methods B* **39**, 318.
Tang, Y. S., Jingping, Z., Heasman, K. C., and Sealey, B. J. (1989). *Solid State Comm.* **72**, 991.
Tang, Y. S., Heasman, K. C., Gillin, W. P., and Sealy, B. J. (1989). *Appl. Phys. Lett.* **55**, 432.
Thilderquist, A., Ahn, S., Michel, J., Ngiam, S., Kohlenbrander, K. D., and Kimerling, L. C. (1996). *Mat. Res. Soc. Symp. Proc.* **405**, 265.
Thonke, K., Hermann, H. U., and Schneider, J. (1988). *J. Phys. C* **21**, 5881.
Tsimperidis, I., Gregorkiewicz, T., and Ammerlaan, C. A. J. (1995). *Mater. Sci. Forum* **196–201**, 591.
Utzig, J. (1989). *J. Appl. Phys.* **65**, 3868.
Widdershoven, F. P., and Naus, J. P. M. (1989). *Mat. Sci. Eng.* **B4**, 71.
Xie, Y. H., Fitzgerald, E. A., and Mii, Y. J. (1991). *J. Appl. Phys.* **70**, 3223.
Yang, Y. S. (1991). *Physica Lett. A* **155**, 219.
Zheng, B., Michel, J., Ren, F. Y. G., Kimerling, L. C., Jacobson, D. C., and Poate, J. M. (1994). *Appl. Phys. Lett.* **64**, 2842.
Zheng, B. (1996). Ph.D. thesis, Massachussetts Institute of Technology.

CHAPTER 5

Silicon and Germanium Nanoparticles

*Yoshihiko Kanemitsu**

INSTITUTE OF PHYSICS
UNIVERSITY OF TSUKUBA
TSUKUBA, IBARAKI 305, JAPAN

I. INTRODUCTION	157
II. FABRICATION OF SILICON (Si) AND GERMANIUM Ge NANOPARTICLES	158
1. Si Nanoclusters: Organic Synthesis	158
2. Isolated Si Nanocrystals: Decomposition of Silane Gas	162
3. Porous Si and Ge: Electrochemical Etching	163
4. Si and Ge Nanocrystals in SiO_2 Matrices: Co-sputtering and Ion Implantation	166
5. Ge Nanocrystals: Chemical Methods	168
III. PHOTOLUMINESCENCE MECHANISM	170
1. Size-dependence of the PL Peak Energy	171
2. Resonantly Excited Luminescence Spectrum	174
3. Three Region Model	177
4. Photoluminescence Dynamics	185
IV. UNIQUE OPTICAL PHENOMENA	189
1. Nonlinear Optical Properties of π-Si	189
2. Tuning of Luminescence Wavelength	194
V. SUMMARY	200
References	202

I. Introduction

In today's tailor-made materials age, our researchers are producing more complex structures that have unique physical properties. Nanometer-sized crystalline semiconductors, often called nanocrystals, are a subset of the notable examples of complex structures. Nanocrystals are intermediate materials between small molecules and bulk crystalline limits. Optical and electronic properties of nanocrystals have attracted much attention, because they exhibit new quantum phenomena and have potential application for

*Present address: Graduate School of Material Science, Nara Institute of Science and Technology, Ikoma, Nara 630-01, Japan.

becoming novel and future devices (Yoffe, 1993; Ogawa and Kanemitsu, 1995). In particular, the discovery of efficient visible photoluminescence (PL) from Si (Takagi et al., 1990) and Ge (Maeda et al., 1991) nanocrystals has initiated numerous studies on luminescence properties of nanocrystals made from indirect-gap group IV semiconductors.

Bulk crystalline Si (c-Si) and Ge (c-Ge) exhibit very weak luminescence due to the indirect bandgap in their electronic structures. Therefore, Si and Ge have not been useful materials for the manufacturing of active optical devices. Recently, low-dimensional Si nanostructures have been artificially fabricated to obtain efficient luminescence at room temperature. It is expected that spatial confinement in low-dimensional nanostuctures causes the relaxation of the wave vector k-selection rule, the blue shift of the bandgap energy, and the enhancement of the oscillator strength in the lowest optical transition. In fact, efficient light emission has been observed from nanocrystals as zero-dimensional (0D) quantum dots, chain-like polymers as one-dimensional (1D) quantum wires, and two-dimensional (2D) quantum wells (QWs) (Kanemitsu, Kondo, and Takeda, 1994). The discovery of room-temperature visible luminescence from 0D Si nanocrystals opens new possibilities for the use of group IV indirect-gap semiconductors as materials for optoelectronic applications such as light-emitting diodes and display devices. However, the mechanism of visible luminescence from Si and Ge nanocrystals is still being disputed. The main issue discussed in this chapter is the mechanism for visible luminescence in Si nanocrystals.

In Section II we describe some typical fabrication methods for light-emitting Si and Ge nanoparticles. Efficient visible luminescence is observed after surface oxidation. Luminescence appears at a particular wavelength region in surface-oxidized Si and Ge nanocrystals. In Section III we present a model for the mechanism of luminescence in surface-oxidized Si and Ge nanocrystals. Topics discussed in Section IV include nonlinear optical properties and the tuning of visible luminescence in porous Si (π-Si) and Si nanocrystals.

II. Fabrication of Silicon (Si) and Germanium (Ge) Nanoparticles

1. Si NANOCLUSTERS: ORGANIC SYNTHESIS

Small Si nanoclusters have been usually produced by pulsed laser evaporation methods (Bloomfield et al., 1985; Jarrold, 1990). Recent laser-evaporation methods combined with mass spectrometer produce size-selected Si clusters. However, since small clusters are short-lived intermediate species and there are few experimental studies of the optical properties of small clusters, we have little information on the optical properties of vapor-phase small clusters. We need to contrive new experimental procedures for the study of the detailed structure and properties of Si clusters. For example, vapor-phase Si clusters deposited onto a solid matrix are prepared for the determination of the structure of small Si clusters (Honea et al., 1993). Despite the perfection of a pure cluster in the gas phase, it is a highly defective system from the point of view of semiconductor physics. In pure systems, surface states appear within the bandgap and degrade the electrical and optical properties of semiconductor nanocrystals and clusters. Passivation of the surface states is very important in semiconductor physics.

On the other hand, modern organic synthesis and purification techniques can produce new Si materials with controlled molecular weight (Miller and Michl, 1989; Kyushin et al., 1994). These techniques have some advantages over the widely used laser-evaporation methods. For example, the size and shape of Si clusters can be exactly controlled and synthesized clusters have no dangling bonds. The study of chemically-synthesized Si clusters helps to understand the electronic properties of vapor-phase Si clusters and to provide new information for the ongoing discussion on the origin of the strong visible luminescence of nanometer-sized Si materials. Here, we discuss the optical properties of Si_8 clusters and the size dependence of the luminescence properties of chemically synthesized Si materials.

Optical absorption and PL spectra of Si_8 clusters with different structures at room temperature are summarized in Fig. 1. Si backbone structures are illustrated in this figure. Synthetic preparation, purification, and characterization methods are described in literature (Kyushin et al., 1994). The dangling bonds are terminated by organic substituents. In each structure, the optical absorption and luminescence properties are not sensitive to the kind of organic termination. Optical absorption and luminescence properties of these clusters reflect the electronic structures of the central Si skeleton structures. Saturatedly bonded Si_8 clusters terminated by an organic substituent have no dangling bonds and become a new model material for small Si clusters.

In the chain and ladder Si_8 structures, absorption spectra are observed in the ultraviolet spectral region. However, in cubic cluster, a small absorption band around 2.5 eV and a large one around 3.5 eV are observed (Kanemitsu et al., 1994b). Absorption edge energy of the cubic structure is very small compared with those of the chain and ladder structures. In the chain structure, a sharp and strong PL band is observed at the absorption edge. In the ladder structure, a broad PL band is observed with a large Stokes shift in the visible spectral region. However, in the cubic structure, PL is observed only at low temperatures. Photoluminescence intensity in the cubic cluster abruptly decreases with increasing temperature above 40 K, while PL intensity in the chain and ladder clusters gradually decreases with increasing

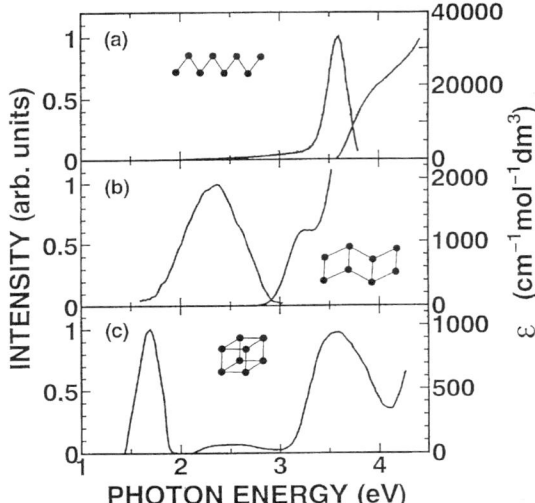

FIG. 1. Optical absorption and PL of chemically synthesized Si_8 clusters at room temperature: (a) chain, (b) ladder, and (c) cubic structures. The optical absorption and luminescence spectra are very sensitive to the Si backbone structure. The lowest absorption peak energy in the cubic structure is low compared with those in the chain and ladder structures. (From Kanemitsu et al. (1995). Phys. Rev. B **51**, 13103.)

temperature. Photoluminescence decay properties of the cubic Si_8 cluster are entirely different from those of the Si_8 chain and Si_8 sheet. Lifetimes of the chain and ladder Si_8 structures are 110 and 700 ps, respectively, at room temperature, while the lifetime of the cubic structures is 3.1 ms at low temperatures (Kanemitsu et al., 1995b). Optical properties of the very small Si clusters are very sensitive to the Si backbone structure.

Figure 2 summarizes the size dependence of the peak energy of the lowest absorption band. A absorption spectrum strongly depends on the size of Si-skeleton chains or sheets (Miller and Michl, 1989; Kanemitsu et al., 1992, 1993c). This suggests the quantum size effect on the lowest exciton energy. Gap energy is sensitive to the shape of clusters, rather than the size of clusters. In order to clarify the electronic structures of Si nanocrystals, we need to chemically fabricate very large Si materials having the number of Si atoms $N = 1000$. However, in this approach from molecules to nanostructures, it is difficult to produce c-Si polymers. In the disordered polymers, the excitons are strongly localized and the noncrystalline nature dominates the optical properties (Kanemitsu et al., 1995a). Therefore, there is a large gap between the Si clusters fabricated from molecules and Si nanocrystals fabricated from bulk c-Si. Detailed experimental studies of the optical

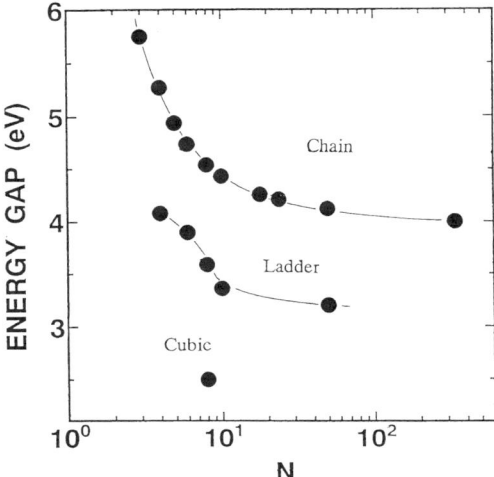

FIG. 2. The absorption peak energy of the lowest excitation state as a function of the number N of Si atoms in chain, ladder and cubic clusters. The solid lines are guides to the eye.

properties of chemically synthesized Si materials (small Si clusters, Si based polymers, siloxenes, etc.) help to understand the microscopic PL mechanisms in Si nanocrystals and other Si materials, because the local Si structures can be controlled in chemically synthesized Si materials. Synthesis of the crystalline 2D polymers and the control of 2D delocalization of excitons would realize wavelength-tunable luminescence in the visible region.

Band structures of sp^3 hybridized Si polymers have been theoretically studied (Takeda and Shiraishi, 1989; Takeda, 1994). The structure of the lowest unoccupied conduction band state depends on the Si backbone structure, while the highest occupied valence band is strongly delocalized in various c-Si polymers, A 1D Si chain [linear trans-polysilanes $(SiH_2)_n$] shows a 3.89 eV direct bandgap at the Brillouin zone Γ point. A 2D Si sheet [planar polysilyne $(SiH)_n$] has two characteristic bandgaps; an indirect one of 2.48 eV and a direct one of 2.63 eV. This 2D Si sheet has a band structure similar to the indirect gap one of 3D bulk c-Si. However, the addition of oxygen atoms to Si polymers causes a strong modification of the band structures (Takeda, 1994). In Si nanostructures, the orbital mixing of Si σ electrons with O nonbonding electrons (σ-n mixing) plays a critical role in the electronic structures. *Ab initio* calculations suggest that band structures SiO_xH_y and GeO_xH_y materials depend strongly on the spatial position of oxygen atoms. Since Si and SiO_2 and their heterostructures are very important in Si-based microelectronics, detailed and systematic studies of chemically synthesized SiO_xH_y materials are needed for a deeper understanding of the electronic properties of Si nanostructures.

2. ISOLATED Si NANOCRYSTALS: DECOMPOSITION OF SILANE GAS

The room-temperature visible PL from Si nanocrystals was first observed by Takagi et al. (1990) using microwave plasma decomposition of SiH_4 (silane) gas. Silicon nanocrystals were produced by means of microwave plasma decomposition of SiH_4 with a H_2 and/or Ar gas mixture in a resonant cavity. Altering the mixture ratio of gases and the microwave power, the size of Si nanocrystals can be varied from 2.5 to 20 nm in diameter. Silicon nanocrystals were blown out through a nozzle to a vacuum chamber and then were deposited on a quartz substrate at room temperature. Transmission electron microscopy (TEM) and X-ray diffraction studies showed Si(111) lattice fringes of 0.31 nm spacing (i.e., single crystals of the diamond structure). After oxidation, Si nanocrystals were covered by SiO_2 matrices and strong visible PL was observed in the red spectral region.

Si nanocrystals have been produced alternatively by laser breakdown of SiH_4 gas (Kanemitsu et al., 1993a). Pure SiH_4 gas was introduced into the vacuum chamber and kept at low pressure. Average diameter of the Si nanocrystals was controlled by this pressure. High power laser pulses were focused into the chamber and a spark was observed at the focal point. Silicon nanocrystals were deposited on quartz or Ge wafer substrates. Strong visible PL was observed after surface oxidation at room temperature. Interplane spacing of the TEM images of the Si core was consistent with the (111) planes of the diamond structure of bulk c-Si. This c-Si core was surrounded by a ~ 1.6-nm-thick Si oxide layer. Fourier-transform infrared (FTIR) spectroscopy and X-ray photoemission spectroscopy (XPS) indicated an a-SiO_2 layer (including a small amount of SiO_x). These surface-oxidized Si nanocrystals show efficient red PL at room temperature as shown in Fig. 3.

Nanometer-sized Si single crystallites have also been made by the pyrolysis of disilane at 860°C in 1.4 atm He pressure (Littau et al., 1993; Brus et al., 1995). These crystallites can be oxidized to create a 1.2 nm SiO_2 layer, collected as an ethylene glycol colloid and size separated using high pressure liquid chromatography. The surface-oxidized nanocrystals show red luminescence with a 5% quantum efficiency at room temperature (Brus et al., 1995), as shown in Fig. 3. Photoluminescence spectra in oxidized Si nanocrystals prepared by these three different methods are similar to each other and are usually observed in the red and infrared spectral region.

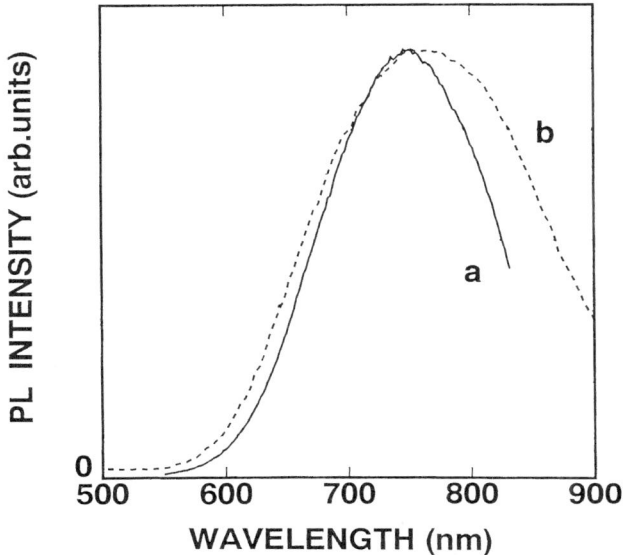

FIG. 3. Photoluminescence spectra of surface-oxidized Si nanocrystals at room temperature: (a) 3.7 nm Si nanocrystals prepared from silane gas (from Kanemitsu et al. (1993a)) and (b) 2 nm Si nanocrystals prepared from disilane gas. (From Littau et al. (1993)). Broad luminescence spectra are observed in the red and infrared spectral region.

3. POROUS Si AND Ge: ELECTROCHEMICAL ETCHING

Porous Si was discovered in 1956 (Uhlir, 1956). The porous layer is created by electrochemical dissolution (anodization or etching) in HF-based electrolytes. Figure 4 shows the schematic of π-Si fabrication cell and the specific processes involved in anodic etching at and close to the pore (Lehmann et al., 1992). It is based on the fact that holes are necessary for the electrochemical dissolution process of Si. Holes arriving at the Si:HF solution interface etch the Si lattice by forming SiF_4 molecules. Etching creates a rough surface. Since the bandgap in π-Si increases compared with bulk c-Si due to the quantum confinement effect, holes need the additional energy E_q to penetrate into the porous layer (lower-right sketch in Fig. 4). If E_q is larger than the bias, porous layer becomes depleted of holes and further dissolution is stopped. Since E_q is a function of the size of the nanocrystals, one can conclude that an increase in the formation bias will result in an increase in bandgap energy and a decrease in crystallite size in the porous layer. This process is self-adjusting. Quantum confinement effect

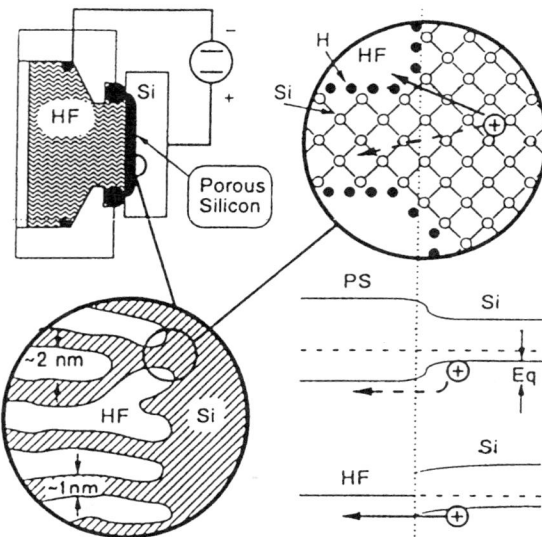

FIG. 4. Schematic of π-Si fabrication (upper left), the chemical process in anodic etching near the pore tip (lower left and upper right) and the band diagram for silicon-electrolyte transition at the pore tip and between the bulk and π-Si (lower right). (From Lehmann et al. (1992). Mater. Res Soc. Proc. **256**, 3.)

limits the size of the nanocrystals (Lehmann and Gosele, 1991). TEM observations indicate that π-Si is a mixture of c-Si spheres or wires and an amorphous phase. Crystalline Si spheres with various diameters are observed and the typical diameter of the Si spheres is several nanometers.

Light-assisted electrochemical etching techniques produce the π-Si layer on n-type Si wafers. By changing the illumination wavelength during electrochemical etching, we can also control PL wavelength from n-type π-Si layers (Mimura et al., 1995). With increasing photon energy, the crystallite size becomes smaller. These light illumination effects support the idea that the quantum size effect and self-adjusting effect are important in the formation process of π-Si nanostructures (van Buuren et al., 1995). Moreover, electrochemical etching under linearly polarized light, induces luminescence anisotropy in the sample plane (Polisski et al., 1996). Both the size and shape of nanocrystals can be controlled by light-assisted anodization of n-type Si wafers. As-prepared π-Si samples show a variety of colors in luminescence. However, after air exposure, π-Si usually shows red luminescence like oxidized Si nanocrystals. This fact implies that oxygen-modified surface states play an important role in the red PL process.

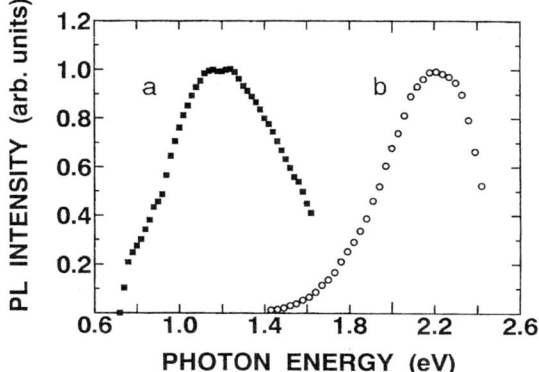

FIG. 5. Photoluminescence spectra of as-prepared π-Ge and oxidized π-Ge samples under 488 nm excitation. The luminescence peak energies of the as-prepared π-Ge and oxidized π-Ge are ~1.17 eV and ~2.15 eV, respectively. The luminescence intensity of the oxidized π-Ge is much larger than that of the as-prepared π-Ge. (From Miyazaki *et al.* (1995). *Thin Solid Films* **255**, 99.)

Porous Ge (π-Ge) and porous SiC (π-SiC) films were also prepared by electrochemical etching of Ge and SiC crystalline wafers in HF solution. Figure 5 shows PL spectra in as-prepared π-Ge and oxidized π-Ge (Miyazaki *et al.*, 1955). As-prepared π-Ge shows at room-temperature a broad luminescence around 1.17 eV. Peak energy of the broad PL is high compared to the bandgap energy of bulk c-Ge, similar to the case of π-Si. After oxidation, a new PL band appears at 2.2 eV at room temperature. Photoluminescence properties of the visible PL in oxidized π-Ge are very similar to oxidized Ge nanocrystals in SiO_2 glasses (Kanemitsu *et al.*, 1992b).

Photoluminescence and electroluminescence (EL) spectra from π-SiC are summarized in Fig. 6. Porous SiC films were prepared from n-type 6H-SiC wafers (Matsumoto *et al.*, 1994a; Mimura *et al.*, 1994). It is well known that the luminescence of indirect-gap SiC is caused by the donor-acceptor pair recombination (D-A luminescence). Porous SiC shows strong PL compared with D-A luminescence in bulk SiC. The geometrical confinement increases D-A pair luminescence strength. The mechanism of the luminescence in π-SiC seems to be complicated. It was reported that doped II–VI nanocrystals can yield both increased luminescence efficiencies and lifetime shortening (Bhargava *et al.*, 1994). These π-SiC and II–VI nanocrystals data also suggest that doped nanocrystals form a new class of light-emitting materials.

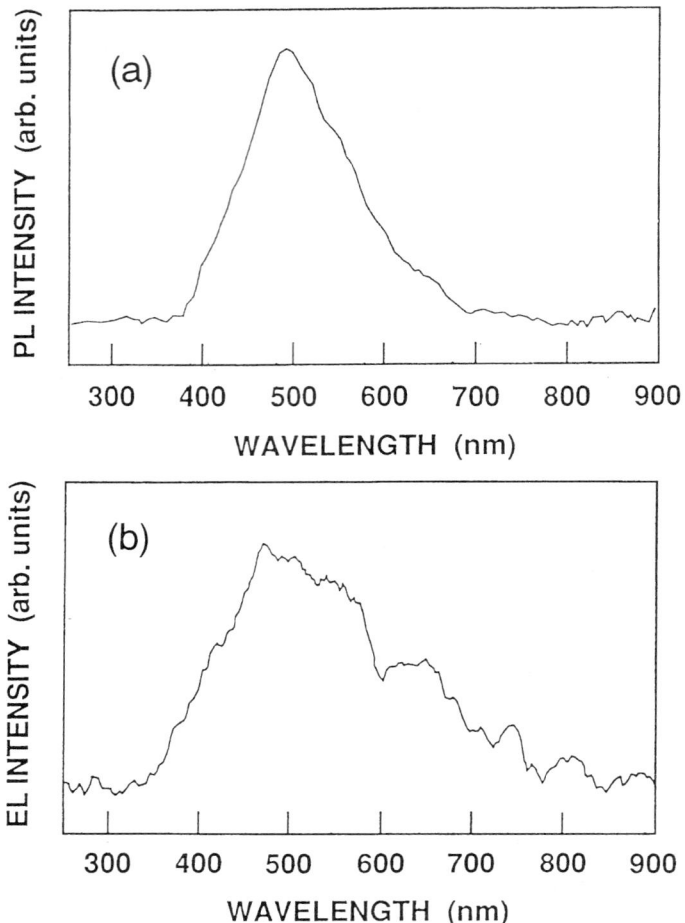

FIG. 6. Luminescence from π-SiC: (a) PL under 325 nm excitation and (b) EL from an ITO/π-SiC junction under a forward voltage 40 V and current of 23 mA. (From Mimura et al. (1994). Appl. Phys. Lett. **65**, 3350.)

4. Si AND Ge NANOCRYSTALS IN SiO_2 MATRICES: CO-SPUTTERING AND ION IMPLANTATION

Group IV semiconductors (Si, Ge or C) nanocrystals in SiO_2 matrices were fabricated using co-sputtering techniques. Visible luminescence from Ge nanocrystals was first observed in samples prepared by the co-sputtering method (Maeda et al., 1991; Kanemitsu et al., 1992b). Ge nanocrystal samples were prepared by a method of radio-frequency magnetron cosputtering of Ge and SiO_2. Thin films of mixture of Ge and SiO_2 were deposited onto Si substrates, and then annealed in an Ar gas atmosphere in order to grow Ge nanocrystals in the SiO_2 glassy matrix and to control the size of Ge nanocrystals. The XPS data showed that both GeO_2 and Ge exist in SiO_2 in the as-deposited glassy state and that most GeO_2 decomposes into Ge nanocrystals after thermal annealing. The crystallinity and size of Ge nanocrystals in the SiO_2 matrix were studied by TEM. XPS and FTIR spectroscopy indicated that amorphous GeO_2 and GeO (a-GeO_2 and a-GeO) were formed at the surface of Ge nanocrystals (Okamoto and Kanemitsu, 1996).

Figure 7 shows the PL spectra from Ge nanocrystals at room temperature under 488 nm laser excitation and PL excitation spectra at the PL peak

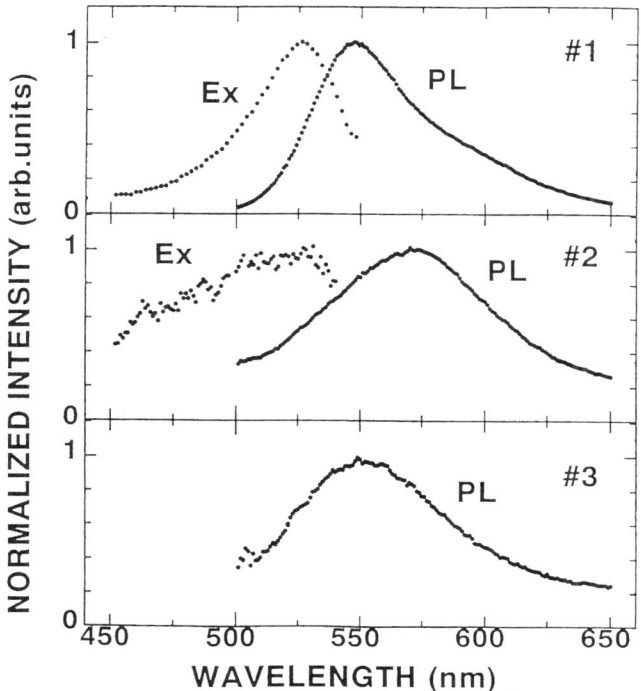

FIG. 7. Photoluminescence spectra under 488 laser excitation and PL excitation spectra at the luminescence peak wavelength in three different samples. The average diameters in #1, #2, and #3 samples are 3.7, 6.8, and 13.8 nm, respectively. No significant size dependence of the luminescence spectrum is observed. The PL intensity increases with decreasing crystal size. (From Kanemitsu et al. (1992). Appl. Phys. Lett. **61**, 2187.)

wavelength (Kanemitsu et al., 1992). Broad PL spectra with a peak around 2.2 ~ 2.3 eV are observed. Photoluminescence peak energy does not depend on the average diameter of Ge nanocrystals in the sample. The luminescence properties of Ge nanocrystals in SiO_2 are similar to those of oxidized π-Ge. Moreover, luminescence spectra of Si nanocrystals in SiO_2 in the red spectral region are similar to those of π-Si.

Ion implantation of Si and Ge in SiO_2 also creates supersaturated Si and Ge solid solutions. In addition, vacuum annealing of ion-implanted samples produces Si and Ge nanocrystals (e.g., Shimizu-Iwayama et al., 1993; Atwater et al., 1994; Komoda et al., 1995). Silicon dioxide grown by thermal oxidation of Si is an extremely pure and well-characterized material, and it is expected that the Si-SiO_2 and Ge-SiO_2 interface has a low density of interface states. Luminescence from Si and Ge nanocrystals embedded in pure SiO_2 matrices has advantages over π-Si for device applications. However, it is very difficult to control the PL wavelength in ion-implanted Si samples, and weak red PL is usually observed. Recently, continuously tunable PL from Si nanocrystals in SiO_2 matrices was reported (Fischer et al., 1995). Annealing temperature and gas ambient are very important factors for the luminescence tunability and the enhancement of luminescence intensity. Tuning of the PL wavelength is obtained by using a forming gas ambient (N_2 and H_2). However, after dry oxidization of ion-implanted samples, luminescence is observed in the red and blue spectral region and there is no correlation between the crystalline size and luminescence peak energy.

5. Ge Nanocrystals: Chemical Methods

Light-emitting Ge nanocrystals in SiO_2 glasses have been prepared by chemical methods such as a sol-gel process and chemical reactions in solid solutions. Germanium dioxide in $Ge_xSi_{1-x}O$ is thermodynamically unstable in the presence of hydrogen and precipitates out as elemental Ge. Using this chemical reaction, Ge nanocrystals were prepared. $Ge_xSi_{1-x}O_2$ alloys can be formed oxidizing Ge_xSi_{1-x} alloys grown epitaxially on a (001) Si substrate. These samples were annealed in a flowing forming gas (N_2 and H_2). The particle size of Ge nanocrystals can be controlled by changing the annealing temperature and time. Figure 8 shows PL spectra of unannealed and annealed samples (Paine et al., 1993). Photoluminescence intensity decreases with increasing annealing time and increasing average particle size. Photoluminescence wavelength is independent of the crystal size. Only small crystallites less than a critical diameter contribute to the luminescence.

Ge nanocrystals in SiO_2 glasses were also prepared by a sol-gel process

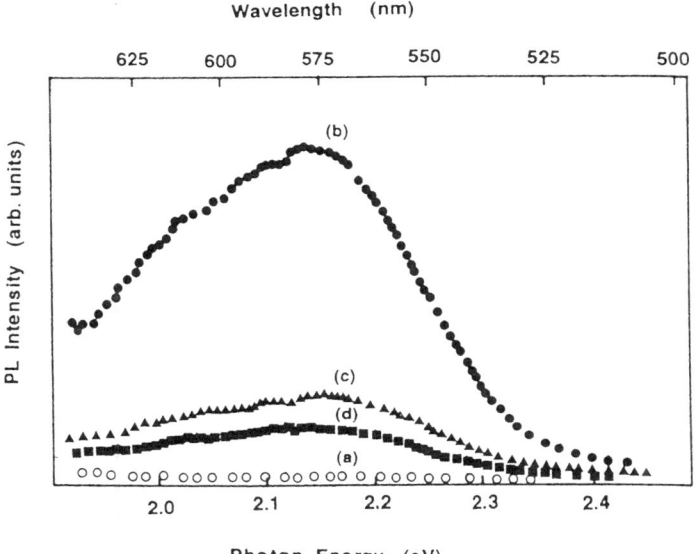

FIG. 8. Photoluminescence from Ge nanocrystals in SiO$_2$ glasses under 488 nm excitation: (a) the as-oxidized sample and after annealing for (b) 10, (c), 30, (d) 60 min at 750°C in H$_2$:N$_2$ gas. No significant size dependence of the luminescence spectrum is observed in a range between 6 and 9 nm. The PL intensity increases with decreasing crystal size. (From Paine et al. (1993). Appl. Phys. Lett. **62**, 2842.)

(Nogami and Abe, 1994). The glass synthesized through hydrolysis of Si(OC$_2$H$_5$)$_4$ and GeCl$_4$ was heated in the presence of hydrogen, whereby Ge ions were reduced to precipitate Ge nanocrystals. The Ge crystal size were controlled by changing the annealing temperature in N$_2$:H$_2$ gas. Figure 9 shows PL spectra in such Ge nanocrystals. A significant size dependence of the PL peak energy is not observed. The intensity depends on crystallite size. Only smaller crystals less than 5 nm in size contribute to visible PL.

Ge nanocrystals have been prepared by many different methods. However, the PL from Ge nanocrystals in SiO$_2$ glasses and oxidized π-Ge, shows the same broad spectrum in the green-yellow region. The PL intensity is very sensitive to size, but PL spectrum is independent of crystal size. Moreover, the structure of Ge nanocrystals depends on the sample preparation condition. For example, new tetragonal Ge films were prepared by a cluster-beam deposition method (Sato et al., 1995). We need to clarify if there is any correlation between the optical properties and the structure of Ge nanocrystals.

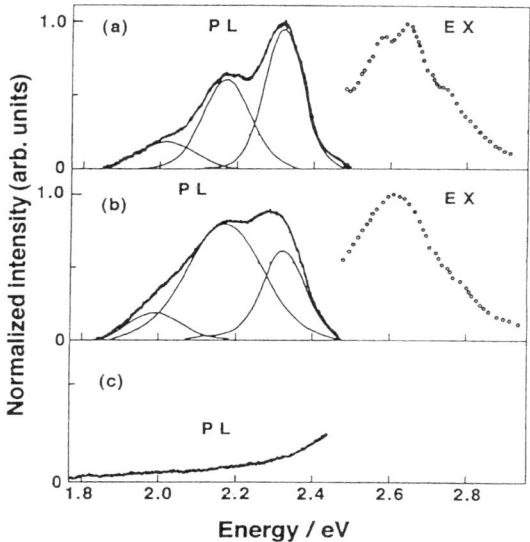

FIG. 9. Photoluminescence from Ge nanocrystals in SiO$_2$ glasses prepared by a sol-gel process. Annealing time and temperature in H$_2$:N$_2$ gas (a) 700°C, 2 h, (b) 700°C, 5 h, and (c) 800°C 3 h. Luminescence originates in Ge nanocrystals with a size of 5 nm or less. (From Nogami and Abe (1994). *Appl. Phys. Lett.* **65**, 2545).

In this section, we described typical fabrication procedures of light-emitting Si and Ge nanocrystals. In particular, visible luminescence spectra of Ge nanocrystals were discussed, because the reader can find further details of the visible luminescence spectra of Si nanocrystals and π-Si in other chapters in this book. There are numerous reports on the fabrication of light-emitting nanocrystals, and other different techniques have been reported in many papers and books, as, for example, Materials Research Society Symposia Proceedings. Strong visible luminescence from Si and Ge nanocrystals is usually observed after surface oxidation. Visible luminescence appears at a particular wavelength region in surface-oxidized Si and Ge nanocrystals. In Section III we shall discuss the mechanism of visible luminescence from oxidized Si nanocrystals.

III. Photoluminescence Mechanism

Efficient visible luminescence from Si nanocrystals and π-Si has initiated remarkable theoretical and experimental studies on low-dimensional Si materials (Takagi *et al.*, 1990; Canham, 1990). Low-dimensionality plays an essential role in the optical transition of semiconductor nanostructures. Theoretical studies show the widening of the bandgap energy and the enhancement of the oscillator strength with decreasing size of wires or dots. The quantum confinement effect plays a role in efficient visible luminescence in Si nanocrystals. However, drastic size reduction to a few nanometers is needed for the observation of strong visible light emission. A large fraction of the Si atoms appears on the surface. Then, it is naturally considered that the surface states on the Si nanocrystals play an active role. By considering the role of the surface, the currently proposed models are roughly classed in two categories: (1) "pure" quantum confinement model (quantum confinement effects in Si nanocrystals without surface states effects), and (2) surface localization model (quantum (quantum confinement effects in Si nanocrystals with surface states effect). In this section, we discuss the mechanism of red luminescence from surface-oxidized Si nanocrystals.

1. SIZE DEPENDENCE OF THE PL PEAK ENERGY

Figure 10 shows optical absorption spectra in π-Si, c-Si, and hydrogenated amorphous Si (a-Si:H). The absorption spectrum of Si nanocrystals is similar to that of π-Si. The exponential absorption tail (Urbach tail) in Si nanocrystals is observed in the near-infrared and visible region (Kanemitsu

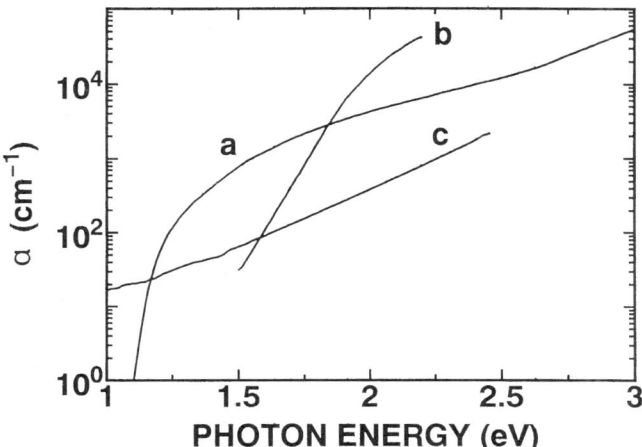

FIG. 10. Optical absorption spectra of (a) indirect-gap c-Si, (b) a-Si:H, and (c) π-Si. Si nanocrystals and π-Si show an exponential absorption spectrum in the near-infrared and visible spectral region. The data are reproduced from Kanemitsu (1996).

et al., 1993b; Koch *et al.*, 1993). In direct-gap semiconductor nanocrystals, the peak of the lowest absorption band is clearly observed even in samples having broad crystallite size distributions. However, in Si nanocrystals, the conduction band has a large density of quantized states. Phonon-assisted optical transitions near the lowest level cause overlapping of the optical transitions. Then, a featureless and continuum absorption spectrum is observed in Si nanocrystals (Hybertsen, 1994; Brus *et al.*, 1995).

Absorption spectra of nanocrystal samples are entirely different from those of bulk c-Si and a-Si:H. The exponential absorption tail (Urbach energy $E_\alpha \sim 250$ meV in Si nanocrystals) is much larger than in a-Si:H ($E_\alpha \sim 40$–50 meV). The large Urbach tail reflects the nanoscopic disordered nature of Si nanocrystals, such as a distribution of nanocrystal sizes and shapes and variations in surface roughness and surface structures. This large Urbach energy causes complicated PL behavior in Si nanocrystals, as will be discussed later. Moreover, it is found that the blue shift of the optical absorption spectrum occurs with a decrease in the average diameter of Si nanocrystals. The blue shift of the absorption spectrum indicates that the quantum confinement effect plays a key role in the absorption process (Kanemitsu *et al.*, 1993b). Moreover, PL excitation spectra depend on the monitored PL wavelength. The blue shift of the PL excitation spectrum occurs with a decrease of the monitored PL wavelength (e.g., Lockwood, 1994). This also implies that the absorption process would be size-dependent in Si nanocrystals.

Size-dependence of the PL peak energy is extensively studied in surface-oxidized Si nanocrystals fabricated by the various techniques described in Section II. Figure 11 summarizes the size dependence of theoretically calculated bandgap energies (Proot *et al.*, 1992; Takagahara and Takeda, 1992; Wang and Zunger, 1994) and of observed PL peak energies (Takagi *et al.*, 1990; Kanemitsu *et al.*, 1993a, 1997; Schuppler *et al.*, 1994). We can observe the size-dependence of PL peak energy experimentally. However, the size dependence of PL peak energy is very small compared with that of the absorption spectrum and theoretical calculations (Kanemitsu *et al.*, 1993; Lockwood, 1994). In particular, there is a large difference between the PL peak energy and the theoretical bandgap energy in small nanocrystals. The above results imply that some energy-loss occurs in the electronic processes from the absorption to light emission or a large Stokes shift is induced by the lattice relaxation in excitonic states. The carriers or excitons relax from the higher-energy absorption state to the lower-energy emission state. The above PL characteristics and the blue shift of the absorption spectrum in oxidized Si nanocrystals imply that the site for photogeneration of carriers is different from that for radiative recombination of carriers.

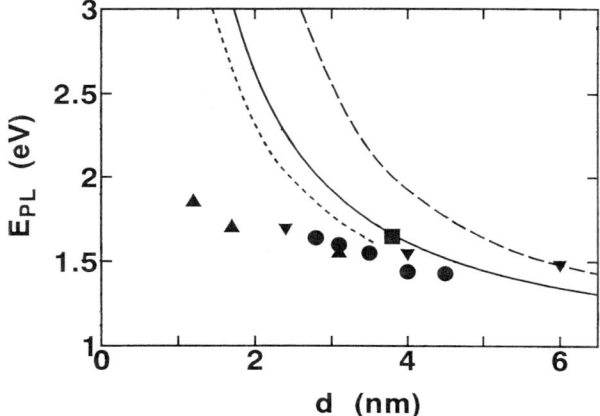

FIG. 11. Size dependence of the PL peak energy in surface-oxidized Si nanocrystals at room temperature. The experimental results are reproduced from Takagi et al. (1990) (●), Kanemitsu et al. (1993) (■), Schuppler et al. (1994) (▲), and Kanemitsu et al. (1997) (▼). Theoretical curves for the band-gap energy are shown by solid (Proot et al., 1992), broken (Takagahara and Takeda, 1992), and dashed (Wang and Zunger, 1994) lines. The size dependence of the luminescence peak energy is very small compared with that of the theoretically calculated bandgap energy.

Alternatively, in as-prepared n-type π-Si without air exposure, the 2.5 nm size samples show green PL near 2.3 eV (Mimura et al., 1995). The observed PL peak energy in H-terminated Si nanocrystals (as-prepared π-Si) depends on the nanocrystals size, and the observed size-dependence is in good agreement with the calculated size-dependence of the exciton energy of Si nanocrystals in Fig. 11 (Kanemitsu and Okamoto, 1997a). Moreover, Ito et al. (1995) have measured the luminescence spectra from the individual nanocrystals of as-prepared π-Si below the probe tip of the scanning tunneling microscope (STM), and found that the measured peak shift with size of the nanocrystal is consistent with the shift of the eneгy gap predicted on the basis of a quantum confinement model. These results suggest that the size-sensitive PL in H-terminated Si nanocrystals comes from mostly delocalized crystalline core states in Si nanocrystals. However, after long-time air exposure, the PL spectra of any of the samples are similar to each other and appear in the red spectral region. No significant size dependence of PL peak energy was observed in π-Si after air exposure. This observation is consistent with the size dependence of PL peak energy in surface-oxidized Si nanocrystals. The surface of as-prepared Si nanocrystals is mainly covered with hydrogen during etching, but after air exposure, the surface is naturally oxidized. Surface oxidation of nanocrystals changes the PL wavelength. Size dependence of PL peak energy in surface-oxidized Si nanocrystals is different from that in H-terminated Si nanocrystals. In smaller nanocrystals, PL energy in surface-oxidized Si nanocrystals is lower than that in H-terminated Si nanocrystals (e.g., Kanemitsu and Okamoto, 1997a). Furthermore, there are large differences between PL peak energy and theoretical bandgap energy in small oxidized nanocrystals. These results suggest that excitons relax from the higher-energy absorption state to the lower-energy emission states in oxidized Si nanocrystals. In particular, in small nanocrystal exhibiting efficient PL in the visible spectral region, the localization processes of excitons (e.g., self-trapped exciton formation and disorder-induced exciton localization) are important in the electronic process from light absorption to light emission (Kanemitsu et al., 1993a, 1997; Allan et al., 1996). The luminescence spectrum and dynamics of Si nanocrystals are very sensitive to the surface structure of Si nanocrystals, particularly with regard to the amount of oxygen and hydrogen on surfaces.

With large surface-to-volume ratios in semiconductor nanocrystals, surface effects become more enhanced on decreasing the size of nanocrystals. The nanocrystal surface acts as the source of novel states other than a boundary. In particular, oxygen atoms critically modify the electronic structure of Si nanostructures (Kanemitsu et al., 1993a). The surface plays a role in the luminescence other than passivation (reduction of nonradiative recombination centers on the crystallite surface). The oxygen-modified surface region within Si nanocrystals affects the luminescence process. Electronic properties of the near-surface region in the nanocrystal are different from those of the c-Si core in the nanocrystal, as will be discussed later.

2. Resonantly Excited Luminescence Spectrum

Si nanocrystals, π-Si, and Si clusters show broad PL spectra in the infrared and visible region. Under ultraviolet (UV) and blue laser excitation, these samples show a featureless and broad PL spectrum. This full luminescence contains contributions from all nanocrystals in the samples, and is inhomogeneously broadened. However, for excitation energies within the inhomogeneously broadened PL band, we observed well-resolved fine structures in luminescence spectra at low temperature (fluorescence line narrowing). Under resonance excitation, fine structure due to coupling to transverse optic (TO) phonons and transverse acoustic (TA) phonons was reported in π-Si (Suemoto et al., 1993, 1994; Calcott et al., 1993; Rosenbauer *et al.*, 1995; Kanemitsu, 1996; Kanemitsu *et al.*, 1997). Polarization memory effect in the luminescence is enhanced by the resonant recitation condition (Kovalev *et al.*, 1996). The resonantly excited luminescence spectrum provides detailed information on the luminescence mechanism of Si nanocrystals and π-Si.

By reducing the excitation energy below ~ 2 eV, step-like phonon structures due to TO phonons are clearly observed in as-prepared π-Si at low temperatures, as shown in Fig. 12. The porous layer is thin ($\sim 1-2 \mu$m thick) to allow uniform optical excitation within the layer and a reduction of the sample inhomogeneity in the depth direction. The step-like structures are more clearly observed with a decrease of the excitation photon energy. Spacing between steps is very close in energy to the 57 meV TO phonons in c-Si (Calcott *et al.*, 1993). Although the quantitative explanation of the step-like PL spectrum is still in dispute, the step-like structure in π-Si suggests that the phonon-assisted absorption process determines the fine structure in the spectrum, rather than the phonon-assisted light emission process (Suemoto *et al.*, 1993, 1994).

On the other hand, the phonon-related structures are not clearly observed in surface-oxidized Si nanocrystals. Figure 13 shows PL spectra in surface-oxidized Si nanocrystals and π-Si at the low-energy side of the laser line (1.959 eV) at 2 K under the same experimental conditions (Kanemitsu,

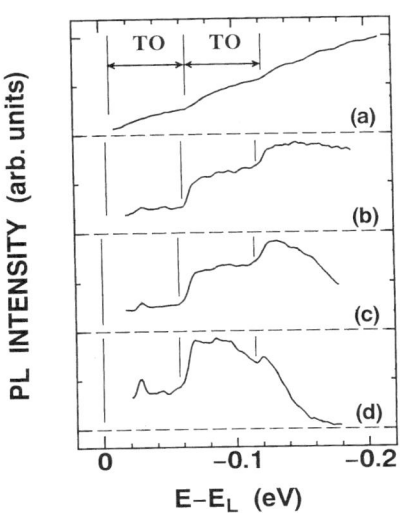

FIG. 12. Resonantly excited PL spectra of freshly prepared π-Si at the low-energy side of the laser line (a) 1.959, (b) 1.724, (c) 1.658, (d) 1.597 eV at 2 K. TO-phonon related structures are clearly observed. The data are produced from Kanemitsu *et al.* (1997).

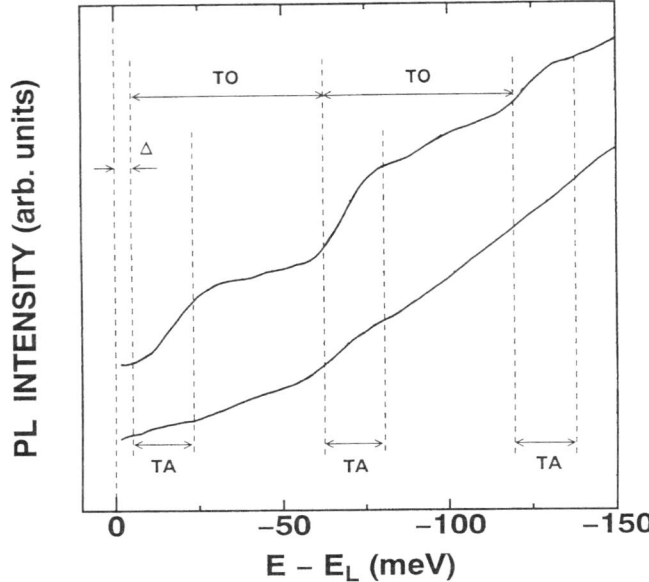

Fig. 13. Resonantly excited PL spectra at the lower-energy side of the laser line (1.959 eV) of surface-oxidized Si nanocrystals (lower part) and freshly prepared π-Si (upper part). Phonon-related structures in the luminescence spectra are observed in π-Si, but not in oxidized Si nanocrystals. (From Kanemitsu (1996). *Phys. Rev. B* **53**, 13515.)

1996). The broad PL spectrum with no fine structure suggests that the potential fluctuation in surface states in oxidized Si nanocrystals is much larger than that of H-terminated Si nanocrystals. It is considered that the disordered potential of oxidized surface states causes the shallow localized states.

Surface oxidation of thin π-Si samples occurs naturally by being kept in air for one year or more. The phonon-related structures are observed in both as-prepared and oxidized thick π-Si samples (∼40 μm thick) (Calcott et al., 1993; Rosenbauer et al., 1995). Porous Si samples are inhomogeneous in the sense that they have broad distributions of the crystallite size and shape, surface roughness, and a fluctuation of surface stoichiometry. In particular, the vertical inhomogeneity in the porous layer plays an important role in determining the PL properties of π-Si. This inhomogeneity is very important in interpreting excitation wavelength dependence of the PL spectrum and the resonantly excited PL spectrum of π-Si.

Vertical inhomogeneity in the porous layer is a consequence of the π-Si formation mechanism. The top of the layer has been in the etching solution longer than the middle and bottom. The c-Si size is then smaller at the top of the layer. After air exposure, the top of the layer is easily oxidized, but the middle and bottom are not easily oxidized. It is considered that the size and surface structure of nanocrystals in the top of the layer are completely different from those in the bottom of the layer. For excitation energies lower than ~ 2 eV, the middle and bottom of the porous layer are excited. H-terminated Si nanocrystals in the bottom of the porous layer contribute to the visible PL. In fact, thick π-Si samples were used in previous work to obtain clear phonon-related structure in the PL spectra in naturally oxidized π-Si. However, in thin porous layer samples ($\sim 1\,\mu$m), the phonon-related structures are only observed in as-prepared π-Si but not in oxidized π-Si (Kanemitsu and Okamoto, 1997b). These experimental observations in thin porous layer samples are consistent with those in SiO_2-capped Si nanocrystals. Phonon-related structure depend strongly on the surface structure and the sample inhomogeneity. H-terminated Si nanocrystals show their crystalline nature, while the oxidized Si nanocrystals show their disordered nature.

With a decrease in size of Si nanocrystals, new structure appears in the resonantly excited PL spectra at low temperatures (Kanemitsu et al., 1996b). The PL peak structures of Si clusters in SiO_2 matrices are different from the step-like structures in π-Si. In Si clusters in SiO_2 matrices, spacing between the PL peaks of gaussian profiles is ~ 135–140 meV and this energy is almost equal to the local vibration energy of the Si–O–Si stretch mode ($\sim 1100\,cm^{-1}$). Since the Si–O bond is polar, the coupling of excitons and stretching vibrations of surface species increases with localization of excitons in smaller dimensions. The peak structures suggest that coupling of localized excitons and surface oxide vibrations are important in the luminescence process. Similar structures are also observed in surface oxidized Ge nanocrystals (Okamoto and Kanemitsu, 1996). The fine structure in the PL spectra under resonance excitations is very sensitive to the surface structure and nanocrystal size. Since nanocrystals have an unusually large surface-to-volume ratio, polar surface bond effects are clearly observed in the optical properties in nonpolar group IV semiconductor nanocrystals.

3. THREE REGION MODEL

The surface-oxidized Si and Ge nanocrystals exhibit stable visible luminescence. While the PL peak wavelength scarcely depends on the average crystallite size $\langle D_{core} \rangle$, the PL intensity of the oxidized Si and Ge nanocrystals depends strongly on the crystallite size (Kanemitsu et al., 1992b;

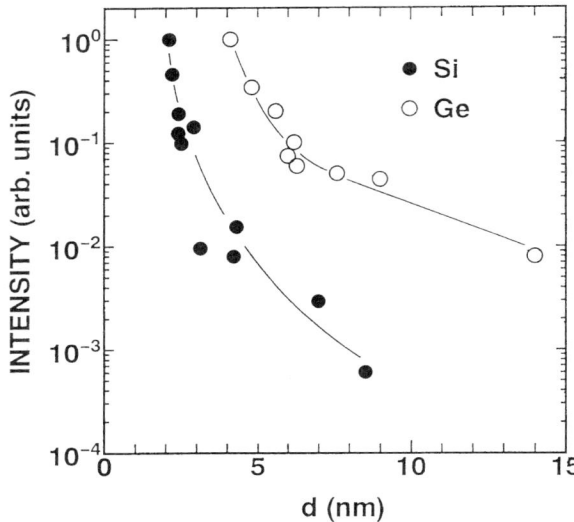

FIG. 14. Size dependence of the luminescence intensity in (a) oxidized Si nanocrystals (red luminescence) and (b) oxidized Ge nanocrystals (yellow luminescence). With a decrease of the crystallite size, the luminescence intensity increases. The data of Si nanocrystals are reproduced from Ruckschloss et al. (1993).

Kanemitsu et al., 1993a; Ruckschloss et al., 1993). Size dependence of the PL intensities is shown in Fig. 14. Size dependence of the PL intensity is very clear compared with that of the PL peak energy. Since the samples have a broad distribution of crystallize size, we consider the effect of size distribution on the PL intensity. The PL intensity increases with the number of oxidized Si and Ge nanocrystals that have a small D_{core}; the PL intensity is approximately proportional to the volume fraction of the oxidized Si nanocrystals with $D_{core} \leqslant \sim 7\text{--}9$ nm (Kanemitsu et al., 1993a) and that of the oxidized Ge nanocrystals with $D_{core} \leqslant \sim 4$ nm (Kanemitsu et al., 1992b), indicating that only the smaller nanocrystals contribute efficiently to strong visible PL. This finding is consistent with the blue shift of the absorption spectrum in nanocrystals. The bandgap widening due to the quantum confinement effect plays an essential role in absorption and luminescence processes in Si and Ge nanocrystals. Here we discuss the PL mechanism of surface oxidized Si nanocrystals.

To evaluate the possible origins of the strong PL in the orange-red spectral region for Si nanocrystals, we consider the electronic and optical properties of the crystalline core. For core sizes of the order of a nanometer, the bandgap energy depends sensitively on core size. Theoretical calculations predict that the lowest (1s) exciton energy of the c-Si core changes from 1.3 eV for 10 nm to 2.0 eV for 4 nm diameter, as described previously. This is inconsistent with size-independent PL peak energy, as shown in Fig. 11. Moreover, the electronic properties of a c-Si core more than several nanometers in diameter is thought to be similar to those of *bulk* Si. The indirect-to-direct transition in the bandgap may not occur even in 3-nm sized samples (Hybertsen, 1994).

The outer surface SiO_2 layer does not itself contribute to strong visible PL because the bandgap energy is out of visible range (>8 eV). Because size-independent PL peak energy implies that visible PL processes have an origin other than the quantum confinement state in the crystalline core, we stress the importance of the interfacial region between the crystalline core and surface oxide layer. We have proposed a three-region model for oxidized nanocrystals composed of (a) a crystalline core with diameter D_{core}, (b) an outer oxide layer, and (c) an interfacial layer between (a) and (b). The a-SiO_2 surface regions form a high-potential barrier to confine the exciton inside. In this model, the interfacial region (an oxygen-modified surface state on the nanocrystals) plays the most important role in the PL process, as illustrated in Fig. 15. In Si nanocrystals, the strong visible PL is observed after the surface is covered by Si oxides. It is considered that the oxidized surface state becomes the localized state exhibiting strong and stable luminescence in the red spectral region.

The composition ratio of the interfacial region should be intermediate between that of the oxide layer (Si:O $= 1:2$) and that of the crystalline core (Si:O $= 1:0$). In incompletely oxidized Si layers, oxygen atoms play important roles in the electronic structure. To understand the most essential feature of the electronic properties of this thin region, we use the result of *ab initio* calculations for one-sided oxidized Si planar sheets. The most characteristic feature is that these Si layers have a direct allowed exciton transition of ~ 1.6 eV at the Γ point (Takeda, 1994). These calculated transition energies are nearly equal to the observed PL peak energies. Transition energy in the interfacial region is almost independent of $\langle D_{core} \rangle$, whereas that in the crystalline core region depends on $\langle D_{core} \rangle$.

When the transition energy of the interfacial layer is lower than that of the crystalline core, excitons or carriers are subject to a confinement potential and are confined on the interfacial spherical-shell region. The condition for exciton confinement is roughly determined by the relative magnitude of the transition energy of each region. Theoretical calculations (e.g., Takagahara and Takeda, 1992; Wang and Zunger, 1994) indicate that the energy gaps of the c-Si core of $D_{core} \leqslant 5-7$ nm are higher than the transition energies of the interfacial layers. As a result, when $D_{core} \leqslant 5-7$ nm for Si nanocrystals, the exciton is confined in the interfacial shell region. A schematic diagram of the energy gap is drawn in Fig. 15(*b*). The exciton

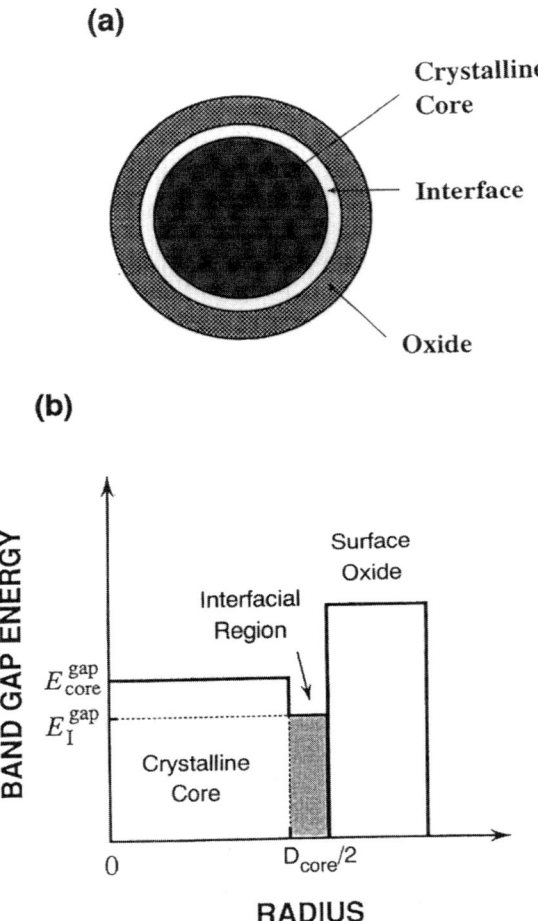

FIG. 15. (a) Schematic view of oxidized silicon nanocrystals. The oxidized Si nanocrystals consists of three regions: the core, interfacial, and outer oxide regions. (b) Energy-gap diagram of the three-region model. The transition energies of the interfacial Si and Ge layers E_I^{gap} are $\sim 1.6\,\text{eV}$ and $\sim 2.4\,\text{eV}$, respectively, including the exciton binding energy. The energy gap of the crystalline core E_{core}^{gap} increases with decreasing core diameter. In Si nanocrystals of $D_{core} \leqslant 5$–$7\,\text{nm}$ and Ge nanocrystals of $D_{core} \leqslant 3$–$4\,\text{nm}$, the exciton is confined in the interfacial region surrounded by the crystalline core and the surface oxide layer.

confinement (or exciton localization) in the interfacial layer is a possible origin of the strong 1.65-eV PL from oxidized Si nanocrystals and the 2.25-eV PL from oxidized Ge nanocrystals. Theoretical evaluations of $D_{core} \leqslant 5$–7 nm for Si nanocrystals and $D_{core} \leqslant 3$–4 nm for Ge nanocrystals agree well with the experimental findings.

We considered the planar c-Si sheet for the calculation of the interface state. However, in Si nanocrystals, spherical curvature of the interface, strains between Si and SiO_2, surface roughness, and variations in surface stoichiometry cause a potential fluctuation of the interface states. Then, broad PL spectra are observed near 1.65 eV. With decreasing size of nanocrystals, a disordered potential of interface states critically affects the electronic structure of interface states and the dynamics of excitons. Moreover, it is likely that the transition energy of the interface state would be size-dependent in very small nanocrystals. The size dependence of energy levels of defect and impurity states is an interesting and important issue. However, it is considered that the size dependence of the transition energy of the interface state is very small compared with that of the crystalline core state, and is negligible in nanocrystals with diameter of 4 nm or more.

According to our three-region model, the process of radiative recombination in surface-oxidized nanocrystals proceeds as follows. The photogeneration of excitons mainly occurs within the c-Si core. Some of the excitons in the core are localized in the interfacial layer (localized states on nanocrystal surface). Strong PL is then caused by the radiative recombination of excitons confined in the interfacial layer. This exciton localization picture explains why PL intensity depends on the size of the c-Si core whereas the PL peak energy is not sensitive to the core size. The one-monolayer c-Si sheet as the interface state is an oversimplified picture. However, recent *ab initio* calculations show that excitons are localized at the c-Si/SiO_2 interface (Kageshima 1996; Kageshima and Shiraishi, 1997) and support our arguments that the oxygen-modified interface state is one candidate for the origin of strong 1.65-eV luminescence.

In the above model, we consider that the active interface state is formed between c-Si and the SiO_2 layer. If the interface state plays an essential role in efficient PL in nanoscale Si/SiO_2 systems, the 1.65 eV luminescence should be observed in 2D Si/SiO_2 wells and 1D Si/SiO_2 wires. The PL properties of nanoscale Si/SiO_2 does not depend on the dimensionality of c-Si. Room-temperature visible PL has been reported in 2D Si quantum wells and 1D wires. Takahashi *et al.* (1995) observed efficient red PL from well-defined 2D Si structures formed on a SIMOX wafer under 488 nm laser excitation. Thin single c-Si film is sandwiched between SiO_2 layers, as shown in Fig. 16(*a*). Figure 16(*b*) shows PL spectra for 2D Si layers of various thickness. They found that all spectra have a peak around 1.65 eV. The peak

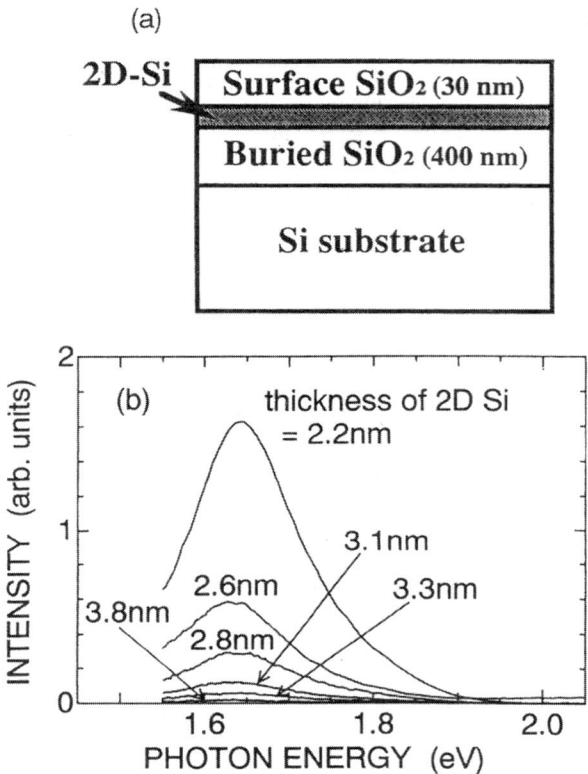

FIG. 16. (a) Schematic cross-sectional view of the two-dimensional Si quantum well structures formed on a SIMOX (separation by implanted oxygen) wafer. (b) Photoluminescence spectra from two-dimensional Si quantum wells of different thickness. (From Takahashi et al. (1995). *Jpn. J. Appl. Phys.* **34**, 950.)

energy of the PL spectra is almost independent of the Si thickness and this fact excludes the luminescence mechanism originating from direct recombination in a pseudodirect 2D Si band. However, the PL intensity depends strongly on the 2D Si thickness. Luminescence properties of 2D Si quantum wells are very similar to those in 0D Si nanocrystals. The photocarrier generation occurs in the 2D Si layer and luminescence comes from the interface recombination centers. Thickness-dependent PL intensity and thickness-independent PL peak energy in surface-oxidized Si 2D Si quantum wells can be well explained by the three region model for 0D Si nanocrystals.

However, in very thin wells (<2 nm), the PL spectrum is asymmetric and depends strongly on well thickness (Okamoto and Kanemitsu, 1997). Figure

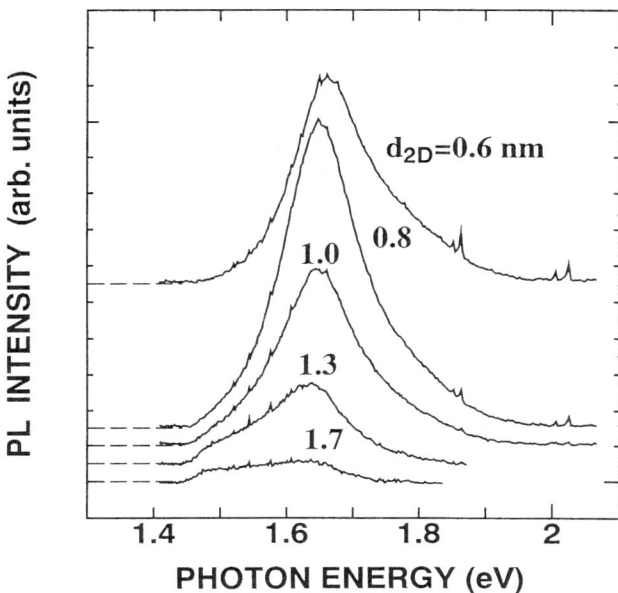

FIG. 17. Photoluminescence spectra for various Si well thickness of Si/SiO$_2$ quantum wells under 2.54 eV excitation at 2 K. The data are reproduced from Okamoto and Kanemitsu (1997).

17 shows PL spectra of 2D Si single quantum wells of various thickness under 2.540-eV excitation at 2 K. In the PL spectrum of 1.7 nm-thick 2D Si, for example, there is a shoulder around 1.5 eV in addition to a main PL peak around 1.65 eV. With a decrease of thickness, the shoulder around 1.5 eV disappears. Instead, a new PL shoulder clearly appears around 1.8 eV. This is indicative of the existence of two different radiative recombination bands.

The asymmetric PL spectra in the red and infrared spectral region can be fitted by two gaussian bands; the weak PL band (denoted as Q) and the strong PL band (denoted as I) (Okamoto and Kanemitsu, 1997). The PL peak energy of the I band is almost independent of the well thickness. In contrast, peak energy of the Q band shifts to higher energy with a decrease of the Si well thickness. Thickness dependence of the PL peak energies of the I and Q bands is plotted in Fig. 18. Peak energy of the main I band does not depend on the well thickness. Similar size-independence of the PL peak energy has been observed in SiO$_2$-capped Si nanocrystals (see, Fig. 11). Experimental observation of the size-insensitive PL in both 0D and 2D Si/SiO$_2$ systems suggests that the main luminescence originates from the interface state between c-Si and SiO$_2$ layers.

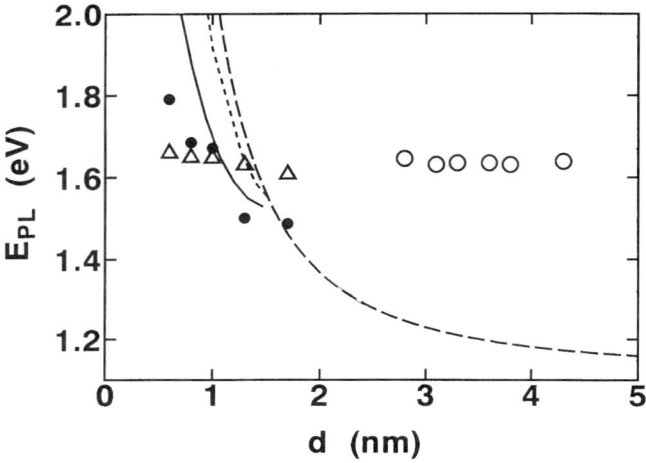

FIG. 18. Thickness dependence of the PL peak energy in Si/SiO$_2$ quantum wells. The PL peak energies in thin Si wells (the Q (●) and I (△) bands) and in thick Si wells (○) are reproduced from Okamoto and Kanemitsu (1997) and Takahashi et al. (1995), respectively. The solid (Kageshima, 1996), dotted (Zhang and Zunger, 1993) and broken (Takahashi et al., 1995) lines show the calculated bandgap energy of 2D Si wells.

Theoretical calculations of the thickness dependence of the bandgap energy for 2D Si wells (Zhang and Zunger, 1993; Kageshima, 1996) are denoted by the lines in Fig. 18. Q band peak energy is roughly consistent with theoretical calculations based on the quantum confinement model. Moreover, resonantly excited PL spectra in the Q band show the TO-phonon-related fine structure and luminescence anisotropy due to the 2D structure. These results imply that Q band is caused by radiative recombination in the Si well. Size-dependence of the PL peak energy in c-Si/SiO$_2$ quantum wells is different from that in a-Si/SiO$_2$ quantum wells (Lockwood et al., 1996). This may be because the coherent length of the wavefunctions and wave vector selection rule in c-Si are different from those in a-Si. Moreover, we speculate that atomic configurations at the interface between c-Si and SiO$_2$ are different from those between a-Si and SiO$_2$. Strains at the interface, interface roughness, and stoichiometry of the interface region affect the electronic structures of very thin 2D wells.

In Si nanostructures with a SiO$_2$ surface layer, both quantum confinement and surface state effects determine the optical properties. Widening of the bandgap energy occurs with a decrease of the size of nanostructures. Silicon oxide passivation reduces the formation of the surface dangling bonds, which act as nonradiative recombination centers. This low defects density is probably one of the reasons for the high luminescence efficiency of surface-oxidized Si nanostructures. Furthermore, Si oxides produce the active surface states for light emission. Oxygen-modified states within Si nanostructures play an essential role in the luminescence process at room temperature. The proposed three region model gives a reason as to why efficient room temperature PL appears at a particular wavelength region in oxidized Si and Ge nanostructures. The co-operation of quantum confinement and surface state effects causes strong visible PL in surface-oxidized Si and Ge nanostructures.

On the other hand, the observed PL peak energy in H-terminated Si nanocrystals (as-prepared π-Si) depends on the nanocrystal size, and observed size-dependence is in good agreement with the size-dependence of the exciton energy predicted on the basis of a quantum confinement model. In large nanocrystals, size-dependent PL spectrum is observed in the infrared spectral region. In infrared spectral region, the size-dependence of the PL peak energy and resonantly excited PL spectra of surface-oxidized Si nanocrystals are very similar to those of H-terminated Si nanocrystals (as-prepared π-Si) (Kanemitsu and Okamoto, 1997a; Kanemitsu et al., 1997). The infrared PL properties are not sensitive to the surface structure. Quantum confinement effect plays a dominant role in the infrared PL in Si nanostructures.

4. Photoluminescence Dynamics

In surface-oxidized Si nanocrystals, surface effects as well as quantum confinement effects produce unique optical and electronic properties. These two effects also complicate the mechanism of broad visible PL in Si nanocrystals. For example, PL spectrum and dynamics depend strongly on the temperature and excitation conditions. With a decrease of temperature, PL peak wavelength becomes shorter. At low temperatures below about 100 K, the PL peak wavelength is not sensitive to temperature. Moreover, the PL intensity gradually increases with decreasing temperature until about 100 K and then decreases at low temperatures below 100 K. These temperature dependences of the PL peak wavelength and intensity imply that the microscopic processes determining PL properties at low temperatures are different from those at high temperatures (Vial et al., 1992; Suemoto et al., 1994; Kanemitsu, 1996). Moreover, time-resolved PL studies in Si nanocrystals indicate that the PL decay exhibits nonexponential behavior and the recombination processes are complex. Here, we discuss the PL decay dynamics in the 10^{-6} to 10^{-2} sec. time regions in surface-oxidized Si nanocrystals.

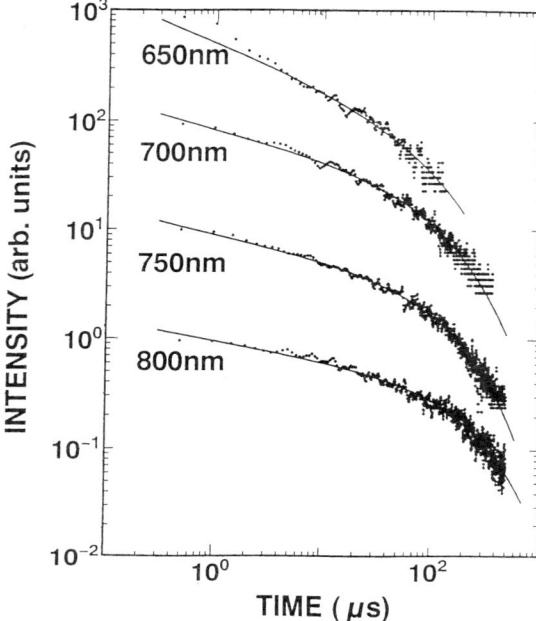

FIG. 19. Double logarithmic plot of the decay curves at different wavelengths in the broad red luminescence spectrum at room temperature. The solid lines are theoretical curves given by stretched exponential functions. (From Kanemitsu (1994). *Phys. Rev. B* **49**, 16845.)

Figure 19 shows PL decay profiles at different wavelengths in the broad PL band at room temperature. The PL decay profiles are nonexponential at temperatures between 10 and 300 K, and are well described by a stretched exponential function (Kanemitsu, 1993):

$$I(t) = I_0(\tau/t)^{1-\beta} \exp[-(t/\tau)^\beta] \tag{1}$$

where τ is an effective decay time, β is a constant between 0 and 1, and I_0 is a constant. Least-squares-fitting of the data gives the value of τ. This stretched exponential decay is usually observed in the PL decay and transport properties of disordered systems.

Temperature and photon-energy dependence of the PL decay rate τ^{-1} are summarized in Fig. 20. At high temperatures, the PL decay time τ depends strongly on both the monitored PL photon energy and temperature (Kanemitsu, 1996). An exponential energy dependence of the PL decay rate τ^{-1} is observed. However, at low temperatures below about 100 K, the PL decay rate is not sensitive to the PL phonon energy. The dominant physical

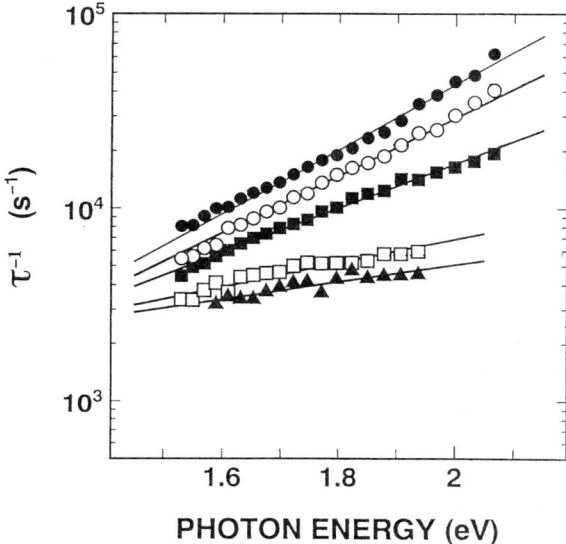

FIG. 20. Effective luminescence decay rate τ^{-1} as function of the monitored photon energy at 280, 220, 150, 80, and 30 K. The exponential slope increases with increasing temperature. At low temperatures, decay rate is almost independent of the luminescence energy. At high temperatures, the exponential energy dependence of the decay rate is clearly observed. (From Kanemitsu (1996). *Phys. Rev. B* **53**, 13515.)

process controlling PL dynamics at high temperatures is different from that at low temperatures.

In isolated Si nanocrystals, it is considered that the diffusion of carriers or excitons and energy transfer between Si nanocrystals are not important in luminescence processes. If the decay of excited electrons and holes within isolated nanocrystals is determined by competing radiative τ_R and nonradiative τ_{NR} processes, the PL lifetime τ_{PL} and PL quantum yield η are given by $\tau_{PL}^{-1} = \tau_R^{-1} + \tau_{NR}^{-1}$ and $\eta = \tau_R^{-1}/(\tau_R^{-1} + \tau_{NR}^{-1})$. The ratio η/τ_{PL} gives the temperature dependence of the radiative decay rate. Temperature dependence of the radiative decay rate is of the Arrhenius type, and the activation energy of the radiative decay rate is about 71 meV in oxidized Si nanocrystals (Kanemitsu, 1996). The activation energy of 71 meV is close to the maximum in the phonon density-of-states of bulk c-Si at 61 meV. Therefore, for the thermal activation process in the radiative decay rate, we consider phonon-limited luminescence processes in oxidized Si nanocrystals.

While the luminescence data of Si nanocrystals are similar to those of π-Si in many ways, there are some important differences. In π-Si, thermal activation energy for the radiative decay rate is 3–17 meV (t'Hooft et al., 1992; Suemoto et al., 1994) and singlet-triplet splitting energy of the exciton states is about 5–10 meV (Calcott et al., 1993). Activation energy in oxidized Si nanocrystals (about 71 meV) is large compared with the activation energy and splitting energy in π-Si. Moreover, clear phonon structure in the PL spectra in oxidized Si nanocrystals was not observed at low temperatures under resonance excitation (Kanemitsu, 1996), as discussed in Section III.2. The broad PL spectrum and nonexponential PL decay imply the existence of shallow tail states below the band edge in oxidized Si nanocrystals. The potential fluctuation of surface states in oxidized Si nanocrystals is much larger than that in hydrides-terminated Si nanocrystals. It is considered that the broad energy distribution of the radiative and nonradiative recombination centers in the crystallite mainly determines the luminescence properties in oxidized Si nanocrystals.

At low temperatures radiative lifetime becomes very long. Photoluminescence properties at low temperatures will be affected by the very long nonradiative relaxation processes. For nonradiative recombination process, we consider the tunneling (Vial et al., 1992) and thermally activated transport processes of carriers to nonradiative recombination centers. Both the tunneling and thermally-activated nonradiative recombination rates depend on the barrier height. Suemoto et al. (1994) estimated that the barrier height for nonradiative recombination processes E_a is ~ 0.2 eV for PL lifetimes between microseconds and milliseconds in Si nanocrystals. This value is one order of magnitude lower than the band offset between Si and SiO_2. Si nanocrystals are isolated from each other in samples. Thus, it is considered that carrier escape from one nanocrystal to another nanocrystal does not occur. The small barrier height (several hundreds of millielectron volts) implies that the carriers are localized in a disordered potential of interface states between the c-Si core and the SiO_2 surface layer. The shallow tail state is formed by the disordered potential due to interface roughness and variations in interface structures. The tail below the band edge is formed with a small state density. The temperature dependence of PL spectrum and dynamics suggests that PL decay properties at high temperatures are controlled by the thermally activated process. The cascade of carriers from higher states to lower states makes the lifetime of higher states shorter, and then the exponential slope is observed. At low temperatures, PL decay rate is not sensitive to PL energy and PL peak energy is not sensitive to temperature. Because radiative lifetime is very long at temperatures below 100 K, the slow tunneling process affects PL properties. Long PL lifetime at low temperatures is mainly determined by the tunneling process. The thermally activated or tunneling processes of carriers in tail states determines the PL properties of oxidized Si nanocrystals. Many experimental results can be understood in terms of the localization of carriers or excitons in a disorder potential of interface states.

Surface-oxidized Si nanocrystals with a disorder potential of interface states show complicated luminescence properties. Si nanocrystals are unique materials whose properties differ from those of amorphous and crystalline solids, and they are a nanoscopic disordered system in the sense that it has a distribution of crystalline sizes, shapes and variations in the surface structure and surface stoichiometry. The nanoscopic inhomogeneity will produce the unique optical properties of oxidized Si nanocrystals and π-Si. Luminescence properties of inhomogeneous systems can be controlled by changing the excitation conditions such as laser wavelength, laser pulse width, incident laser fluence, measurement temperature, and ambient conditions.

IV. Unique Optical Phenomena

1. NONLINEAR OPTICAL PROPERTIES OF π-Si

Since the discovery of efficient visible PL from Si, Ge, and SiC nanocrystals, there have been numerous reports describing the visible luminescence properties of nanocrystals made from indirect-gap group IV semiconductors. In particular, π-Si is receiving widespread interest motivated by potential applications as light-emitting devices, because the PL efficiency of π-Si is very high and a very large porous layer is obtained. In inhomogeneous systems, nonlinear optical measurements provide detailed information on electronic structure. The study of nonlinear optical properties allows a new approach to understanding the origin and mechanism of efficient visible luminescence in Si nanocrystals and π-Si. We discuss here the third-order nonlinear-optical susceptibility $|\chi^{(3)}(-3\omega;\omega,\omega,\omega)|$ spectra in naturally oxidized π-Si films by the third harmonic generation (THG) Maker fringe method for a third harmonic photon energy region from 540 to 720 nm, which covers the visible luminescence region (Kanemitsu et al., 1995).

Figure 21 shows a typical example of the Maker fringe pattern, that is the THG intensities as a function of incident angle of the fundamental laser beam for a 42 μm-thick π-Si film. At large angles, a fringe pattern is clearly observed. Optical harmonic generation from a multilayered sample was treated by Bethune (1989) using a transfer matrix technique. Using this technique, we calculated THG Maker fringe pattern from a multilayered sample structure composed of a naturally oxidized π-Si film and two SiO_2 glasses. A calculated fringe pattern is shown in Fig. 21. The periods of fringe

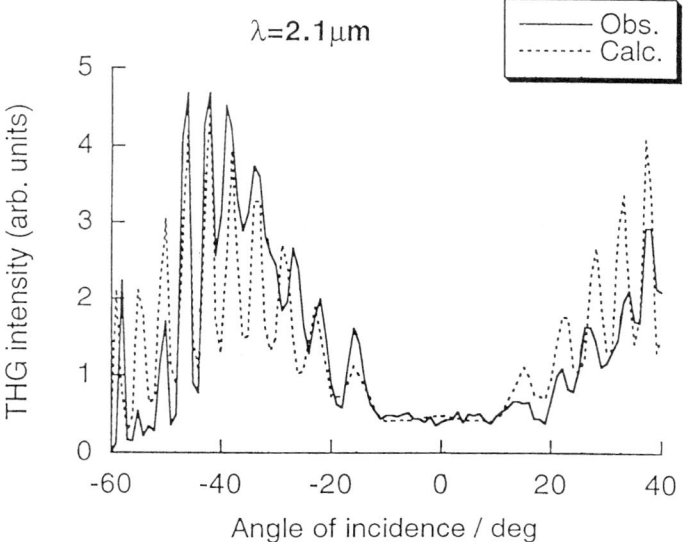

FIG. 21. Typical third-harmonic intensity pattern for a π-Si film as a function of angle of incidence of the fundamental pulsed laser light on the sample (solid line). The broken line is the theoretically calculated fringe pattern. (From Kanemitsu *et al.* (1995). *Phys. Rev. B* **52**, 10752.)

patterns in the experiment are reproduced by theoretical calculations and we can evaluate the value of $|\chi^{(3)}|$ at a given wavelength.

The third-order nonlinear optical susceptibility $|\chi^{(3)}|$ spectrum as a function of the THG wavelength is summarized in Fig. 22. The value of $|\chi^{(3)}|$ is not sensitive to the pump laser wavelength. The value of $|\chi^{(3)}|$ is about 0.5×10^{-12} esu in this region. We find that π-Si has a large value of $\chi^{(3)}$, although the absorption coefficient is very small and the refraction index ($n \sim 1.5$) is low in the near-infrared wavelength region.

In direct-gap semiconductor nanocrystals such as CuCl or CuBr quantum dots, the $|\chi^{(3)}|$ spectrum exhibits sharp peaks at exciton states near the absorption edge. Resonant enhancement of $|\chi^{(3)}|$ due to exciton confinement is clearly observed in the luminescence wavelength region (near band edge) (Ogawa and Kanemitsu, 1995). In π-Si, no sign of enhancement effect in $|\chi^{(3)}|$ is observed in the luminescence spectral region. This fact provides important information on the optical absorption and luminescence processes in π-Si. Two possibilities can be pointed out. First, the indirect-gap nature is important in optical transitions in Si nanocrystals. Porous Si has the indirect-gap semiconductor nature and the lowest optical transition is the phonon-assisted optical transition near the absorption edge. Then, it is

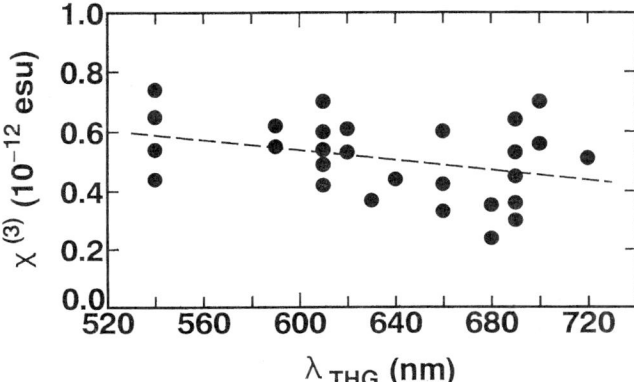

FIG. 22. Third-order nonlinear optical susceptibility $|\chi^{(3)}|$ spectrum of π-Si as a function of the wavelength of third harmonic generation. There is no enhancement of $|\chi^{(3)}|$ in the luminescence wavelength region. (From Kanemitsu et al. (1995). Phys. Rev. B **52**, 10752.)

speculated that enhancement of $|\chi^{(3)}|$ is not observed near the indirect-transition region. Second, we can consider that there is no resonant enhancement due to the band edge in this wavelength region and the bandgap is above this THG photon energy region. The red PL with a peak of ~ 700 nm is the light-emission from localized state with a small density, rather than the band edge emission.

Porous Si shows strong visible luminescence under wavelengths shorter than ~ 500 nm. Furthermore, pump-and-probe experiments show that large photoinduced absorption signals related to the photocarrier generation (imaginary part of $\chi^{(3)}$) are observed, but these signals are also insensitive to pump wavelengths longer than 550 nm (Matsumoto et al., 1994c). Wavelength dependence of the nonlinear absorption coefficient α_{NL} is summarized in Fig. 23. Nonlinear absorption α_{NL} in the π-Si film is clearly observed at laser wavelength shorter than about 500 nm (about 2.4 eV). The 2.4-eV photon energy is approximately equal to the calculated bandgap energy of a 2-nm nanocrystal (e.g., Wang and Zunger, 1994). This implies that carriers are generated in Si nanocrystals and the nonlinear absorption is caused by these photogenerated carriers. This result is consistent with the above THG Maker fringe result. Moreover, induced absorption change at probe wavelengths of 600–800 nm is clearly observed when excitation wavelength is shorter than 600 nm (Fauchet, 1995). These results are explained by a model in which photocarriers generated at the excitation laser wavelength are localized at the lower-energy states and then localized carriers cause a large absorption change in the luminescence wavelength region. Nonlinear optical measurements imply that the effective band edge

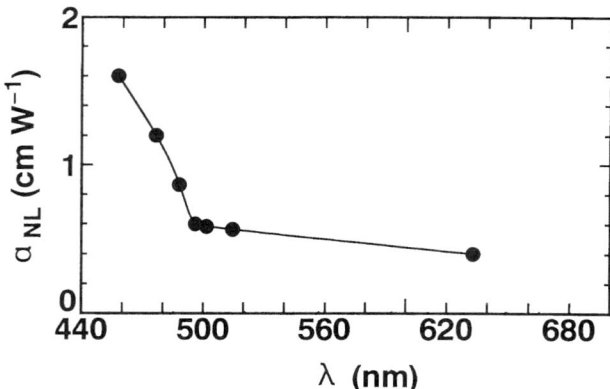

FIG. 23. Photoinduced absorption intensity α_{NL} in π-Si as a function of the excitation wavelength. A nonlinear absorption coefficient is clearly observed at laser wavelengths shorter than about 500 nm.

exists above the luminescence peak energy. It is considered that partially oxidized π-Si has an indirect-gap semiconductor nature in the optical absorption process and efficient room-temperature PL comes from lower-energy localized states with small density of states, rather than band edge emission.

Multiphoton-excitation luminescence experiments reveal that a particular surface state plays a role in the visible PL process (Chin et al., 1995; Diener et al., 1995). Visible PL from π-Si is induced by infrared multiphoton excitation at room temperature. Figure 24 summarizes red PL spectra excited by picosecond laser pulses at 0.532, 1.06, 1.3, and 4.9 μm (Chin et al., 1995). The PL spectra are similar to each other and to the red PL spectrum under UV excitation above the bandgap energy. The emission of a single photon at 700 nm requires the absorption of 1, 2, 3, and 7 photons at 0.532, 1.06, 1.3, and 4.9 μm, respectively, in the excitation process. However, the observed power-dependence is weak compared with the above estimation. This implies that the carriers are excited into the quasicontinuum Urbach state. Chin et al. (1995) pointed out that the excitation spectrum is correlated well with the stretching vibrational spectrum of the surface SiH_x species. On the other hand, in partially oxidized π-Si, the visible luminescence is observed after high-level vibration excitation in the Si–O absorption band (Diener et al., 1995). The radiative center is an isolated, Si–O related, atomic configuration on the surface of the Si particles. Surface vibrational modes are directly involved in the excitation of room-temperature PL in π-Si. Nonlinear absorption and luminescence experiments can reveal the microscopic mechanism of luminescence in semiconductor nanocrystals.

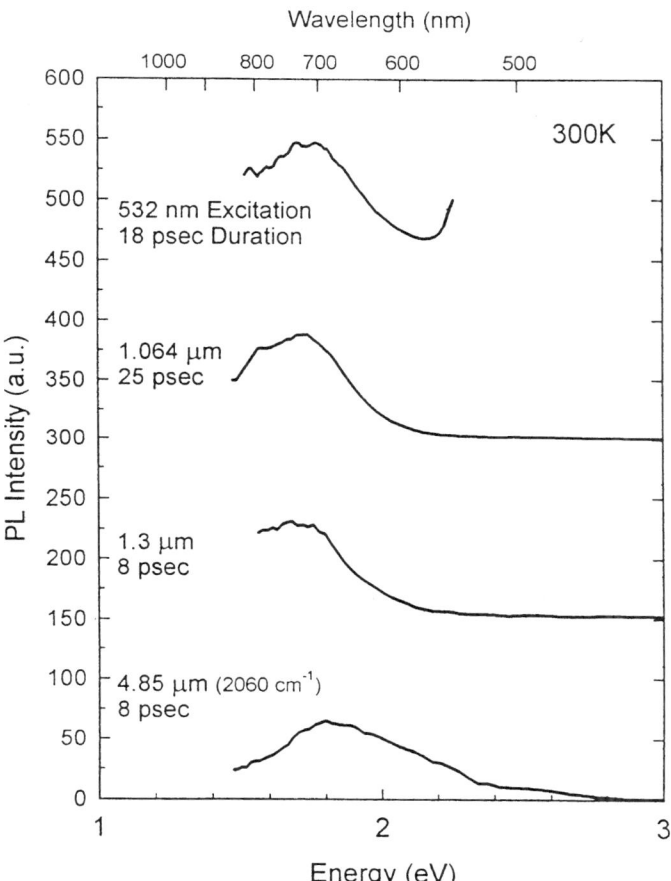

FIG. 24. Photoluminescence spectra of π-Si observed by excitation at various infrared and visible wavelengths. The spectra are similar to each other. Visible PL is observed by infrared multiphoton excitation at room temperature. (From Chin et al. (1995). *Science* **270**, 776.)

Moreover, using nonlinear absorption phenomena, we have demonstrated some all-optical logic and switching experiments needed for promotion of Si-based all-optical integrated circuits (Matsumoto et al., 1994b, 1994c). We observed the optical nonlinear absorption and optical bistability in π-Si at room temperature (Matsumoto et al., 1994). However, the response time of the photoinduced absorption change and bistable loop is about 0.5–1 ms. This relaxation time is longer than the luminescence lifetime at room temperature. Rather, the relaxation time is nearly equal to the carrier diffusion lifetime determined from time-of-flight photocurrent measurements (Fajfer et al., 1995). These slow and wavelength-dependent nonlinear phenomena can be explained by a nonuniform space charge effect (Kanemitsu, 1995). The study of nonlinear optical properties opens a new approach to discover unique optical properties other than visible luminescence and to understand the electronic states in semiconductor nanocrystals.

2. Tuning of Luminescence Wavelength

In the previous section, it was shown that surface-oxidized Si nanocrystals and π-Si exhibit stable red luminescence at room temperature. Many researchers have tried to control the PL wavelength in Si nanocrystals and π-Si by changing the nanocrystal size, surface structure, and excitation condition. Here, we discuss the excitation power dependence of both red and blue PL in π-Si.

In the case of weak continuous-wave laser excitation, π-Si samples show the usual red luminescence. However, using high-power 240-nm 500-fs laser pulses at a repetition rate of 81 MHz, the luminescence peak varies from red to blue. Figure 25(a) shows the PL spectra at different excitation intensities. At low intensities, the red peak only is observed. At high intensities, a new blue PL peak appears. At intermediate intensities, the PL spectrum, clearly shows both the red and blue PL peaks. Figure 25(b) shows the red and blue PL intensities as a function of the 240-nm 500-fs laser-excitation intensity. Red PL is observed at less than $100\,mW/cm^2$ excitation intensity and increases sublinearly with increasing excitation intensity. The red luminescence intensity saturates at around $500\,mW/cm^2$. However, the intensity of blue luminescence increases linearly with increasing excitation intensity. These nonlinear PL behaviors can be explained by a change of the recombination route from lower energy states to the higher ones due to population saturation (Kanemitsu et al., 1996a). These results suggest that there are two luminescent states in π-Si, in which the PL intensity depends on excitation conditions such as pulse energy and pulse duration.

The above power-dependence of PL intensity can be quantitatively explained by a simple three-level model consisting of the ground, higher excited and lower excited states. To fit the experimental results, we determined $\gamma_{cs} \sim 10^{12}\,s^{-1}$, $\gamma_C \sim 10^{10}\,s^{-1}$, $\gamma_s \sim 10^8\,s^{-1}$, where the coefficient γ_{cs} is the nonradiative decay rate from higher states to lower states, and γ_C and γ_S are the decay rates from higher state to the ground state and from the lower state to the ground state, respectively. Solid lines in Fig. 25(b) obtained with these parameters describe the nonlinear luminescence behaviors well. The above calculation shows that there are two different lumines-

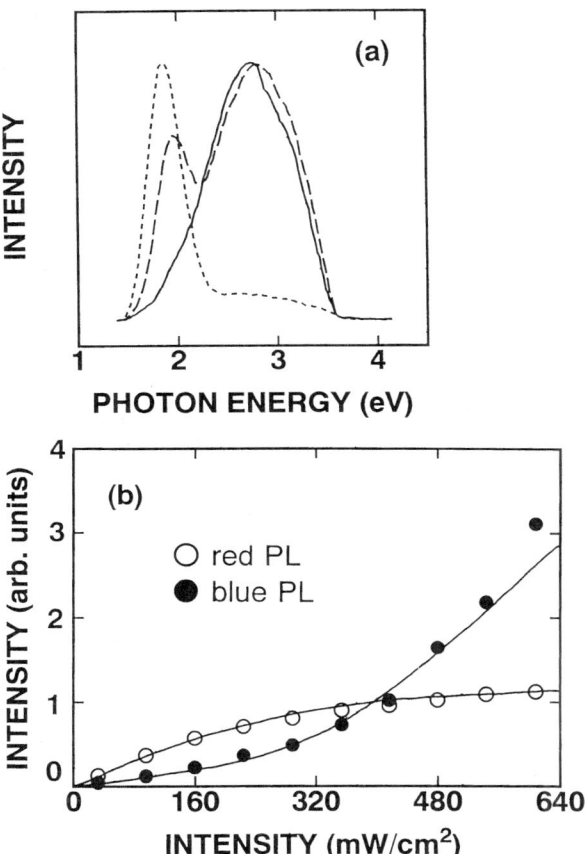

FIG. 25. (a) Luminescence spectra under high-power and high-repetition-rate femtosecond laser excitation at 30 mW/cm² (dashed line), 450 mW/cm² (broken line) and 700 mW/cm² (solid line). (b) Blue and red luminescence intensities as a function of excitation intensity. The solid lines represent the theoretically fitted curves. We can control the luminescence wavelength by changing the excitation conditions. The data are reproduced from Kanemitsu et al. (1996a).

cence states in π-Si in the visible spectral region. Blue PL is caused by the change of carrier recombination routes due to the population saturation of lower-lying surface localized states. A similar shift cycle in PL spectra between green and red was obtained by changing the solutions (Li et al., 1993). As-prepared samples in HF solution showed green PL. When the sample was rinsed in water and blow dried, the PL spectrum changed from green to red. Returning the samples to HF solution caused the green PL to appear again. The spectral shift between blue-green and red is a very interesting phenomenon and also confirms that there are different luminescence states in π-Si. The luminescence spectrum from two different states is overlapped, because π-Si has a distribution of crystalline sizes and shapes and a fluctuation of the surface structure and stoichiometry. By changing excitation conditions and surface structures, we can change luminescence wavelengths.

The strong blue PL is also observed in π-Si oxidized at high temperatures. Thermal oxidation processes produce a higher quality oxide at the Si surface for microelectronics technology. For optoelectronic device applications and the ongoing discussion on the mechanism of visible PL, we need to control surface structure and preserve crystalline cores with diameters of several nanometers. High quality SiO_2 formed on the surface of the c-Si core provides: (1) mechanical and chemical stability, (2) a reduction of nonradiative recombination centers, and (3) optically transparent windows for the study of the c-Si core state.

The surface structure of π-Si has been modified by rapid-thermal-oxidization (RTO) processes (Petrova-Koch et al., 1992; Kovalev et al., 1994). Porous Si samples were processed in a RTO apparatus. In our experiments (Kanemitsu et al., 1994), the heating rate was 200°C/s and samples were kept for 35 sec. at the oxidation temperature T_{ox} ranging from 480 to 1200°C under 1 atm O_2 pressure. Cooling rate was 100°C/s. Figure 26 shows PL spectra from oxidized π-Si under 325 nm excitation at room temperature as a function of T_{ox}. In π-Si oxidized at low T_{ox} (<800°C), the PL peak is around 750 nm. The spectral width of the broad PL is about 0.3 eV (FWHM). Both the PL peak and PL width in oxidized π-Si are nearly equal to those in the naturally oxidized π-Si. At higher T_{ox} above 800°C, the strong blue PL near 400 nm appears, while red PL near 750 nm disappears. FTIR spectroscopy provides information about the internal surface of π-Si. During RTO processes, the composition of internal surface changes from Si hydrides and Si oxyhydrides (Si:O:H) to a-SiO_2 for $T_{ox} > 800°C$. The PL peak energy changes from about 1.65 eV ($T_{ox} < 800°C$) to 3 eV ($T_{ox} > 800°C$). It is found that there is a good relationship between the PL spectrum and surface-termination condition (Kanemitsu et al., 1994).

Both the temperature dependence of blue PL intensity and decay dynamics of the blue PL are entirely different from those of the red PL (Kanemitsu et al., 1994). Initial decay of the blue PL in the picosecond time region is approximately given by a single exponential function. The long nonexponential decay of the blue PL is not observed, but that of the red PL is clearly observed. The mechanism of the blue PL is different from that of the red PL originating from the long-lived near-surface state. It is speculated that after high-temperature oxidation, there hardly exists any Si crystallite larger than 2 nm in the sample. The small c-Si core has hybrid electronic

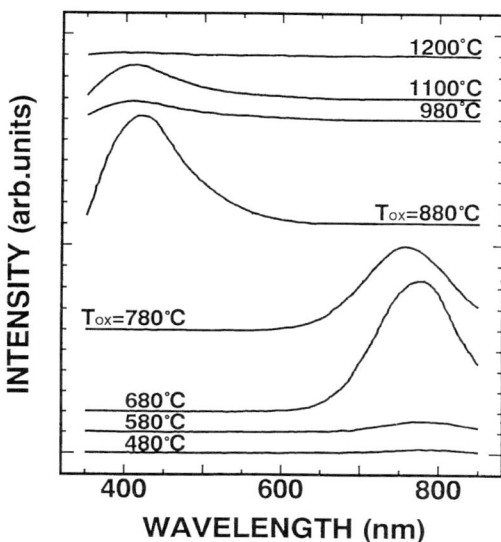

FIG. 26. Luminescence spectra from oxidized π-Si under 325 nm excitation as a function of oxidation temperature T_{ox}. The luminescence spectrum drastically changes at temperatures higher than $T_{ox} = 800°C$. There is a correlation between the luminescence peak wavelength and the surface structure of nanocrystalls. (From Kanemitsu et al. (1994). *Phys. Rev. B* **49**, 14732.)

properties between the molecular and crystalline states. We cannot distinguish surface state from core state in a very small crystallite like a molecular cluster. In very small Si nanoparticles, the average Si coordination is far from 4 coordination of bulk-like Si materials, and almost all of the Si atoms are terminated by H or O or both. Then the concept of the interior and surface states is not available for molecular-like systems. The small Si cluster acts similar to Si-rich oxides (nonstoichiometric oxides, SiO_x).

The blue PL of π-Si and RTO π-Si shows complicated behaviors. Porous Si after air exposure shows a high value of the linear polarization of blue-green PL (Andrianov et al., 1993), and the degree of luminescence polarization of π-Si is higher than that of RTO π-Si (Kanemitsu et al., 1996a). The picosecond lifetime of blue PL in π-Si depends strongly on monitored wavelength, where blue PL in π-Si is observed under high excitation as shown in Fig. 25. The lifetime decreases with decreasing monitored PL wavelength (e.g., 100 ps at 2.4 eV and 300 ps at 1.5 eV). It seems that this wavelength dependence of the PL lifetime implies a size dependence of the radiative lifetime from crystalline core state. On the other hand, the lifetime of the blue PL in RTO π-Si does not depend on the monitored wavelength. Wavelength dependence of the PL lifetime in π-Si is completely different from that in RTO π-Si. The various and complicated optical response of blue PL reflects complicated Si structures. Optical properties are very sensitive to the local atomic configuration of small Si clusters (Kyushin et al., 1994). Thus, we consider that very small Si clusters with different atomic configurations play a role in the strong and complicated luminescent process in the blue-green spectral region.

Oxidized π-Si (RTO π-Si and π-Si after air exposure) shows efficient luminescence in the red and blue spectral region. Continuously tunable luminescence has not been observed in oxidized π-Si and surface-oxidized Si nanocrystals. However, it is accepted that during the etching process, quantum confinement effect determines the crystallite size in π-Si. Therefore, by changing the illumination wavelength during electrochemical etching, we tried to control the photoluminescence wavelength from an n-type π-Si layer without air exposure (Mimura et al., 1995). Figure 27 shows PL spectra in as-prepared π-Si as a function of the light wavelength in light-assisted electrochemical etching. With increasing photon energy, the crystallite size becomes smaller. The as-prepared π-Si shows a variety of colors in luminescence, but the luminescence is very weak and unstable.

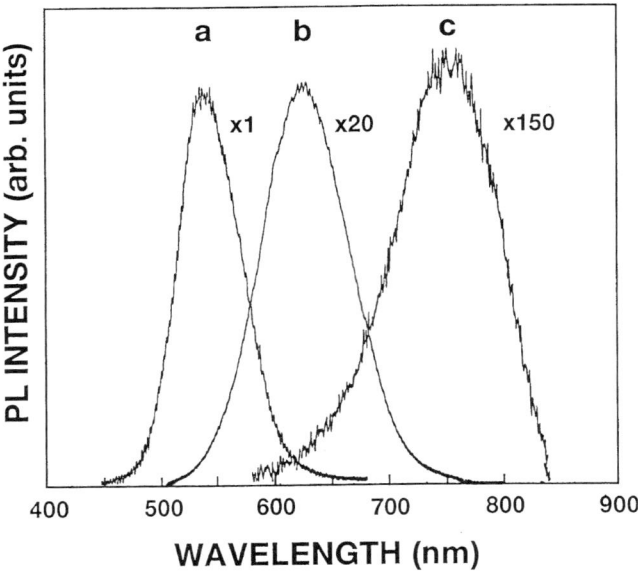

FIG. 27. Photoluminescence spectra from n-type π-Si prepared by light-assisted electrochemical etching: cutoff wavelength during etching (a) 390 nm, (b) 440 nm, and (c) 490 nm. We can control the crystal size and PL wavelength by changing the incident wavelength during electrochemical etching. (From Mimura et al. (1995). Mater. Res. Soc. Proc. **358**, 635.)

Similar size-dependent luminescence spectra in as-prepared π-Si have been reported by Ito et al. (1995) using a scanning tunneling microscopy (STM) technique. After air exposure, π-Si shows the red luminescence like oxidized Si nanocrystals. Band structures of hydrogen-terminated Si nanocrystals are different from those of oxidized Si nanocrystals. Tunable luminescence in as-prepared π-Si without air exposure is related to crystal size. Moreover, EL of π-Si after air exposure still exhibits green luminescence similar to the PL of π-Si before air exposure. Thus, green EL is caused by the mostly delocalized region in π-Si. Similar green luminescence due to delocalized states is observed in p-type π-Si in solutions (Dubin et al., 1994). It is speculated that the green luminescence is due to radiative recombination in the delocalized state in π-Si. Moreover, voltage tunable EL is observed in π-Si in contact with solutions (Bsiesy et al., 1993). The applied voltage allows a selective excitation of a group of nanocrystals. The EL tunability is related to nanoscopic inhomogeneity and the size distribution of nanocrystals. The study of tuning of luminescence and EL mechanism is very important for the understanding of quantum confinement effects in Si nanocrystals.

Co-sputtering and ion implantation techniques create supersaturated Si solid solutions and precipitation of Si in a SiO_2 glassy matrix. After various annealing treatments, we can observe a broad PL band. Almost all samples show PL in a particular wavelength region; red and blue PL are observed like π-Si. Fisher et al. (1996) reported continuous tuning of the PL from 3.0 eV to 1.6 eV with successive annealing cycles in a forming gas (H_2 and N_2), as shown in Fig. 28. With increasing nanocrystal size, PL peak energy shifts to the lower energy. The observed shift from 3.0 to 1.6 eV is a size-related phenomenon. Tunable PL under the forming gas conditions shows that interface of Si nanocrystals or clusters is different from that for thermal oxidation.

From tunable luminescence experiments, we note that the role of hydrogen-passivation is very important. We need to clarify the effect of hydrogen passivation in tunable luminescence in Si nanocrystals and clusters. The rainbow color PL is observed in as-prepared π-Si and Si nanocrystals annealed under a $H_2:N_2$ gas, rather than Si nanocrystals covered by SiO_2 thin layer. The ordering of nanocrystals depends on bonding of surface atoms. Hydrogen-terminated π-Si shows a crystalline nature, while oxygen-terminated π-Si shows atomic disorder within the nanoparticles (Tsang et al., 1992). Hydrogen termination is very important to understand the size-related PL mechanism. Surface oxidation causes novel electronic states in nanocrystals. Strong visible luminescence comes from the radiative recombination of excitons localized in oxygen-modified surface states. However, surface oxidation also produces nonradiative recombination centers or defects at the Si/SiO_2 interface. Hydrogen-termination also reduces

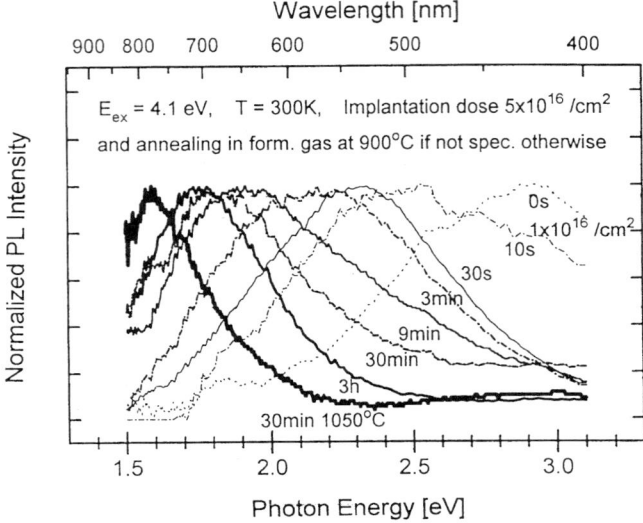

FIG. 28. Photoluminescence spectra of Si implanted SiO$_2$ showing the continuous tuning from 3.0 eV to 1.6 eV with Si precipitation in a H$_2$:N$_2$ gas atmosphere. The redshift of PL wavelength is a size-related effect. (From Fischer et al. (1996). Thin Solid Film **276**, 100.)

surface recombination centers (Yablonovitch et al., 1986), and changes bandgap energy of the surface state. Thus, thermal annealing under a forming gas (N$_2$:H$_2$) enhances the PL intensity. Since Si and SiO$_2$ and their heterostructures are very important in Si-based microelectronics, detailed and systematic studies of Si nanostructures with SiO$_x$H$_y$ surface structures are needed for a deeper understanding of the electronic properties of Si nanostructures. In particular, we should try to prepare fully hydrogen-terminated Si nanocrystals and investigate their optical properties.

V. Summary

In this chapter, we presented an overview of current experimental studies of the optical properties of Si and Ge nanocrystals. The quantum confinement effect causes widening of bandgap energy and enhancement of oscillator strength in nanocrystals. In Si nanocrystals with diameters of several nanometers, the band structure remains indirect as in bulk c-Si. Luminescence from π-Si and Si nanocrystals is very sensitive to surface chemistry of the crystallites, particularly with regard to relative amounts of oxygen and hydrogen on the surface. Surface effect plays an active role other than in the reduction of nonradiative recombination centers. In surface-oxidized Si nanocrystals, novel electronic structure of the surface state is caused by oxygen interconnecting the crystallite core and the outer surface layer. Size dependence of optical properties of surface-oxidized nanocrystals shows that the visible luminescence comes from tail states below the band edge, while the absorption process would be size-dependent in Si nanocrystals. It is concluded that the cooperation of the quantum confinement effect and surface states in oxidized nanocrystals causes efficient visible luminescence at room temperature. However, the observed PL peak energy in H-terminated Si nanocrystals (as-prepared π-Si) is sensitive to nanocrystal size, and the TO phonon-related structure is clearly observed under resonant excitation. H-terminated Si nanocrystals show their crystalline nature, while oxidized Si nanocrystals show their disordered nature.

Si and Ge nanocrystals samples are inhomogeneous in that they have broad distribution of crystallite sizes and shapes and the surface roughness and surface stoichiometry. This structural inhomogeneity on a nanometer scale causes unique optical and electronic properties. Photoluminescence and EL spectra are controlled by a change in excitation conditions such as laser wavelength, laser power, laser pulse, or applied voltage. For continuous tuning of the luminescence wavelength and the enhancement of PL intensity in Si and Ge nanocrystals, we need to know the positional correlation between structure and luminescence itself with nanoscale resolution. It is pointed out that the electronic structure of hydrogen-terminated Si nanocrystals is different from that of surface-oxidized Si nanocrystals. We should try to produce fully hydrogen-terminated Si and Ge nanocrystals and clusters and investigate their optical properties. Recently, a new type of Si-based nanostructure has been studied: Si/SiO_2 (Takahashi et al., 1995; Lockwood et al., 1996), Si/CaF_2 (Vervoot et al., 1995), Si/Si:O:H (Tsu et al., 1995) quantum well structures. The fabrication of these new nanostructures and the development of new techniques (scanning microprobe techniques, single-dot spectroscopy, etc.) will clarify the details of the electronic properties of Si and Ge nanocrystals.

Acknowledgments

The author would like to thank S. Okamoto, K. Suzuki, H. Uto, N. Shimizu, Y. Masumoto, H. Mimura, T. Matsumoto, T. Ogawa, K. Takeda, K. Shiraishi, M. Kondo, A. Mito, S. Oda, H. Matsumoto, and S. Kyushin for collaborations and F. Koch, V. Petrova-Koch, and T. Komoda for discussions. He also thanks many of the authors cited in this review for discussions.

References

Allan, G., Delerue, C., and Lannoo, M. (1996). *Phys. Rev. Lett.* **76**, 2961.
Andrianov, A. V. Kovalev, D. I., Zinov'ev, N. N., and Yaroshetskii, I. D. (1993). *JETP Lett.* **58**, 427.
Atwater, H. A., Shcheglov, K. V., Wong, S. S., Vahala, K. J., Flagan, R. C., Brongersma, M. L., and Polman, A. (1994). *Mater. Res. Soc. Symp. Proc.* **316**, 409.
Bhargava, R. N., Gallagher, D. C., Hong, X., and Nurmikko, A. (1994). *Phys. Rev. Lett.* **72**, 416.
Bethune, D. S. (1989). *J. Opt. Soc. Am. B* **6**, 910.
Bloomfield, A., Freeman, R. R., and Brown, W. L. (1985). *Phys. Rev. Lett.* **54**, 2246.
Brus, L. (1994). *J. Phys. Chem.* **98**, 3575.
Brus, L. E., Szajowski, P. F., Wilson, W. L., Harris, T. D., Schuppler, S., and Citrin, P. H. (1995). *J. Am. Chem. Soc.* **117**, 2915.
Bsiesy, A., Muller, F., Ligeon, M., Gaspard, F., Herino, R., Romestain, R., and Vial, J. C. (1993). *Phys. Rev. Lett.* **71**, 637.
Calcott, P. D. J., Nash, K. J., Canham, L. T., Kane, M. J., and Brumhead, D. (1993). *J. Phys. Condens. Matter.* **5**, L91.
Canham, L. T. (1990). *Appl Phys. Lett.* **57**, 1046.
Chin, R. P., Shen, Y. R., and Petrova-Koch, V. (1995). *Science* **270**, 776.
Diener, J., Ben-Chorin, M., Kovalev, D. I., Ganichev, S. D., and Koch, F. (1995). *Phys. Rev. B* **52**, R8617.
Dubin, V. M., Ozanam, F., and Chazalviel, J. N. (1994). *Phys. Rev. B* **50**, 14867.
Fauchet, P. M. (1995). *Mater. Res. Soc. Proc.* **358**, 525.
Fejfar, A., Pelant, I., Sipek, E., Kocka, J., Juska, G., Matsumoto, T., and Kanemitsu, Y. (1995). *Appl. Phys. Lett.* **66**, 1098.
Fischer, T., Petrova-Koch, V., Shcheglov, K., Brandt, M. S., and Koch, F. (1996). *Thin Solid Films* **276**, 100.
Honea, E. C., Ogura, A., Murray, C. A., Raghavachai, K., Sprenger, W. O., Jarrold, M. F., and Brown, W. L. (1993). *Nature* **366**, 42.
'tHooft, G. W., Kessner, Y. A. R. R., Rikken, G. L. J. A., and Venhuizen, A. H. J. (1992). *Appl. Phys. Lett.* **61**, 2344.
Hybertsen, M. S. (1994). *Phys. Rev. Lett.* **72**, 1514.
Ito, K., Ohyama, S., Uehara, Y., and Ushioda, S. (1995). *Appl. Phys. Lett.* **67**, 2536.
Jarrold, M. F. (1991). *Science* **252**, 1085.
Kageshima, H. (1996). *Surf. Sci.* **357–358**, 312.
Kageshima, H., and Shiraishi, K. (1997). To be published.
Kanemitsu, Y. (1993). *Phys. Rev. B* **48**, 12357.
Kanemitsu, Y. (1994). *Phys. Rev. B* **49**, 16845.
Kanemitsu, Y. (1995). *Phys. Rep.* **263**, 1.
Kanemitsu, Y. (1996). *Phys. Rev. B* **53**, 13515.
Kanemitsu, Y., Kondo, M., and Takeda, K. (1994). *Light Emission from Novel Silicon Materials.* The Physical Society of Japan, Tokyo.
Kanemitsu, Y., Suzuki, K., Nakayoshi, Y., and Masumoto, Y. (1992a). *Phys. Rev. B* **46**, 3916.
Kanemitsu, Y., Uto, H., Masumoto, Y., and Maeda, Y. (1992b). *Appl. Phys. Lett.* **61**, 2187.
Kanemitsu, Y., Ogawa, T., Shiraishi, K., and Takeda, K. (1993a). *Phys. Rev. B* **48**, 4883.
Kanemitsu, Y., Uto, H., Masumoto, Y., Matsumoto, T., Futagi, T., and Mimura, H. (1993b). *Phys. Rev. B* **48**, 2827.
Kanemitsu, Y., Suzuki, K., Masumoto, Y., Komatsu, T., Sato, K., Kyushin, S., and Matsumoto, H. (1993c). *Solid State Commun.* **86**, 545.

Kanemitsu, Y., Futagi, T., Matsumoto, T., and Mimura, H. (1994a). *Phys. Rev. B* **49**, 14732.
Kanemitsu, Y., Suzuki, K., Kondo, K., and Matsumoto, H. (1994b). *Solid State Commun.* **89**, 619.
Kanemitsu, Y., Suzuki, K., Kyushin, S., and Matsumoto, H. (1995a). *Phys. Rev. B* **51**, 13103.
Kanemitsu, Y., Suzuki, K., Kondo, K., Kyushin, S., and Matsumoto, H. (1995b). *Phys. Rev. B* **51**, 10666.
Kanemitsu, Y., Okamoto, S., and Mito, A. (1995c). *Phys. Rev. B* **52**, 10752.
Kanemitsu, Y., Matsumoto, T., and Mimura, H. (1996a). *J. Non-Cryst. Solids* **198–200**, 977.
Kanemitsu, Y., Shimizu, N., Komoda, T, Hemment, P. L. F., and Sealy, B. J. (1996b). *Phys. Rev. B* **54**, R14329.
Kanemitsu, Y., and Okamoto, S. (1997a). *Mater. Sci. & Eng. B*, in press.
Kanemitsu, Y., and Okamoto, S. (1997b). *Phys. Rev. B* **56**, in press.
Kanemitsu, Y., Okamoto, S., Otobe, M., and Oda, S. (1997). *Phys. Rev. B* **55**, R7375.
Koch, F.. Petrova-Koch, V., and Muschik, T. (1993). *J. Jumin.* **57**, 271.
Komoda, T., Kelly, J. P., Nejim, A., Homewood, K. P. Hemment, P. L. F., and Sealey, B. J. (1995). *Mater. Res. Soc. Symp. Proc.* **358**, 163.
Kovalev, D., Averboukh, B., Ben-Chorin, M., Koch, F., Efros, Al. L., and Rosen, M. (1996). *Phys. Rev. Lett.* **77**, 2089.
Kovalev, D., Ben-Chorin, M., Koch, F., Efros, Al. L., Rosen, M., Gippius, N. A., and Tikhodeev, S. G. (1995). *Appl. Phys. Lett.* **67**, 1585.
Kovalev, D. I., Yaroshetzkii, I. D., Muschik, T., Petrova-Koch, V., and Koch, F. (1994). *Appl. Phys. Lett.* **64**, 214.
Kyushin, S., Matsumoto, H., Kanemitsu, Y., and Goto, M. (1994). In *Light Emission from Novel Silicon Materials* (eds. Y. Kanemitsu, M. Kondo, and K. Takeda). Physical Society of Japan, Tokyo, p. 46.
Lehmann, V., Cerva, H., and Gosele, U. (1992). *Mater. Res. Soc. Symp. Proc.* **256**, 3.
Lehmann, V., and Gosele, U. (1991). *Appl. Phys. Lett.* **58**, 856.
Li, K. H., Tsai, C., Sarathy, J., and Campbell, J. C. (1993). *Appl. Phys. Lett.* **62**, 3192.
Littau, K. A., Szajowski, P. F., Muller, A. J., Kortan, R. F., and Brus, L. E. (1993). *J. Phys. Chem.* **97**, 1224.
Lockwood, D. J. (1994). *Solid State Commun.* **92**, 101.
Lockwood, D. J., Lu, Z. H., and Baribeau, J. M. (1996). *Phys. Rev. Lett.* **76**, 539.
Maeda, Y., Tsukamoto, N., Yazawa, Y., Kanemitsu, Y., and Masumoto, Y. (1991). *Appl. Phys. Lett.* **59**, 3168.
Matsumoto, T., Futagi, T., Mimura, H., and Kanemitsu, Y. (1993). *Phys. Rev. B* **47**, 13876.
Matsumoto, T., Takahashi, J., Tamaki, T., Futagi, T., Mimura, H., and Kanemitsu, Y. (1994a). *Appl. Phys. Lett.* **64**, 226.
Matsumoto, T., Hasegawa, N., Tamaki, T., Ueda, K., Futagi, T., Mimura, H., and Kanemitsu, Y. (1994b). *Jpn. J. Appl. Phys.* **33**, L35.
Matsumoto, T., Futagi, T., Hasegawa, N., Mimura, H., and Kanemitsu, Y. (1994c) In *Semiconductor Silicon 1994* (eds. H. R. Huff, W. Bergholz, and K. Sumino). *Electrochemical Society Proceedings*, vol. 94-10. Electrochemical Society, Pennington, p. 545.
Miller, R. D., and Michl, J. (1989). *Chem. Rev.* **89**, 1359.
Mimura, H., Matsumoto, T., and Kanemitsu, Y. (1994). *Appl. Phys. Lett.* **65**, 3350.
Mimura, H., Matsumoto, T., and Kanemitsu, Y. (1995). *Mater. Res. Soc. Symp. Proc.* **358**, 635.
Miyazaki, S., Sakamoto, K., Shiba, K., and Hirose, M. (1995). *Thin Solid Films* **255**, 99.
Nogami. M., and Abe, Y. (1994). *Appl. Phys. Lett.* **65**, 2545.
Ogawa, T., and Kanemitsu, Y. (1995). *Optical Properties of Low-Dimensional Materials.* World Scientific, Singapore.
Okamoto, S., and Kanemitsu, Y. (1996). *Phys. Rev. B* **54**, 16421.

Okamoto, S., and Kanemitsu, Y. (1997). To be published.
Paine, D. C., Caragianis, C., Kim, T. Y., Shigesato, Y., and Ishihara, T. (1993). *Appl. Phys. Lett* **62**, 2842.
Petrova-Koch, V., Muschik, T., Kux, A., Meyer, B. K., Koch, F., and Lehmann, V. (1992). *Appl. Phys. Lett.* **61**, 943.
Polisski, G., Averboukh, B., Kovalev, D., Ben-Chorin, M., and Koch, F. (1996). *Thin Solid Films* **276**, 235.
Proot, J. P., Delerue, C., and Allan, G. (1992). *Appl. Phys. Lett.* **61**, 1948.
Rosenbauer, M., Finkbeiner, S., Bustarret, E., Weber, J., and Stutzmann, M. (1995). *Phys. Rev. B* **51**, 10539.
Ruckschloss, M., Landkammer, B., and Veprek, S. (1993). *Appl. Phys. Lett.* **63**, 1474.
Sato, S., Nozaki, S., Morisaki, H., and Iwase, M. (1995). *Appl. Phys. Lett.* **66**, 3176.
Schuppler, S., Friedman, S. L., Marcus, M. A., Adler, D. L., Xie, Y. H., Ross, F. M., Harris, T. D., Brown, W. L., Chabal, Y. J., Brus, L. E., and Citrin, P. H. (1994). *Phys. Rev. Lett.* **72**, 2648.
Shimizu-Iwayama, T., Ohshima, M., Niimi, T., Nakao, S., Saitoh, K., Fujita, T., and Itoh, N. (1993). *J. Phys. Condens. Matter* **5**, L375.
Suemoto, T., Tanaka, K., Nakajima, A., and Itakura, T. (1993). *Phys. Rev. Lett.* **70**, 3659.
Suemoto, T., Tanaka, K., and Nakajima, A. (1994). In *Light Emission from Novel Silicon Materials* (eds. Y. Kanemitsu, M. Kondo, and K. Takeda). The Physical Society of Japan, Tokyo, p. 190.
Takagahara, T., and Takeda, K. (1992). *Phys. Rev. B* **46**, 15578.
Takagi, H., Ogawa, H., Yamazaki, Y., Ishizaki, A., and Nakagiri, T. (1990). *Appl. Phys. Lett.* **56**, 2379.
Takahashi, Y., Furuta, T., Ono, Y., Ishiyama, T., and Tabe, M. (1995). *Jpn. J. Appl. Phys.* **34**, 950.
Takeda, K. (1994). In *Light Emission from Novel Silicon Materials* (eds. Y. Kanemitsu,, M. Kondo, and K. Takeda). The Physical Society of Japan, Tokyo, p. 1.
Takeda, K., and Shiraishi, K. (1989). *Phys. Rev. B* **39**, 11028.
Tsang, J. C., Tischler, M. A., and Collins, R. T. (1992). *Appl. Phys. Lett.* **60**, 2279.
Tsu, R., Morais, J., and Bowhill, A. (1995). *Mater. Res. Soc. Symp. Proc.* **358**, 825.
Uhlir, A. (1956). *Bell Syst. Tech. J.* **35**, 333.
van Buuren, T., Tiedje, T., Patitsas, S. N., and Jones, C. D. W. (1995). In *22nd International Conference on The Physics of Semiconductors* (ed. D. J. Lockwood). World Scientific, Singapore, p. 2125.
Vial, J. C., Bsiesy, A., Gaspard, F., Herino, R., Ligeon, M., Muller, F., Romestain, R., and Macfarlane, R. M. (1992). *Phys. Rev. B* **45**, 14171.
Vervoot, L., Bassani, F., Mihalcescu, I., Vial, J. C., and Arnaud dAvitaya, F. (1995). *Phys. Stat. Sol. (b)* **190**, 133.
Wang, L. W., and Zunger, A. (1994). *J. Chem. Phys.* **100**, 2394.
Wang, X. (1994). In *Porous Silicon* (eds. Z. C. Zeng, and R. Tsu). World Scientific, Singapore, p. 77.
Yablonovitch, E., Allara, D. L., Chang, C. C., Gmitter, T., and Bright, T. B. (1986). *Phys. Rev. Lett.* **57**, 249.
Yoffe, A. D. (1993). *Adv. Phys.* **42**, 173.
Zang, S. B., and Zunger, A. (1993). *Appl. Phys. Lett.* **63**, 1399.

CHAPTER 6

Porous Silicon: Photoluminescence and Electroluminescent Devices

Philippe M. Fauchet

DEPARTMENT OF ELECTRICAL ENGINEERING
THE INSTITUTE OF OPTICS
DEPARTMENT OF PHYSICS AND ASTRONOMY, AND LABORATORY FOR LASER ENERGETICS
UNIVERSITY OF ROCHESTER
ROCHESTER NY 14627

I. INTRODUCTION	206
1. *Si Light Emission*	206
2. *Porous Si*	207
II. PROPERTIES OF THE PL BANDS	210
1. *The "Red" Band*	210
2. *The "Blue" Band*	212
3. *The "Infrared" Bands*	213
4. *The Extrinsic Luminescence Bands*	216
III. ORIGIN OF THE INTRINSIC PL BANDS	218
1. *Quantum Confinement and the Red PL Band*	218
2. *Silicon Oxide and the Blue PL Band*	222
3. *Recrystallization, Dangling Bonds, and the Infrared PL Bands*	224
IV. PURE QUANTUM CONFINEMENT AND SURFACE STATES: A CRITICAL DISCUSSION	226
V. NONOPTICAL PROPERTIES	233
1. *Introduction*	233
2. *Electrical Properties*	233
3. *Structural Properties*	236
VI. ELECTROLUMINESCENT DEVICES	238
1. *General Survey*	238
2. *LED Lifetime*	239
3. *Power Efficiency*	240
4. *Response Time*	241
5. *Spectral Coverage*	243
6. *Compatibility with Microelectronics*	244
VI. CONCLUSIONS AND OUTLOOK	246
References	247

I. Introduction

1. Si Light Emission

Radiative recombination of an electron with a hole across the bandgap of semiconductors produces luminescence. The emitted photons have an energy equal to the bandgap energy and a momentum that is negligible. Thus, the electron and the hole must be located at the same point in the Brillouin zone, which is the case in direct gap semiconductors such as GaAs. Under these conditions, the radiative recombination rate is large and the radiative lifetime is short (typically of the order of a few nanoseconds for a dipole-allowed transition). To obtain a large luminescence efficiency, the nonradiative recombination rate must be smaller than the radiative recombination rate. The efficiency is defined by

$$\eta = \tau_{nonrad}/(\tau_{nonrad} + \tau_{rad}) \tag{1}$$

where η is the quantum efficiency, τ_{rad} is the radiative lifetime, and τ_{nonrad} is the nonradiative lifetime. Nonradiative recombination occurs at the surface (surface recombination) and in the bulk (Auger recombination; recombination at defects). To minimize it, the surface should be well passivated and the bulk must be free of defects that act as radiation killers. The efficiency of direct gap III–V semiconductors such as GaAs exceeds 1% at room temperature and 10% at cryogenic temperature.

Silicon (Si) is an indirect gap semiconductor, in which electrons and holes are found at different locations in the Brillouin zone. Therefore, recombination by emission of a single photon is not possible. Photon emission is possible only if another particle, such as a phonon, capable of carrying a large momentum is involved. In this case, both energy and momentum can be conserved in the radiative transition. The participation of a third particle in addition to the electron and the hole makes the rate of the process substantially lower and the radiative lifetime is typically in the millisecond regime. Thus, the efficiency drops by several orders of magnitude, even for high purity materials and good surface passivation. At room temperature, the efficiency of crystalline Si (c-Si) is usually of the order of 10^{-4} to 10^{-5}%, which makes it unattractive for light emitting devices (LEDs) (Iyer and Xie, 1993). Several attempts have been made to improve the luminescence efficiency of Si and Si-based alloys. They include the use of Si-Ge alloys and superlattices (Sturm et al.,1993) and isoelectronic impurities (Bradfield, Brown, and Hall, 1989) in a manner similar to N in GaP. The strategy in these attempts can be divided into two classes: increasing the radiative rate or decreasing the nonradiative rate. The former can be achieved by "bandgap engineering" or "defect engineering" which essentially consists of "convincing" the electron and the hole that they do not need a phonon to recombine radiatvely. This can be accomplished through zone folding in thin superlattices or with isoelectronic impurities whose energy level extends throughout the Brillouin zone. The latter consists of confining the electron and the hole to a small volume where the probability of finding a nonradiative center is equal to zero. This can be accomplished in Ge-rich regions in SiGe and with isoelectronic impurities. These approaches have led to very good photoluminescence (PL) efficiencies (up to 10% in some cases) and good electroluminescence (EL) efficiency in some structures (approaching 1%), but only at cryogenic temperatures. At room temperature, these numbers unfortunately drop by several orders of magnitude. A thorough discussion of these approaches can be found in other chapters in this volume.

2. POROUS Si

Porous Si (π-Si) is a material that has been known for nearly 40 years (Uhlir, 1956). It is usually formed by electrochemical etching in an HF solution under an anodic current. It has found some use in microelectronics, especially in the silicon-on-insulator (SOI) technology, as π-Si becomes a good insulator after oxidation (Bomchil et al., 1988). In the 1980s, several studies of the optical properties of π-Si were published and PL in the deep red/near infrared was detected at cryogenic temperatures (Pickering et al., 1984). In 1990, Canham reported that when π-Si is further etched in HF for hours after preparation, it emits bright red light under illumination with blue or UV light (Canham, 1990). Figure 1 shows the evolution of the PL taken from Canham's paper. Prolonged etching in HF increases the porosity and thus produces smaller columnar structures, which leads to brighter PL at shorter wavelengths. When the dimension of the columns decreases below 5 nm, theory indicates that the bandgap widens by quantum confinement in the conduction and valence bands (Sanders and Chang, 1992) and thus smaller columns produce larger bandgaps. In Canham's original model, the luminescence was attributed to band-to-band recombination across the increased bandgap. Interestingly, at the same time Canham was publishing his seminal paper, Lehmann and Gösele independently showed experimentally that the bandgap of π-Si is larger than that of c-Si and attributed this increase to quantum confinement as well (Lehmann and Gösele, 1991).

To produce π-Si, the Si wafer (anode) and a metal cathode are immersed in an aqueous solution containing HF. In the presence of holes, HF dissolves Si (Smith and Collins, 1992). To produce π-Si, the HF concentration is usually kept between 10 and 25% by weight, alcohol is added to improve the penetration of the solution into the pores and to minimize

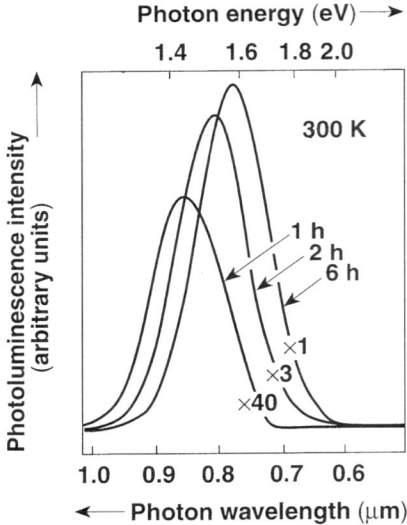

FIG. 1. Photoluminescence spectra of π-Si formed on a p^- Si wafer and subsequently immersed in 40% aqueous HF for 1, 2, and 6 h. (From Canham, 1990.)

hydrogen bubble formation, and a constant current density ranging from 1 to 100 mA/cm^2 is maintained. In general, when the current density increases or the HF concentration decreases beyond a critical value, pore formation changes to electropolishing (Smith and Collins, 1992). In this regime, the Si wafer is dissolved uniformly and the material on the wafer surface is not luminescent. The etching rate and the PL spectrum of the porous layer are controlled by many parameters, such as the Si doping level and type, the properties of the near-surface top layer of the wafer, the anodization time and current density, the solution concentration, and the pH. Consider three representative results: (1) The formation rate of light emitting π-Si can be enhanced by producing a gentle damage of the wafer surface by ion bombardment or it can be suppressed by amorphizing a few hundred nanometer thick surface layer (Duttagupta et al., 1995b). (2) By illuminating the wafer during anodization, which results in the generation of photoinjected holes, light emitting π-Si can be made from n-type Si, which does not contain naturally the holes necessary to form π-Si. (3) By changing the anodization conditions or subjecting the π-Si layer to selected processing steps, it is possible to make π-Si emit in a very wide range of wavelengths, as shown in Fig. 2 (Fauchet, 1994).

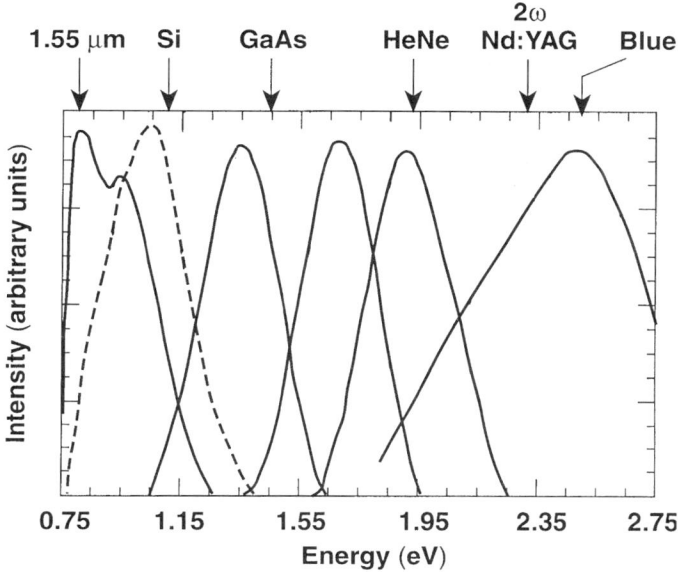

FIG. 2. Normalized PL spectra of six π-Si samples prepared and processed using different conditions. (From Fauchet, 1994.)

The wide range of conditions under which π-Si is produced, the low cost of the equipment, and the striking results that can be achieved and easily measured have encouraged hundreds of groups worldwide to work on π-Si. Since 1990, the understanding of the luminescence mechanisms has improved, although they are not yet fully established, and, more recently, π-Si EL devices have reached important milestones. The purpose of this chapter is to review the properties of π-Si, especially its PL and EL. Section II is a succint review of the optical properties, including a description of the three intrinsic PL bands and important extrinsic PL bands. Section III discusses the origin of the three intrinsic PL bands. Section IV presents an in-depth investigation of whether quantum confinement alone can explain the red PL band. Section V reviews selected nonoptical properties of π-Si, namely electrical transport and mechanical strength. Section VI is devoted to a survey of state-of-the-art EL devices made with π-Si. The discussion emphasizes the stability, efficiency, speed, emission wavelength, and compatibility with microelectronics. Finally, Section VII presents some conclusions and an outlook for future development.

II. Properties of the PL Bands

1. THE "RED" BAND

The PL described in Canham's original report (see Fig. 1) was in the red/near infrared part of the spectrum and is by far the most intensely studied band. The properties of this "red" band are similar over a range of wavelengths from slightly above the bandgap of c-Si to the yellow part of the spectrum. Results obtained with ultrahigh porosity samples that were supercritically dried to maintain their structural integrity indicate that the green PL can belong to the red band as well (Canham et al., 1994). The PL characteristics are: wide spectrum (100 to 500 meV), large efficiency under UV pumping (1–5% at room temperature, $\sim 10\%$ at cryogenic temperature), and long decay time (1–100 μsec at room temperature, up to 10 msec at cryogenic temperature). Figure 3 shows the evolution of a typical π-Si spectrum after pulsed excitation (Fauchet et al., 1994a) and the temperature dependence of the PL lifetime and intensity (Vial et al., 1993). The short-wavelength components usually decay faster than the long-wavelength components but this spectral diffusion with time varies from sample to sample and is even absent in some samples (Koyama and Koshida, 1994). The decay of the full band or that of any individual wavelength component does not follow a simple law and can be fitted very well using a stretched exponential function (Kanemitsu, 1993).

As mentioned above, tuning of the red PL band can be achieved by changing many parameters. The relationship between peak PL wavelength and porosity has already been mentioned. Figure 4 shows how the peak PL wavelength and the PL intensity vary with porosity for p^- wafers (decreasing HF concentration increases the porosity) and for p^+ wafers (increasing the current density increases the porosity) (Duttagupta et al., submitted for publication, 1997). In these measurements, the porosity was measured gravimetrically (Herino et al., 1987). When the porosity increases, the PL peak shifts to shorter wavelenghs and the PL intensity increases dramatically. Figure 5 shows the blue shift of the peak PL wavelength with time and with the shortest wavelength of the broadband light used during anodic etching of the n-type Si wafers (Koyama and Koshida, 1993). It had been known that under conditions of light assistance (LA), a longer anodic etching time produces a shorter PL peak wavelength (Peng, Tsybeskov, and Fauchet, 1993). In Fig. 5, the PL peak wavelength approaches the shortest wavelength used during etching with LA. This result is consistent with the model proposed by Lehmann and Gösele to explain the formation of nanometer-size columnar crystallites ("quantum wires") during anodization (Lehmann and Gösele, 1991). In their model, quantum confinement in the

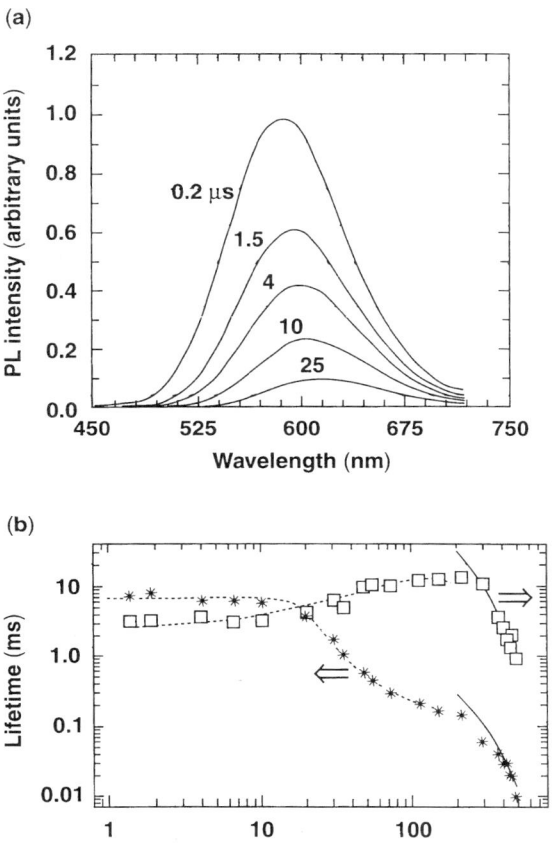

FIG. 3. (a) Time-resolved PL spectra taken during a short ~50 ns time window at selected times after excitation with a picosecond ultraviolet laser pulse. (From Fauchet et al., 1994.) (b) Photoluminescence lifetime and intensity versus temperature for a typical π-Si sample. (From Vial et al., 1993.)

wires produces a potential barrier for hole transport from the bulk to the tip of the wires and as a result etching of the wires stops due to the lack of holes. In the results of Fig. 5, the holes are photoinjected and etching continues until the bandgap of the quantum wires approaches the most energetic photon energy used during etching with LA, when the process self-terminates.

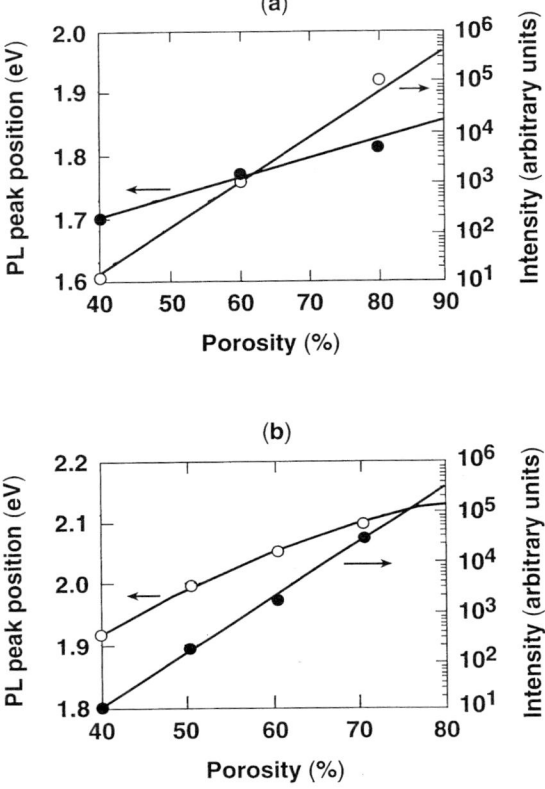

FIG. 4. Photoluminescence peak position and intensity versus porosity for (a) p$^-$ Si and (b) p$^+$ Si. For the same type of samples, the PL peak position shifts to the blue and the PL intensity increases with increasing porosity. (From Duttagupta et al., unpublished.)

2. THE "BLUE" BAND

Blue PL is especially strong in π-Si following thermal oxidation at high temperatures ($T_{ox} \geqslant 1000°C$) (Petrova-Koch et al., 1993). Figure 6 shows the PL spectrum and dynamics in π-Si oxidized at 1100°C for 2 min (Tsybeskov, Vandyshev, and Fauchet, 1994). The broad PL band (FWHM > 0.5 eV) with a peak near 2.6 eV and a decay on a nanosecond time scale are typical of these samples. The decay is nonexponential, as for the red band, but in contrast to the red band no significant wavelength dependence of the blue PL decay is observed from 440 nm to 650 nm, the decay dynamics do not

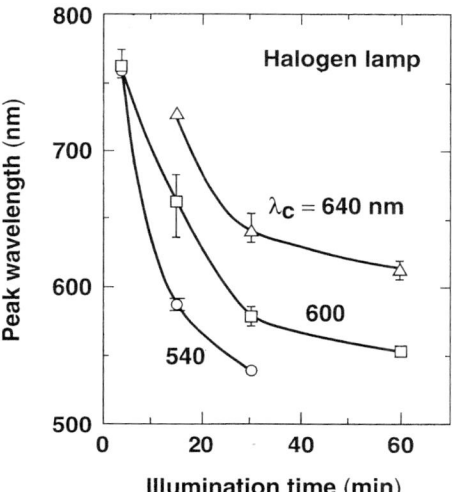

FIG. 5. Dependence of the PL peak wavelength on the anodization time and the shortest wavelength used during broadband illumination. (From Koyama and Koshida, 1993.)

change appreciably when the measurements are performed at cryogenic temperatures, and the decay dynamics are not fitted well by a stretched exponential. The estimated efficiency of the blue PL is greater than 0.1%, one order of magnitude below the efficiency of the red PL. The strong infrared absorption line at 1080 cm^{-1} shown in the inset of Fig. 6 is related to the asymmetric stretching vibrational mode of the Si–O–Si oxygen bridge.

The blue PL is also observed in samples that are oxidized chemically e.g., following immersion in 10% HNO_3 for 2 min prepared by anodization under light assistance or even freshly anodized after handling in air (Tsybeskov, Vandyshev, and Fauchet, 1994). The blue PL dynamics in these samples also show a nonexponential decay with a characteristic time in the 1 ns range. The intensity of the blue PL band appears to be correlated with the strength of the infrared absorption line at 1080 cm^{-1}.

3. THE "INFRARED" BANDS

The properties of the infrared (IR) PL have been studied much less. Several different mechanisms are responsible for the IR PL. Indeed, some of these measurements are consistent with bandgap or near-bandgap PL from

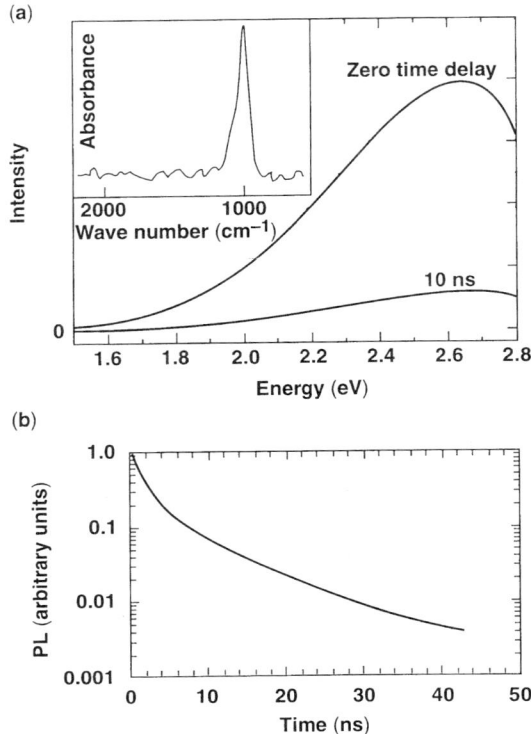

FIG. 6. Photoluminescence spectrum immediately after excitation and 10 ns later, and decay dynamics of the blue band. The IR absorption spectrum of the oxidized π-Si sample is shown in the inset. (From Tsybeskov et al., 1994.)

bulk-like c-Si (Pickering et al., 1984; Perry et al., 1992), while others involve a band that is commonly near 0.8 eV but can be tuned above the bandgap of bulk c-Si (Koch, 1993; Koch et al., 1993; Fauchet et al., 1993; Mauckner et al., 1995). Consider first the PL that appears at or near the bandgap of c-Si. Figure 7 shows the PL spectrum and integrated PL intensity produced by mesoporous Si after annealing in dry oxygen near 950°C for ~1 h (Tsybeskov et al., 1996c). At cryogenic temperature, this PL coincides with the PL spectrum observed in bulk c-Si (Davies, 1989). As the temperature increases toward 300 K, the PL spectrum widens but the PL efficiency remains unchanged near or above 0.1%.

The tunable infrared band can be made more prominent by annealing in ultrahigh vacuum at a temperature as high as ~500°C for 5 min (Fauchet

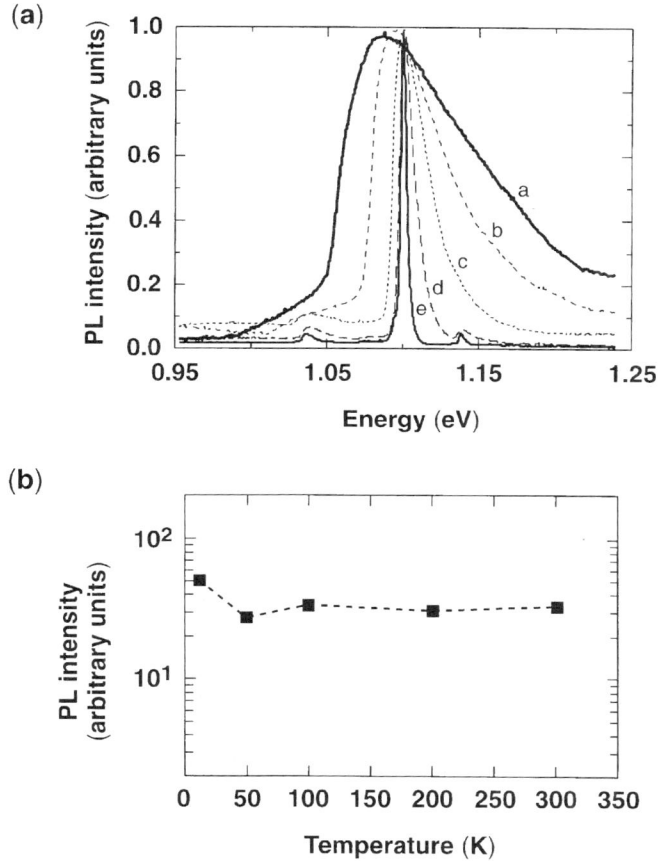

FIG. 7. (a) Photoluminescence spectra of mesoporous silicon annealed in dilute oxygen for 30 min at 950°C. The measurement temperature is (1) 300 K, (2) 200 K, (3) 100 K, (4) 50 K, and (5) 12 K. The spectra are identical to those of c-Si. (From Tsybeskov et al., 1996c.) (b) The integrated PL spectrum is temperature-independent. (From Tsybeskov et al., 1996c.)

et al., 1993). Figure 8 shows the room-temperature PL spectra for a π-Si sample before annealing and after annealing at two temperatures. The broad IR PL peak of the as-prepared sample is small but measurable at room temperature. As the annealing temperature is increased, its relative intensity increases until it dominates the spectrum. The IR PL appears to be most intense after annealing at a temperature where the red PL disappears, that is, when most of the hydrogen that passivates the π-Si surface has been

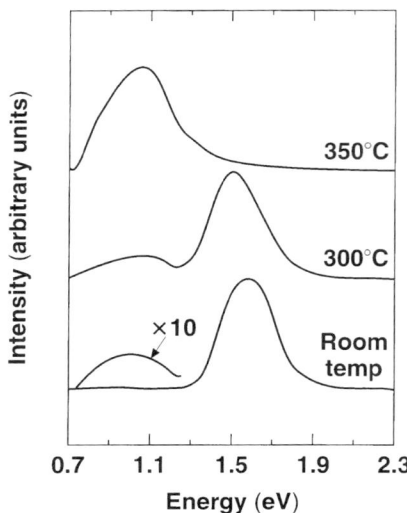

FIG. 8. Normalized room-temperature PL spectra of a π-Si sample before and after annealing at the indicated temperature for 5 min. The annealing and the measurements are performed in ultra high vacuum. (From Fauchet *et al.*, 1993.)

desorbed. In a few samples, after 350°C annealing, the integrated PL intensity measured at room temperature was comparable to the integrated intensity of the initial red PL band and the luminescence extended well past 2 μm. The conditions required for obtaining such samples are not well understood. At 77 K, time-resolved PL measurements indicate that the IR band decays in the range from tens of nanoseconds to ∼ 10 μsec. (Mauckner *et al.*, 1995; Fauchet *et al.*, 1994b), which is faster than the red PL decay with which it coexists. Figure 9 compares the red and IR PL dynamics measured at different temperatures on the same π-Si sample (Mauckner *et al.*, 1995). The facts that the IR and visible PL bands coexist in the same sample and that their decay dynamics and temperature dependence are independent suggest that the two PL bands are produced at distinct sites in π-Si, presumably in different nanocrystallites.

4. The Extrinsic Luminescence Bands

Porous Si can also be used as a host material for luminescent species. Compared to the three intrinsic PL bands, little work has been performed so far and most of it is concerned with Er as the active center. Erbium is a

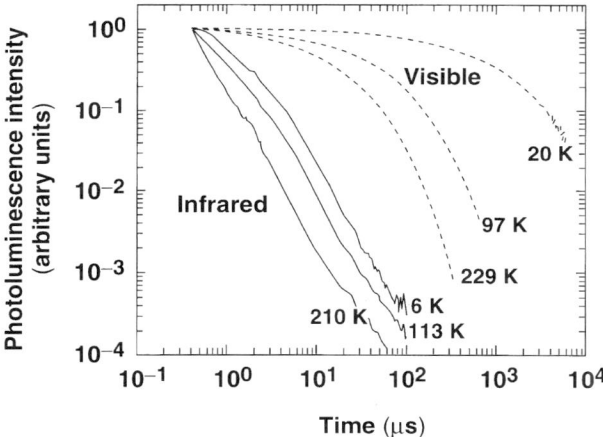

FIG. 9. Comparison of the PL decay dynamics of the red and IR bands in a π-Si sample of 80% porosity. At all measurement temperatures, the IR band decays much faster than the red band. (From Mauckner et al., 1995.)

rare earth element that has an intra-center luminescence line at 1.54 μm, which corresponds to the wavelength of choice for long-distance optical fiber communications. When Er is implanted in c-Si, PL and EL can be obtained at this wavelength (Zheng et al., 1994). The maximum concentration of active Er centers is limited however to near 10^{18} cm^{-3}, beyond which Er-rich precipitates are formed. Porous Si is thought to provide natural barriers to this precipitation because of its nanostructure. In addition, it can contain a large amount of oxygen, which favors the formation of Er-O complexes that enhance the luminescence efficiency (Michel et al., 1991) and it has a larger bandgap than c-Si, which is expected to lead to stronger, more temperature-independent luminescence (Favennec et al., 1993). Several groups have reported intense 1.54 μm PL in π-Si samples where Er had been introduced by a variety of means (Kimura et al., 1994; Dorofeev et al., 1995; Namavar et al., 1995; Taskin et al., 1995).

Porous Si can in fact be used as a host for many different luminescent centers. Using as a host Si-rich Si oxide prepared from π-Si, as in Tsybeskov et al. (1997a), luminescence from centers such as Er, Nd, S, or Se can be obtained. Figure 10 shows the room-temperature PL spectra of three different centers in this π-Si based material. This area of research is still at the very early stages and much work remains to optimize the efficiency of the emission. Note however that EL from Er-doped π-Si has been achieved very recently (Tsybeskov et al., 1997b).

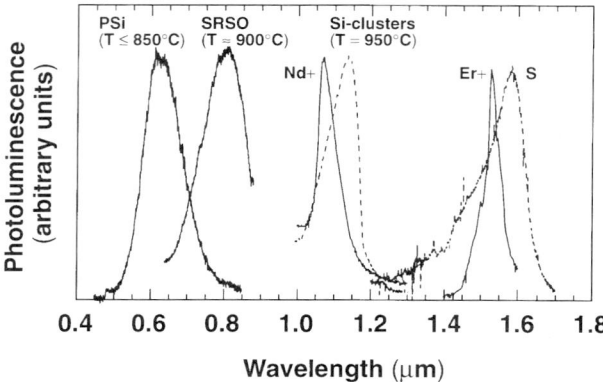

FIG. 10. Room temperature PL in silicon-rich silicon oxide nanocomposites formed by oxidation of π-Si. The PL from three luminescent centers (Nd, Er, and S) is shown, together with three luminescence spectra obtained after oxidation at different temperatures. (From Tsybeskov et al., submitted.)

III. Origin of the Intrinsic PL Bands

1. Quantum Confinement and the Red PL Band

Numerous models have been proposed to explain the strong, visible red PL (Canham, 1995). They can be divided into two categories: the models in which the luminescence occurs in quantum confined c-Si nanostructures and the models in which it does not. As already mentioned, the model proposed by Canham to explain the visible PL (Canham, 1990) and that of Lehmann and Gösele to explain the opening of the bandgap (Lehmann and Gösele, 1991) invoked quantum confinement in nanometer-size c-Si objects. The opening of the bandgap by quantum confinement not only pushes the bandgap to shorter wavelengths but also increases the PL energy, provided that the radiative recombination occurs at or near the bandgap of the Si nanocrystals. Evidence in support of this model is obtained from experiments in which the columnar structures that make π-Si are further thinned by a slow etching step in HF in the open circuit configuration (Canham, 1990) and other experiments in which the PL peak wavelength is controlled by the shortest wavelength used to illuminate the sample during etching (Koyama and Koshida, 1993). As the crystallite sizes decrease, the PL shifts to shorter wavelengths. The second intriguing aspect of the PL in π-Si is its very high efficiency at room temperature ($>1\%$). The efficiency results from a competition between radiative and nonradiative recombination mechanisms (Fauchet *et al.*, 1994b). In c-Si, the radiative recombination rate is very low, because the radiative process involves the participation of a third particle, such as a phonon, capable of providing momentum conservation during recombination. As a result, nonradiative recombination predominates and the quantum efficiency is typically <0.0001% at room temperature. To increase the efficiency, the radiative rate must increase and/or the nonradiative rate must decrease. Both occur in π-Si. The dominant effect is the fact that each photogenerated electron–hole pair is confined to a small crystalline object, containing a few thousand atoms, and whose surface is very well passivated by Si–H or Si–O bonds. In addition, Auger recombination, which limits the recombination lifetime in bulk c-Si except at low excess carrier density, is absent unless more than one electron–hole pair is injected in each crystallite. Ideally, the only recombination pathway is thus radiative and the efficiency could approach 100%. A dramatic demonstration of this is provided by Wilson, Szajowski, and Brus (1993), where it was shown that isolated and perfectly passivated Si quantum dots can luminesce in the red with 50% quantum efficiency at low temperature. Porous Si is not made of perfectly isolated and passivated quantum dots and thus nonradiative recombination takes place, albeit at a reduced rate. It is thus important that, as predicted theoretically (Hybertsen, 1994), the radiative rate also increases compared to that of c-Si. As the motion of electrons and holes is restricted in real space, they sample a wider region in reciprocal space and both phonon-assisted and phonon-less radiative recombination become more efficient. The measured PL decay time of the order of microseconds to milliseconds is indeed faster than the calculated c-Si radiative recombination time.

What experimental evidence exists to support quantum confinement? There is ample evidence for the presence of nanometer-size crystallites. Simple calculations within the effective mass approximation (Takagahara and Takeda, 1992) or more realistic simulations (Proot, Delerue, and Allan, 1992; Wang and Zunger, 1994) predict that the object size leading to a bandgap in the red should be in the 2 to 4 nm range, depending on whether the crystallites are columnar or dot-like. Transmission electron microscope (TEM) (Cullis and Canham, 1991), X-ray (Lehmann *et al.*, 1993), and Raman (Lockwood and Wang, 1995) measurements have demonstrated the presence of crystallites in the correct size range in red luminescing. Figure 11 is a TEM micrograph of highly porous π-Si showing the presence of an interconnected network of columnar crystallites in the nanometer range (Cullis, 1993). Depending on the preparation conditions of π-Si, one finds either columnar or dot-like crystallites. In the Raman spectrum, the broadening and shift of the optic phonon line near $520\,\text{cm}^{-1}$ can be analyzed to yield a crystallite size. Figure 12 shows how the Raman

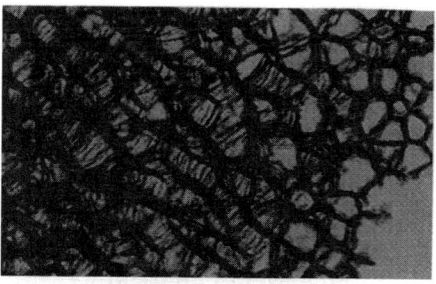
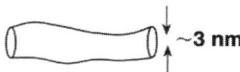

- **Undulating wires of <5-nm diam form a sponge-like material**

- **Crystallinity is preserved**
- **Porosity between 50% and >95%**

Porosity = 92%
(A. Cullis, DRA, UK)

FIG. 11. Transmission electron micrograph of a π-Si sample of 92% porosity. The structure is made of interconnected wires having a diameter of a few nanometers. (From Cullis, 1993.)

spectrum of Si nanocrystallites depends on size (Campbell and Fauchet, 1986) and demonstrate a correlation between PL wavelength and Raman crystal size in π-Si (Lockwood and Wang, 1995). Theory also predicts that the Raman spectra of a spherical crystallite of diameter d and a cylindrical crystallite of the same diameter are distinguishable (Fauchet and Campbell, 1988). Although the evidence is still incomplete, it appears that most Raman spectra of π-Si are consistent with confinement of the optic phonons in a spherical crystallite.

Alternative models were proposed shortly after Canham's report. One model suggested that the PL might originate not from c-Si but from a highly disordered, amorphous Si (a-Si) tissue formed during electrochemistry (George et al., 1992). Strong PL had been reported earlier in a-Si:H alloys containing $\sim 50\%$ of hydrogen (Wolford et al., 1983). Evidence for the presence of a hydrogen-rich a-Si phase in π-Si came from TEM and other measurements that showed a large amount of disordered material in highly luminescent π-Si prepared by stain etching. This model fails, however, because (1) samples with no detectable amorphous phase and yet very strong luminescence have been prepared (Canham et al., 1994), and (2) samples that have been annealed in oxygen at 900°C contain no detectable Si hydride bonds and still luminesce in the same spectral region and with the same efficiency (Petrova-Koch et al., 1992). Another class of models proposed that although the absorption might take place in c-Si nanocrystals, the luminescence takes place in specific bonds or molecules that would be present on or attached to the surface. The most popular of these models was the one in which siloxene was the luminescing species (Brandt et al., 1992). The optical properties of siloxene ($Si_6O_3H_6$) and its numerous related

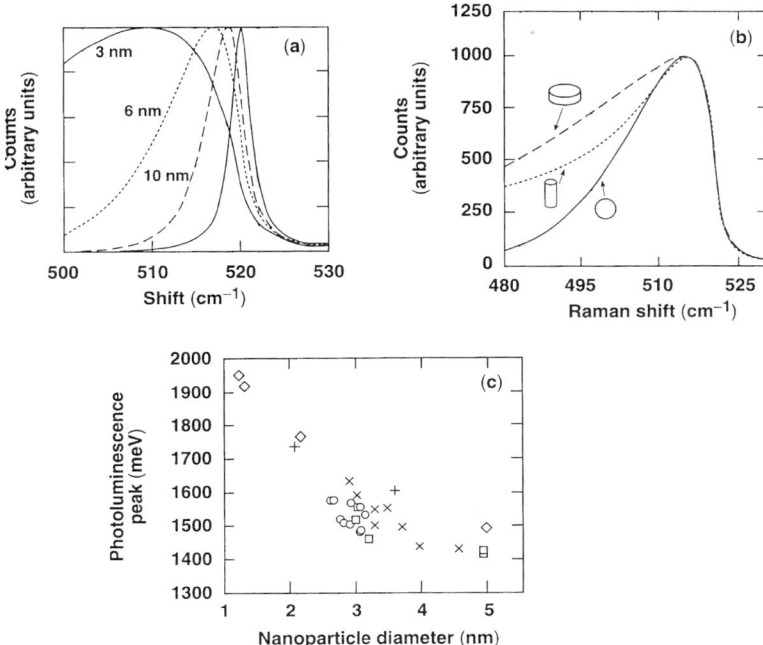

FIG. 12. (a) Calculated Raman spectra for bulk Si and spherical Si nanocrystals of 10 nm, 6 nm, and 3 nm diameter. The shift, broadening and asymmetry change systematically with diameter. (From Fauchet and Campbell, 1988.) (b) Calculated Raman spectra for Si nanocrystallites containing 2000 atoms of different shapes: sphere, cylinder (length = 4 × diameter), and disk (diameter = 4 × length). The difference is only in the low wavenumber tail. (From Fauchet and Campbell, 1988.) (c) Correlation between the PL peak energy and the nanocrystallite size determined by Raman spectroscopy. (From Lockwood and Wang, 1995.)

compounds have indeed many similarities with π-Si. Nevertheless, siloxene is not the luminescing species in π-Si. Samples containing no detectable Si hydride bonds (Petrova-Koch et al., 1992) and samples with no detectable Si oxide bonds (including π-Si kept in HF solution (Wadayama, Yamamoto, and Hatta, 1994)) can be prepared, all of which exhibit strong, red luminescence.

The third class of models involves defects in π-Si. These models are not easily refutable and, in fact, quite possibly explain the PL measured in some samples. One proposal is that the red luminescence is related to defects in Si dioxide, specifically the nonbridging oxygen hole center (NBOHC) (Prokes, 1993). It is well known that SiO_2 can luminesce rather efficiently; in fact, this type of luminescence will be invoked to explain the blue PL. The

NBOHC in glass has been rather extensively studied and shown to produce luminescence near 1.7 to 1.8 eV, the spectral region that is most commonly seen in π-Si. There are two problems with assigning the luminescence in π-Si to this defect: (1) once again, samples containing no detectable oxygen luminesce efficiently, and (2) the red band is tunable from 1.5 eV to past 2.2 eV, which represents too large a tuning range for a specific defect.

Another strong indication that the red PL band is associated with the presence of Si nanometer-size objects comes from the luminescence detected in a wide variety of Si nanostructures. Visible PL has been detected in isolated Si nanocrystals prepared by gas phase techniques (Wilson, Szajowski, and Brus, 1993; Schuppler et al., 1994; Takagi et al., 1990; Saunders et al., 1993), in nanocrystalline Si films (Tamura et al., 1994), in spark-processed Si (Hummel et al., 1993), in lithographically defined Si columns (Fischer et al., 1993; Zaidi, Chu, and Brueck, 1995; Nassiopoulos et al., 1995), in c-Si superlattices (Vervoort et al., 1995), in a-Si superlattices (Lu, Lockwood, and Baribeau, 1995), and in porous a-Si (Bustarret, Ligeon, and Rosenbauer, 1995). All of these different light-emitting forms of Si have in common the fact that at least one dimension of the Si nanostructure is of the order of one to a few nanometers.

2. Si Oxide and the Blue PL Band

A comparison of the infrared absorption and PL spectra of many π-Si samples prepared by different procedures indicates clearly that the intensity of the blue component is directly related to the strength of the infrared absorption near $1080 \, cm^{-1}$ that measures the presence of SiO_2. In fact, in samples where both the blue and red bands are present, it is nearly certain that they are produced at distinct sites in the porous layer. Since the blue band is very broad and overlaps with the red band, it is necessary to exercise caution when analyzing the decay of both bands: in most samples, the small nanosecond component that is sometimes measured in the red band may be due to the tail of the blue band and, vice-versa, the small microsecond component that is sometimes measured in the blue band may be due to the tail of the red band.

The blue PL band may have one of three origins: luminescence from very small c-Si nanocrystals; luminescence from the bulk of the Si dioxide present in samples where blue PL is detected; or luminescence from the surface of the Si dioxide. Strong oxidation implies that a large fraction of the c-Si nanocrystals has been transformed into SiO_2; thus, the average size of any c-Si remnants can be as small as ~1 nm. Theory suggests that for this size, the bandgap should approach 3 eV (Proot, Delerue, and Allan, 1992; Wang and Zunger, 1994) and phonon-less radiative transitions with characteristic lifetimes in the nanosecond, regime should dominate (Hybertsen, 1994). This model, which is consistent with the short PL wavelength and lifetime, has been proposed by Harvey *et al.* (1992), but it cannot explain the chemical quenching experiments performed on oxidized π-Si (Rehm *et al.*, 1995). Because the c-Si remnants are surrounded by an oxide layer that is at least several monolayers thick, the blue PL should be insensitive to the chemical environment. Instead, when blue-emitting oxidized π-Si is exposed to liquid methanol, a reversible quenching of the blue PL is observed (Fig. 13). One can thus exclude the hypothesis that the blue PL would originate from the interior of Si nanocrystallites. It should be noted that when π-Si luminesces in the red and its surface is covered by only few SiO_2 monolayers, exposure to methanol does not quench the luminescence. The protection provided by the SiO_2 layer can explain this lack of quenching. Quenching is observed in red emitting π-Si samples when the surface is passivated with Si–H bonds and the Si nanocrystal is in direct contact with the chemical species.

In the second model, the PL originates in the oxide itself. Si dioxide is known to luminesce efficiently in the visible under appropriate conditions, and it has been shown that blue PL with a nanosecond decay time is observed after excitation by ultraviolet (UV) photons in high-purity, wet synthetic silica (Anedda *et al.*, 1993). The blue PL in oxidized π-Si could then originate from the oxide surrounding the Si remnants or from the

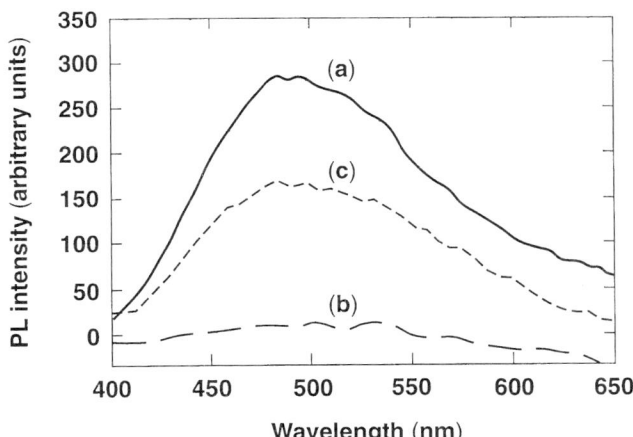

FIG. 13. Photoluminescence spectra of thermally oxidized π-Si: (a) before exposure to methanol, (b) immediately after exposure, and (c) one day later, when most of the methanol has evaporated. (From Rehm *et al.*, 1995.)

SiO_2/Si interface. This model cannot be ruled out, and the observation of a 4.1 eV threshold for the existence of a very long-lived (0.6 sec.) blue PL component may be consistent with this model (Kux, Kovalev, and Koch, 1995).

In the third model, the blue band is not produced by the oxide itself but rather by species present at the surface of the oxide. This model is consistent with the chemical quenching experiments mentioned earlier (Rehm et al., 1995) and with the observation that at least in some cases the intensity of the blue PL increases by more than one order of magnitude after many days of exposure to laboratory conditions, when a large quantity of many different molecular species such as water vapor or carbon may accumulate on the microscopic surface (Loni et al., 1995a). It has been proposed that silanol is the species responsible for the blue luminescence (Tamura et al., 1994) but there remains considerable debate as to its role since blue PL is observed in samples where no silanol bonds are observed (Hummel, 1995).

3. Recrystallization, Dangling Bonds, and the Infrared PL Bands

The PL band near 1.1 μm has all the properties of the PL from c-Si. The insensitivity of the integrated PL intensity to temperature from 10 K to 300 K and the fact that the position and the strength of the zero-phonon and all the phonon lines are identical in π-Si and c-Si are remarkable (Tsybeskov et al., 1996c). The preparation conditions of the samples that display this behavior are consistent with recrystallization of Si in non-stoichiometric, Si-rich Si oxide (Sakamoto et al., 1995). Recrystallization does not necessarily imply melting. However, it is known that the melting point of small semiconductor nanocrystallites is depressed compared to the bulk form. In the case of II–VI nanocrystals, the melting temperature can decrease by more than 1000°C for the smallest clusters (Goldstein, Echer, and Alivisatos, 1992). The effect has not been quantified yet in Si but it has been proposed that, as in II–VI semiconductors, the melting temperature can become extremely low (Goldstein, 1996), well below the temperature at which the π-Si samples were processed. Because the depression of the melting temperature is rooted in the large surface to volume ratio, the magnitude of the effect is expected to depend very strongly on the environment of the nanocrystals. For the samples of Fig. 7, the microcrystals produced during the heat treatment have an average diameter close to 100 nm, which explains the absence of quantum confinement effects. These microcrystals are imbedded in a SiO tissue that provides passivation and carrier confinement (but no quantum confinement). As expected, the processing window to obtain this unusual infrared PL is well-defined (Tsybeskov et al., 1996c).

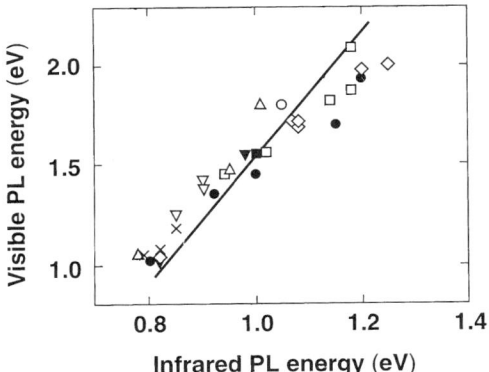

FIG. 14. Compilation of numerous measurements of the red and IR PL peak energies performed in π-Si by different groups. The IR peak energy tracks the red peak energy. The line is the theory of Hill and Whaley (1996). (From Petrova-Koch et al., 1995.)

The tunable infrared band has been associated with dangling bonds (Koch et al., 1993; Mauckner et al., 1995; Petrova-Koch et al., 1995). Figure 14 shows that the peak energies of the red and infrared bands have a linear relationship (Petrova-Koch et al., 1995). This can be explained if the infrared band corresponds to recombination of a near band edge electron at a dangling bond whose absolute energetic position is insensitive to size. The PL shift with size should then be proportional to ΔE_c, the conduction band quantization energy. If the red PL corresponds to recombination of an electron with a hole near their respective band edges, the PL shift with size is expected to be proportional to $\Delta E_c + \Delta E_v$, where ΔE_v is the valence band quantization energy. Theory suggests (Ren and Dow, 1992) and experiments confirm (van Buuren et al., 1993; Rehm et al., 1996) that, in π-Si, $\Delta E_v = 2\Delta E_c$ and thus the infrared band shift with size should be one third of the red band shift, in agreement with the data. Figure 14 also compares recent theoretical predictions (Hill and Whaley, 1996) with the observed relation between IR and red PL. From these calculations. it is concluded that the IR PL is due to recombination of an electron with a trapped hole (and not a hole with a trapped electron) and that the dangling bond trap energy is ~0.3 eV above the bulk Si valence band edge. Dangling bonds are usually associated with nonradiative recombination. However, theoretical investigations (Delerue, Allan, and Lannoo, 1993) of radiative recombination involving dangling bonds suggest that it is possible in nanocrystals and even predict a lifetime that is not inconsistent with the observed decay (Mauckner et al., 1995).

IV. Pure Quantum Confinement and Surface States: A Critical Discussion

In Section III.1, it was concluded that Si nanocrystals in which carriers were quantum confined were responsible for most of the red PL. What was not established in that section is whether the luminescence peak energy coincides with the bandgap energy. If recombination involves free carriers, the PL energy should equal the bandgap minus the binding energy of the exciton. On the other hand, if the PL energy is within the bandgap, one or both carriers are trapped in defect states, most likely on the surface of the nanocrystals. At first glance, the problem is easily resolved by comparing the PL energy and the bandgap in the same sample. The results of such a comparison are inconclusive. The nanocrystals present in π-Si are not all of the same size; a decrease in diameter by 1 nm in the size regime of 2 to 5 nm leads to a blueshift of the bandgap by hundreds of meV. As a result, the larger crystallites dominate the absorption spectrum and determine the absorption edge (von Behren *et al.*, 1995a; von Behren *et al.*, 1995b). In contrast, radiative recombination comes preferentially from the smaller nanocrystals, which are less likely to contain a nonradiative center (Xie *et al.*, 1994). As a result, the PL is dominated by the smaller crystallites. Thus, the PL and absorption measurements do not sample the same fraction of the population and comparing them does not answer the question of the origin of the PL. Figure 15(a) shows an extreme case where the PL spectrum remains unchanged despite the fact that the absorption spectrum is strongly blue shifted when the crystallite size deduced from Raman measurements changes from 9 nm to 2 nm (Kanemitsu *et al.*, 1993). Figure 15(b) shows that the nearly exponential absorption spectrum often observed in π-Si (Koch *et al.*, 1993b) and sometimes attributed to the presence of band-tails, as in a-Si, may be entirely explained by the nanocrystallite size distribution.

When the photogenerated carriers are produced by light tuned well above the bandgap, all the nanocrystals are excited and the PL peak is broad and featureless. If instead the luminescence is excited at low temperature by a laser tuned inside the broad PL spectrum, only the crystallites larger than a critical size are excited. Under these conditions of size-selective excitation, the PL intensity is strongly reduced but, as shown in Fig. 16, the PL spectrum shows discrete steps that coincide with the zone-edge phonons of c-Si that are known to be involved in the bandgap absorption and emission in c-Si (Calcott *et al.*, 1993; 1995). These results not only indicate that Si is the absorbing and luminescing species in π-Si under these excitation conditions but also strongly suggest that the efficient, broadband luminescence observed with well above gap excitation occurs at the Si crystallite

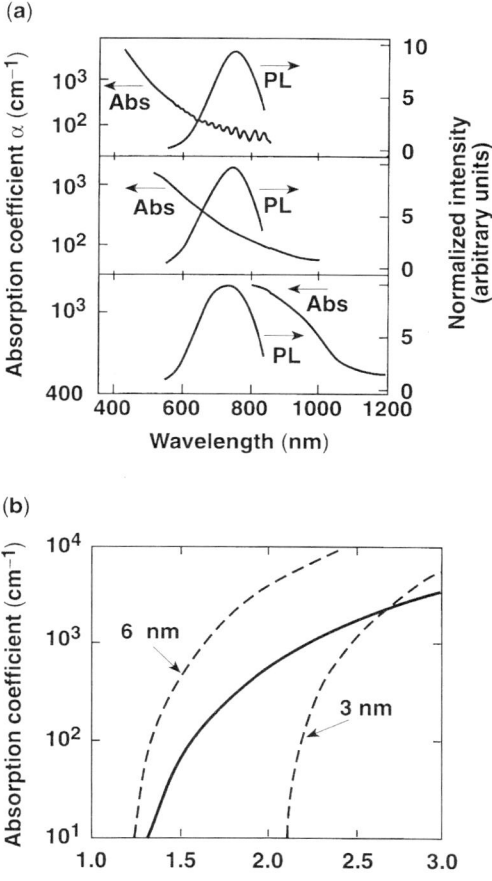

FIG. 15. (a) Absorption and PL spectra measured on π-Si films having an average crystallite size determined by Raman spectroscopy of 2 nm, 3.5 nm, and 9 nm (top to bottom). (From Kanemitsu *et al.*, 1993.) (b) Calculated absorption spectrum for π-Si made of a uniform distribution of nanocrystals in the range of 2 to 6 nm, together with the calculated absorption coefficients for nanocrystals of 3 and 6 nm.

bandgap. This experiment supports the "pure" quantum confinement model. It is worth noting at this point that the distinction between direct and indirect bandgap becomes blurred as the crystallite size decreases. A reasonable criterion is whether the matrix element for an optical transition is larger with or without phonon assistance. With this criterion, Fig. 16

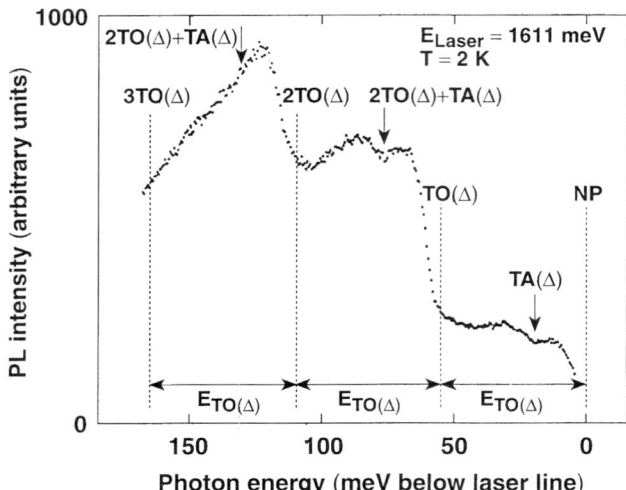

FIG. 16. Photoluminescence spectrum taken at 21 K in π-Si excited with a laser tuned inside the broad red band. The steps in the emission spectrum are correlated with the participation of bulk c-Si phonons in aborption, emission, or both. (From Calcott et al., 1993.)

indicates that π-Si remains an indirect gap semiconductor, in agreement with theoretical predictions (Hybertsen, 1994; Hill and Whaley, 1996).

A totally different kind of experiment, femtosecond photoinduced absorption/bleaching pump-probe measurement, also shows that π-Si remains an indirect gap semiconductor. A femtosecond laser pulse (the "pump") tuned above the bandgap of the semiconductor injects electrons and holes. The change in transmission produced by these photoinjected carriers is then measured with another femtosecond laser pulse (the "probe"), synchronized with the pump pulse and, in the present experiments, having the same photon energy. In the case of a direct gap semiconductor like GaAs, bleaching is observed because the final state is partially occupied, the initial state is partially empty, and the two states are dipole-coupled (Young et al., 1994). In the case of an amorphous semiconductor like a-Si:H (Fauchet et al., 1992) or an indirect gap semiconductor like Si (Fauchet and Nighan, 1986; Bambha et al., 1988), intraband or free carrier absorption dominates and induced absorption is observed. Figure 17 compares the femtosecond response of π-Si with that of GaAs, c-Si, and a-Si:H (Fauchet et al., 1995; von Behren et al., 1996). It is clear that π-Si is not a direct gap semiconductor and furthermore that it behaves in these experiments very much like c-Si and not like a-Si:H. A detailed interpretation of the data of Fig. 17 can be found in von Behren et al. (1996).

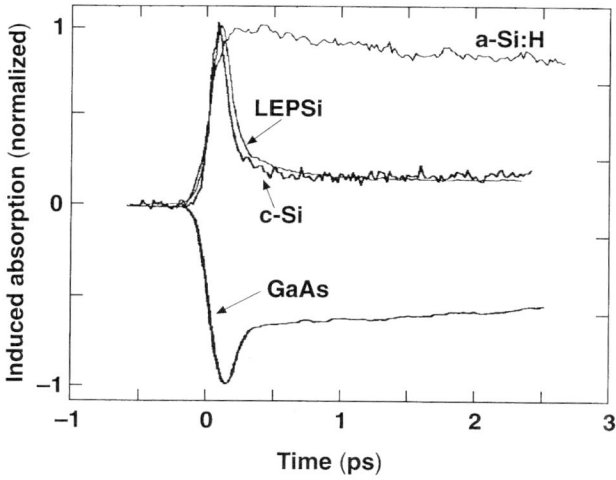

FIG. 17. Normalized photoinduced absorption measured on thin films of a-Si:H, c-Si, π-Si (LEPSi), and GaAs, when the pump and probe pulses are at 2 eV. (From Fauchet et al., 1995.)

If the PL in π-Si not only requires quantum confinement but also involves surface states in the radiative process, changes in the surface chemistry are expected to lead to modifications in the PL properties. This hypothesis was tested by an experiment in which the surface chemistry of identical samples was controlled by various treatments, including annealing in air for 5 min at various temperatures (Tsybeskov and Fauchet, 1994). As Fig. 18 shows, the PL peak energy increases as the ratio of Si–O bonds to Si–H bonds increases. These results obtained with nanocrystallites having the same nominal size demonstrate a correlation between surface chemistry and PL peak wavelength. For a particular surface coverage of the nanocrystallites, a specific set of surface states is expected. When the surface coverage changes, the nature and the energetic position of the surface states present within the bandgap, and thus the PL peak energy, should change. In this model, the wide PL spectrum may result at least in part from recombination involving different surface configurations.

A direct comparison between bandgap and PL energies can also be attempted by other methods. One experiment consists of quenching the PL of π-Si by exposing it to electron or hole traps. This was accomplished by immersion in benzene solution containing a molecule having a known redox potential (Rehm, McLendon, and Fauchet, 1996a; 1996b). For molecules acting as electron (hole) traps, electron (hole) transfer from the Si nanocrystallites to the quencher molecule occurs only if the redox level lies below

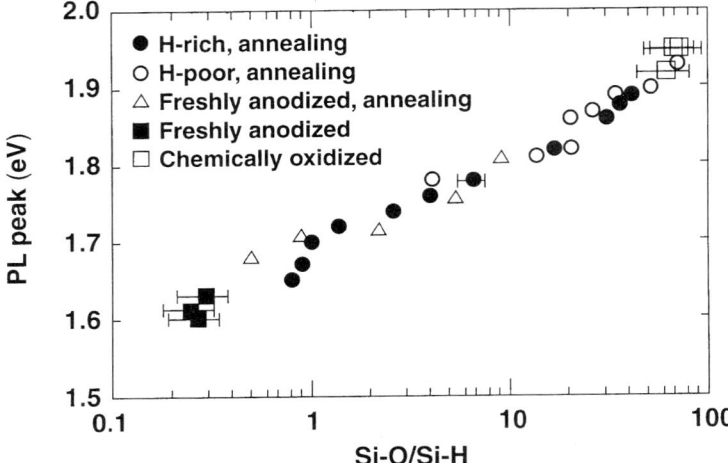

FIG. 18. The PL peak energy increases when the nanocrystallite surface coverage changes from Si–H to Si–O. The ratio of Si–O bonds over Si–H bonds, determined by IR absorption measurements, was not corrected for differences in absorption strength. (From Tsybeskov and Fauchet, 1994.)

(above) the conduction (valence) band edge. When one carrier is transferred to the quencher, the PL intensity decreases dramatically. Figure 19 shows the quenching for electron traps: When the redox potential reaches a value of -1.20 V, a dramatic change in the quenching ability of the adsorbate occurs. For π-Si samples having a peak PL energy of 2 eV, the conduction band edge increased by 0.5 eV and the valence band edge by 1 eV compared to bulk c-Si. Thus, the bandgap of these samples was 2.6 eV, while the PL was at 2 eV. Because the PL itself is used to determine the bandgap, the emitting and absorbing sites should be the same.

This Stokes shift between the bandgap and PL energies is consistent with data obtained by photoelectron spectroscopy (PS) to deduce the bandgap of π-Si (van Buuren *et al.*, 1993; van Buuren *et al.*, 1995). In these experiments, the PS measurements probe a thin (~ 10 nm) layer near the macroscopic surface of the π-Si layer, while the PL probes a ~ 1 μm-thick layer, whereas in the experiment of Fig. 19, the PL itself is used to deduce the bandgap and thus the PL and bandgap are deduced from measurements on the same nanocrystallites. Figure 20 shows that quantum confinement decreases the valence band edge twice as much as it increases the conduction band edge (van Buuren *et al.*, 1993), in agreement with theory (Ren and Dow, 1992). Figure 21 is a compilation of several experimental and

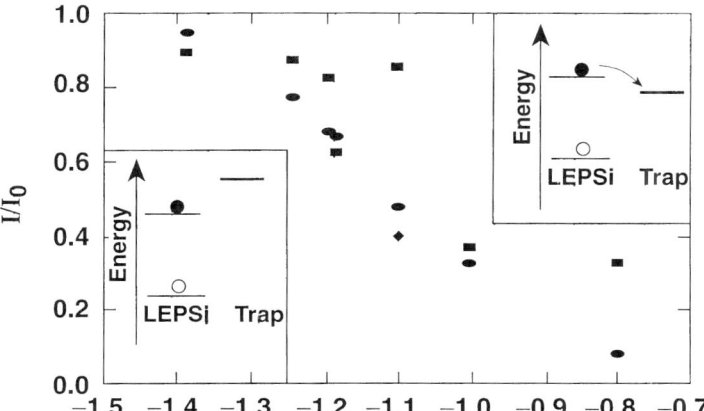

FIG. 19. Ratio of the PL intensity after and before exposure of π-Si to conduction band quenchers (CBQ) versus redox potential of the CBQ. The π-Si samples are prepared from n-type Si with light assistance and emit at 2 eV. The solvent used in the quenching experiments is benzene. The insets show the relative position of the conduction band edge and the electron trap energies when quenching is and is not observed. (From Rehm et al., 1996a.)

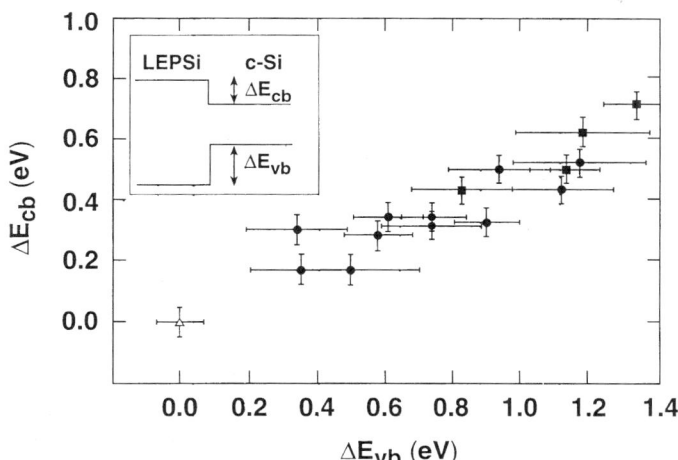

FIG. 20. Increase in the conduction band energy versus increase in the valence band energy for different π-Si samples. The effect of quantization is twice as large in the valence band as in the conduction band. (From van Buuren et al., 1993.)

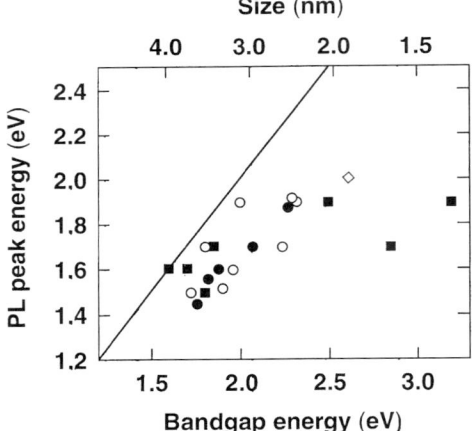

FIG. 21. Photoluminescence peak energy versus bandgap energy in π-Si. The open circles are from van Buuren et al. (1995), the solid diamond from Rehm et al. (1996b), the solid circles from Kux and Ben-Chorin (1995), and the solid squares from Saunders et al. (1993); Schuppler et al. (1995); Littau et al. (1993) for the PL energy and Wang and Zunger (1994) for the bandgap energy. If the PL and bandgap energies were the same, the data would fall on the solid line. An early version of this graph appears in Rehm et al. 1996b.

theoretical results and shows that the Stokes shift between bandgap and PL energies increases with decreasing size. The datum obtained by using the quenching experiment is shown in Fig. 21. Other data are obtained by two other methods: (1) The PL peak energy measured for Si nanocrystallites of a known size (Saunders et al., 1993; Schuppler et al., 1995; Littau et al., 1993) is plotted versus the bandgap energy calculated according to Wang and Zunger (1994) for quantum dots of the same size. (2) The PL peak energy measured in π-Si is plotted versus the bandgap extrapolated from PL excitation spectra of the same samples (Kux and Ben-Chorin, 1995). For sizes above 3 nm, the difference between bandgap and PL energies remains below 300 meV, a value that could be attributed to the exciton binding energy in small crystallites (Ohno, Shiraishi, and Ogawa, 1992); however, below 3 nm, the difference appears to be too large (up to 1 eV and more) to be explained by the expected increase in the free exciton binding energy.

Although considerable experimental uncertainty remains in determining unequivocally the bandgap and PL energies of Si nanocrystallites and despite continuing theoretical discussions (e.g., the calculated dependence of the bandgap on size of Hill and Whaley (1996) is much weaker than that calculated in Proot, Delerue, and Allan (1992); and Wang and Zunger (1994), it seems likely that the luminescence does not coincide with the bandgap and thus involves surface states. If this is true, then the nature of these surface states remains to be elucidated. A step in that direction may have been taken recently, when it was proposed, based on theoretical calculations, that recombination from self trapped excitons at the surface of Si nanocrystallites may explain the very large Stokes shift seen in Fig. 21 (Allan, Delerue, and Lannoo, 1996). This recent theory appears also to explain why, under resonant excitation conditions, no Stokes shift is observed (neglecting the very small exchange splitting (Calcott *et al.*, 1993)).

V. Nonoptical Properties

1. INTRODUCTION

The optical properties of π-Si, in particular its bright luminescence, have been the subject of numerous investigations. In contrast, the nonoptical properties of π-Si have not been so thoroughly studied despite their fundamental and technological importance. Porous Si can be viewed as a network of interconnected quantum wires or a collection of quantum dots imbedded in a matrix such as Si dioxide. The chemical, electrical, and structural properties of these classes of objects are expected to be quite different from those of bulk semiconductors. For example, the melting point of semiconductor nanocrystals is lower than that of their bulk counterpart (Goldstein, Echer, and Alivisatos, 1992) and the magnitude of the decrease should be dependent on the surface environment. Some work has already been performed on the chemical sensitivity of π-Si (Rehm *et al.*, 1995; Lauerhass *et al.*, 1992; Lee, Ha, and Sailor, 1995; Zhang *et al.*, 1995). This section discusses two nonoptical properties of interest for electroluminescent devices made of π-Si. First, the electrical properties of π-Si and π-Si devices are investigated and some numbers derived such as the carrier mobility. Second, the mechanical properties of π-Si are briefly discussed with emphasis on hardness measurements.

2. ELECTRICAL PROPERTIES

Consider a simple metal/π-Si device, where the p-type Si substrate has a resistivity of 5 to 10 Ω.cm, the π-Si layer has a porosity of 80%, and the metal is a thin, semitransparent gold contact layer. A good ohmic contact is made to the backside of the wafer. The current density (J)-voltage (V) characteristic of the device shown in Fig. 22 (Peng *et al.*, 1995; Peng, Hirschman, and Fauchet, 1996) is rectifying and follows, in forward bias, a

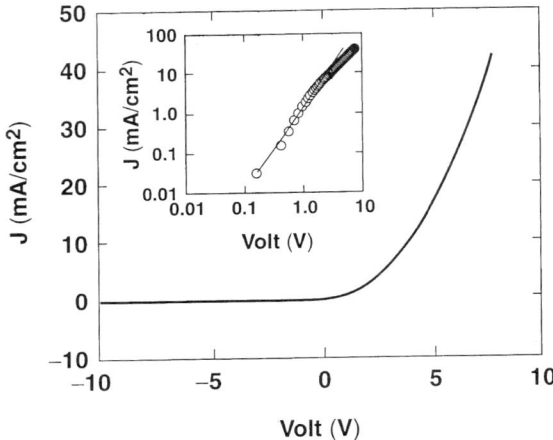

FIG. 22. J-V relation of a Au/π-Si/c-Si device. The inset shows that J increases as V^2 (solid line). (From Peng, Hirschman and Fauchet, 1996.)

power law relationship $J = KV^m$ where $m \approx 2$, which is typical of a space charge limited current. The current is dominated by carrier transport in the high resistivity ($\sim 10^{10}\, \Omega.\text{cm}$) π-Si layer and the device can be modeled as an insulator sandwiched between two conductors, with no measurable band bending at the two interfaces (Peng, Hirschman, and Fauchet, 1996). When the injected carrier concentration exceeds the thermal carrier concentration, these devices no longer follow an ohmic behavior and for double carrier injection, the J-V relationship is given by

$$J = \varepsilon\varepsilon_0 \mu_{\text{eff}} V^2/d^3 \qquad (2)$$

where $\varepsilon\varepsilon_0$ is the dielectric function of π-Si (in F/m), d is the thickness of the π-Si layer, and μ_{eff} is the effective carrier mobility. The dependence of J on both V and d has been measured to satisfy Eq. (2) (Peng, Hirschman, and Fauchet, 1996).

Although Eq. (2) can in principle be used to extract the carrier mobility, a better method consists of measuring the response of the EL to a small alternating current (AC) bias in a device held under a large forward bias (Peng et al., 1995). Since there is no measurable band bending at the metal/π-Si interface, the injected carriers drift through the π-Si layer where they recombine radiatively. The transit time through the device (τ_t) is given by

$$\tau_t = d^2/\mu_{\text{eff}} V \qquad (3)$$

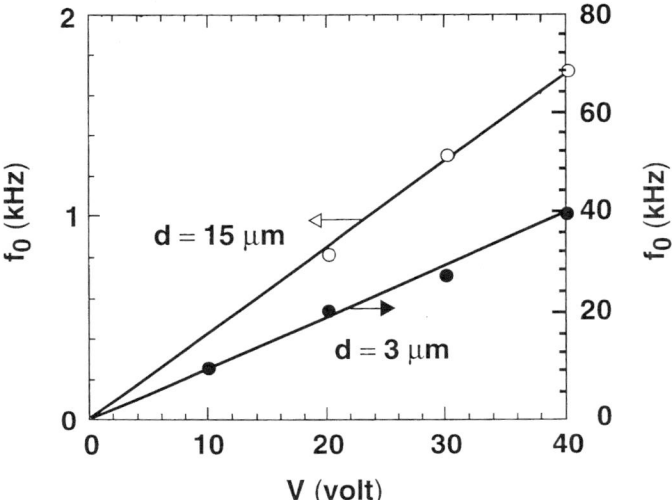

FIG. 23. Scaling of the −3 dB frequency f_0 with the applied voltage and the thickness of the π-Si layer. The solid lines are obtained from Eq. (3) if $\mu = 10^{-4}$ cm^2/V sec. (From Peng and Fauchet, 1996.)

When the modulation frequency of the AC bias becomes large, the electric field reverses itself in a time shorter than the transit time and the magnitude of the AC EL decreases. Figure 23 plots the −3 dB frequency f_0 (the frequency at which the magnitude of the AC EL drops by 50%) versus direct current (DC) bias for two devices with a π-Si layer of 3 and 15 μm, respectively (Peng and Fauchet, 1996). Other measurements have confirmed that f_0 scales with d^{-2} and with V (Peng, Hirschman, and Fauchet, 1996). Thus, f_0 is equal to the inverse of τ_t and the effective mobility obtained from these measurements is 10^{-4} cm^2/Vs.

In an effort toward designing efficient and fast π-Si LEDs, several groups have anodized a c-Si p-n junction (Peng et al., 1995; Peng and Fauchet, 1995; Steiner et al., 1993). It was proposed that after anodization a p-n junction can actually be formed inside the π-Si layer. The J-V and the capacitance-voltage (C-V) characteristics of these devices support this proposal (Peng, Hirschman, and Fauchet, 1996). Figure 24 shows that the J-V curve of such a π-Si LED starts as an exponential at low voltage, with an ideality factor of 2.1. This is in contrast to the J-V curve of the metal/π-Si devices and is consistent with the presence of a junction inside the π-Si layer. In contrast to the metal/π-Si devices, the C-V curves of the device of Fig. 24 are dependent on the value of the reverse bias voltage, which also suggests

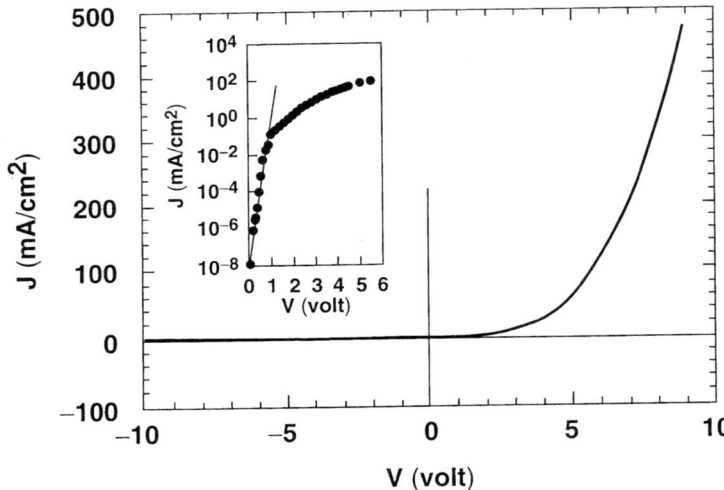

FIG. 24. J-V relation of a π-Si p-n junction device. The inset shows that for $V < 1$ V, $J = J_0 \exp\{eV/nkT\}$ with the ideality factor $n = 2.1$. (From Peng, Hirschman and Fauchet, 1996.)

that a p-n junction is present. Finally, the large deviation from the exponential behavior at larger forward bias in the J-V curve is caused by the carrier diffusion current and the space charge limited current.

3. Structural Properties

The structural properties of π-Si have been investigated by a wide array of techniques such as TEM, atomic force microscopy (AFM), scanning electron microscopy (SEM), X-ray scattering, etc. Red π-Si has a very open structure, made of an interconnected network of thin wires or even of isolated Si dots imbedded in an oxide matrix. The poor mechanical strength of π-Si has been an argument used to reject its use in devices. In this section, the dependence of the microhardness of π-Si on porosity is investigated. Hardness is an important mechanical property of any solid and is usually understood to mean resistance to local deformation (McColm, 1990). It is also a convenient, economical, and locally destructive characterization technique, which can provide useful information regarding the mechanical stability, crystallite size, and surface quality of π-Si (Duttagupta et al., 1997).

The Vickers test for microhardness was performed as a function of porosity and substrate doping in π-Si films that did not undergo any special processing step after anodization. A square based pyramidal indenter

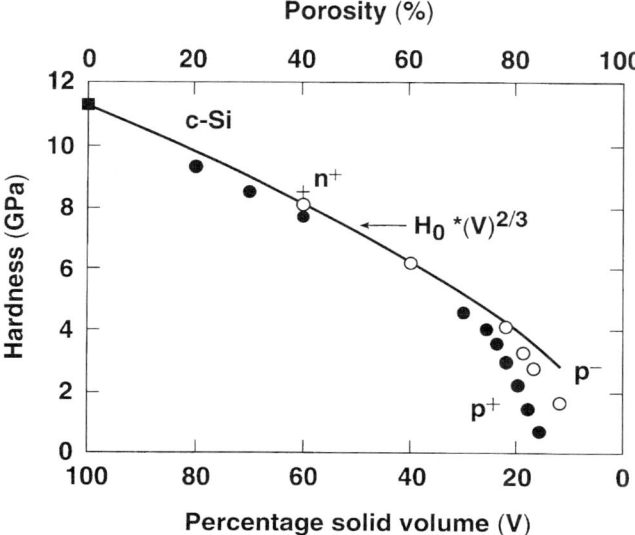

FIG. 25. Hardness versus porosity for π-Si produced by anodization of p⁻, p⁺, and n⁺ substrates. The solid line shows that the hardness decreases as $V^{2/3}$, where $V(\%) = 100-$Porosity(%) is the percentage solid volume. (From Duttagupta et al., submitted.)

produces a diamond shaped indent on the surface. The hardness is determine from the applied load and the diagonal of the indent (Vingsbo et al., 1986). The variation of hardness with percentage solid volume V ($V = 100-P$, where P is the porosity) is shown in Fig. 25 for p⁻ and p⁺ films (the mean hardness for the n⁺ sample was 8.5 GPa). A steady decrease from the hardness of (100) c-Si ($H_0 = 11.5$ GPa) was observed with an increase in porosity. The fitting of the data shows a $(V)^{2/3}$ dependence at lower porosities. At porosities greater than 75–80%, a sharp decrease in hardness is noticeable, and there is a considerable deviation from the $(V)^{2/3}$ proportionality. This observation agrees with small-angle scattering X-ray studies of π-Si that indicated at high porosities a large increase in pore size, as the columns become stand-alone structures and the pores coalesce (Vezin et al., 1992).

The low hardness of high porosity π-Si films (almost an order of magnitude lower than single crystal Si) must be considered in the choice of π-Si and in the design and processing of π-Si devices. It is expected that oxidized films will have a higher porosity. Another possibility would be to form a π-Si nanocomposite by filling the pores with an inert material, as shown by Duttagupta et al., 1997).

VI. Electroluminescent Devices

1. General Survey

Electroluminescence was observed in π-Si shortly after Canham's report of strong room-temperature PL, first in solution during anodic oxidation (Halimaoui et al., 1991) and then in a solid-state device (Richter et al., 1991). Many different device structures have been demonstrated and strong (0.1–1% efficiency), voltage-tunable EL has even been reported under cathodic polarization of n-type π-Si using liquid contacts (Fig. 26) (Bsiesy et al., 1993). This section is exclusively concerned with solid-state devices, because of their technological importance and because liquid contacts seem to be inherently unstable. A typical π-Si LED consists of a transparent or semitransparent contact (metal, ITO or conducting polymers) and a 1 to 10 μm thick π-Si layer on a c-Si substrate (p- or n-type) (Richter et al., 1991; Kalkhoran, Namavar, and Maruska, 1992; Bassous et al., 1992; Koshida and Koyama, 1992; Koshida et al., 1993). Typical threshold conditions for EL have been reported to be $V \sim 10$ V and $J \sim 10$ mA/cm^2 although the best numbers are now ~ 2 V and $\ll 1$ mA/cm^2 (Loni et al., 1995b). Until 1995, π-Si LEDs had a low EL external quantum efficiency ($\leqslant 0.01\%$) and degraded irreversibly in one hour of operation or less. Recent results to be discussed below have renewed the prospect for commercial π-Si LEDs.

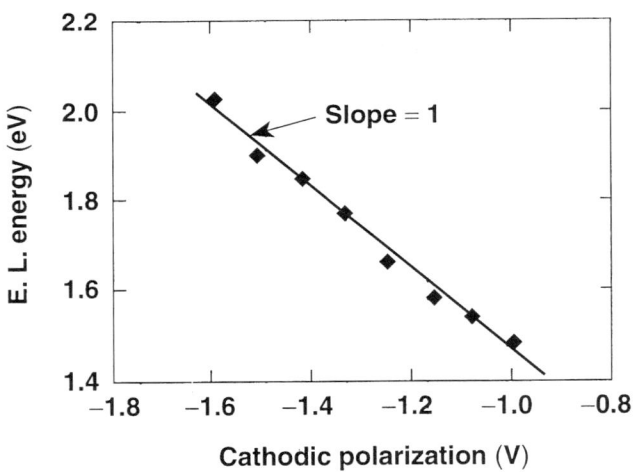

Fig. 26. Tuning of the EL peak energy with cathodic bias in a liquid junction device made of π-Si in contact with a 0.2 M (NH$_4$)$_2$S$_2$O$_8$ solution. (From Bsiesy et al., 1993.)

Below, we address in some detail several specific questions that must be answered satisfactorily before a real technology emerges from π-Si:

1. can π-Si LEDs remain stable for weeks to years?
2. can the power efficiency reach 1%?
3. can they be modulated at high frequencies?
4. can they emit throughout the visible spectrum and/or at appropriate wavelengths in the infrared?
5. can they be compatible with standard microelectronic processing?

2. LED LIFETIME

The stability of most π-Si LEDs is poor: for example, the record continuous-wave efficiency LEDs reported in Loni *et al.* (1995b) degrade within minutes in air and hours in moderate vacuum. The cause of this degradation is that the Si–H bonds that passivate the Si nanocrystal surfaces are very fragile and can be easily broken by exposure to light, ambient air, moderate temperatures, and large electric fields (Robinson *et al.*, 1992; Collins, Tischler, and Stathis, 1992). The temperature of the π-Si layer in a device driven well above EL threshold can reach $\sim 100°C$ and the local electric field can be well in excess of the typical macroscopic field of 10^4 to 10^5 V/cm. Under these conditions, the Si–H bonds can be broken relatively easily, and rapid degradation follows. To improve the stability, the fragile Si–H bonds should be replaced by the stronger Si–O bonds (Tsybeskov, Duttagupta, and Fauchet, 1995) and the devices should be engineered to provide better heat sinking and greater mechanical stability (Tsybeskov *et al.*, 1996a; 1996b). When these improvements are implemented, no degradation is observed after several weeks of operation under ambient conditions.

It has been shown that the PL stability is improved when the π-Si layer is partially oxidized (Tsybeskov *et al.*, 1993). Partial oxidation means that the Si–H bonds have been removed and the surface of the Si nanocrystals is now covered by one or a few SiO_2 monolayers. This material, unlike π-Si passivated by Si hydride bonds or the fully oxidized, blue-emitting π-Si, does not lose its PL when immersed in methanol (Rehm *et al.*, 1995). Because the oxide layer is thin enough, electronic transport from crystallite to crystallite is not compromised. Simple semitransparent metal/partially oxidized π-Si/c-Si LEDs were found to be stable over 100 h of continuous operation (Tsybeskov, Duttagupta, and Fauchet, 1995), but were relatively inefficient (power efficiency $<0.01\%$).

The quality of the oxide and the Si/SiO_2 interface has been improved by using elevated temperatures (Tsybeskov *et al.*, 1996a; 1996b). The temperature treatment takes place in a diluted oxygen atmosphere or in nitrogen to

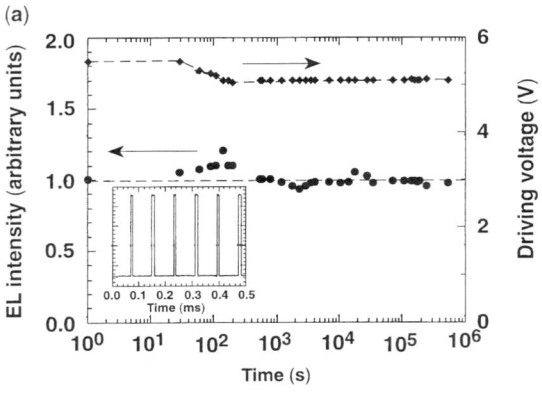

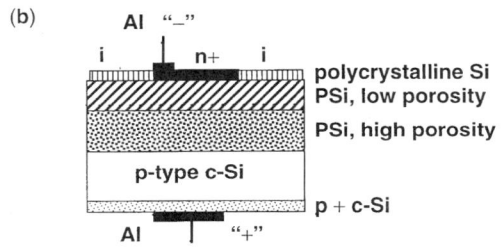

FIG. 27. (a) Stability of the EL and the applied voltage under steady-state pulsed current excitation. (b) Schematic of the device under test in Fig. 27(a). (From Tsybeskov et al., 1996b.)

slow down the conversion of Si into SiO_2 and reduce the oxide thickness to a few monolayers. Furthermore, the device structures have been modified to provide better heat sinking by inserting a low porosity Si layer and a doped μc-Si layer between the π-Si region and the top metal contact. Figure 27 shows the results of stability tests, performed under pulsed drive conditions well above EL threshold, and a schematic of the device. No degradation is seen for several weeks. As discussed below, the power efficiency of the best π-Si LED of this type is near 0.1%. Therefore, the stability of the π-Si LEDs is no longer an obstacle, although commercial devices must show stability over periods of 1,000 to >10,000 h depending on the application.

3. POWER EFFICIENCY

The efficiency of π-Si LEDs quickly increased from $\sim 0.0001\%$ to $\sim 0.01\%$ mostly through better device engineering and processing such as the use of a p-n junction in π-Si (Steiner et al., 1993; Lang, Steiner, and Kozlowski, 1993). Why was the efficiency of these devices still two to three orders of magnitude below the typical quantum efficiency measured in PL? There are two major reasons for this: the difficulty in injecting carriers from the contact into the π-Si and the poor transport properties in π-Si. Making a good quality electrical contact to a highly porous structure such as π-Si is difficult. Evaporation is a line-of-sight technique that produces contacts only with the top nanostructures. Electropolymerization to form an intimate contact has also been successful (Koshida et al., 1993), but the efficiency of the LEDs was not improved. In 1995, the power efficiency of π-Si LEDs was increased to $>0.1\%$ (Loni et al., 1995b; Tsybeskov et al., 1996a; Linnros and Lalic, 1995; Simons et al., 1996) and the internal quantum efficiency, defined as the ratio of the number of photons generated inside the π-Si layer to the number of carriers injected inside the π-Si layer, was estimated to be $>1\%$ [129], which is comparable to the PL efficiency.

The first published report of high efficiency was in the pulsed mode of operation only (Linnros and Lalic, 1995) and there were severe limitations on the repetition rate of the device and its lifetime. A device that was operational in the continuous wave mode of operation and had a power efficiency of 0.1% for a relatively narrow window of processing parameters was reported in Loni et al. (1995b) and Simons et al. (1996). The threshold voltage (2 V) and current density (1 μA/cm^2) of this device are much lower than in all previous reports. The high efficiency could be maintained only for low input electrical power, which means that these LEDs were not bright (the maximum spectrally integrated intensity was $<10\,\mu$W/cm^2), and for a period of time from minutes to 1 h, after which irreversible degradation took place. The recent π-Si LED structures of Fig. 27 also have a power efficiency $\sim 0.1\%$ but they are capable of emitting $\sim 1\,$mW/cm^2 of red light for weeks with no detectable degradation. Electron injection takes place from a heavily doped μc-Si film into a low-porosity Si layer and then into the π-Si layer. Hole injection takes place from the p-type Si wafer directly into the π-Si layer. The thickness of the π-Si layer is kept at $\sim 1\,\mu$m to minimize losses during transport. As will be discussed later, these LEDs are processed under semiconductor fabrication line compatible conditions, including high temperature treatments. Taken together, the results are very promising and suggest that EL can become as efficient as PL provided that greater attention is paid to the device engineering.

4. Response Time

The response time of π-Si LEDs is an important parameter that determines their suitability in applications such as optical interconnects and real-time optical signal processing. The first test consists of measuring the

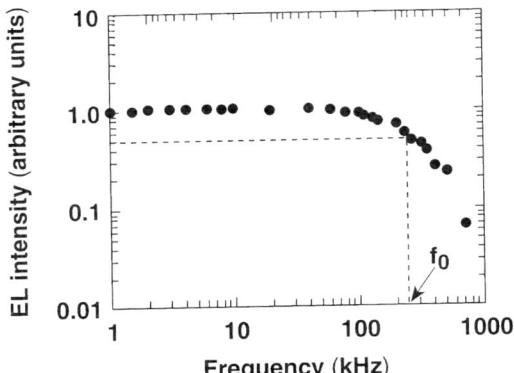

FIG. 28. Amplitude of the AC EL versus modulation frequency for a π-Si p-n junction device. The junction depth is 0.2 μm and the π-Si layer was partially oxidized by a dip in HNO_3. The 3-dB frequency is 230 kHz. (From Peng and Fauchet, 1995.)

frequency response in the small signal limit. The device is held at a large forward bias, well above threshold for detectable EL. A small AC voltage is applied, which modulates the EL. The amplitude of the AC EL is monitored as a function of frequency, as already mentioned in Section V.2. Figure 28 shows that the magnitude of the AC EL signal remains constant up to a critical frequency beyond which it drops quickly. The 3-dB frequency, defined as the frequency at which the AC EL has dropped by a factor of 2 and found in Section V.2 to be the inverse of the transit time, determines the maximum frequency response of the diode in this small-signal analog test. Thus, the response time of the EL is limited by the time it takes carriers to cross the π-Si layer, and not the PL lifetime. By making the π-Si layer thinner than 1 μm or by using a π-Si p-n junction, the 3-dB frequency can approach 1 MHz when it is limited by the luminescence lifetime. Clearly, the small-signal AC modulation speed of these π-Si LEDs cannot approach 1 GHz, unless they can be made using blue-emitting π-Si, whose PL lifetime is ~1 ns. Although blue-emitting π-Si LEDs have been demonstrated, no modulation results have been reported.

Another important mode of operation for LEDs is the pulsed mode of operation, in which a voltage pulse is applied to the device to produce pulsed EL. Figure 29 shows the current pulse and the measured EL on a metal/π-Si/c-Si LED structure (Lazarouk et al., 1996). The 10% to 90% rise and fall times of similar devices are usually below 100 ns. In a very encouraging result obtained using oxidized π-Si with a measured PL lifetime of ~30 ns, the rise and fall times were as short as 20 nsec (Wang et al., 1994).

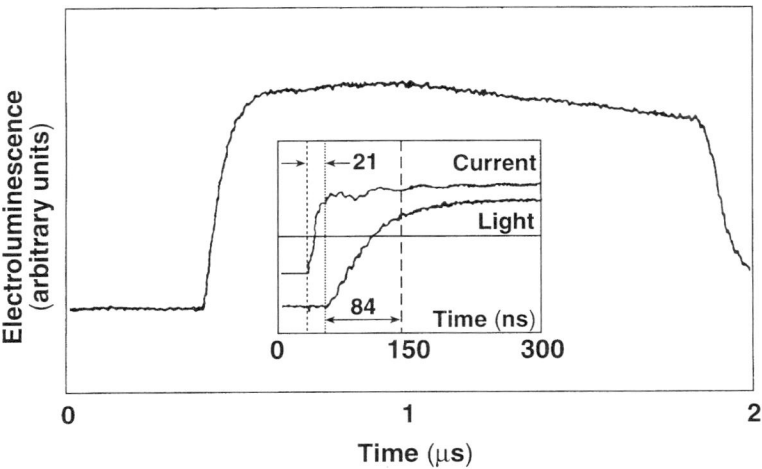

FIG. 29. Response of the EL of an Al/π-Si/c-Si LED driven by a rectangular current pulse of 500 mA. The delay time is 21 ns and the rise time is 84 ns. (From Lazarouk et al., 1996.)

A delay is often observed between the onset of the EL and that of the current pulse. This delay, which was as long as 1 μsec for large bias in Wang et al. (1994), may correspond to the charging of the traps that are present in the oxide or at the oxide/π-Si interface. If this interpretation is correct, it should be possible to decrease and perhaps eliminate this delay by using better processing techniques, as shown in Fig. 29, where the delay was reduced to 21 ns.

5. SPECTRAL COVERAGE

Since the PL can span the spectrum from the infrared to the blue, it is reasonable to expect that π-Si LEDs can cover the same range of wavelengths. As early as 1992, LEDs with peak wavelengths ranging from the deep red to the blue were fabricated (Fig. 30) (Steiner et al., 1993). The EL efficiency of these devices was between 0.005 and 0.01%. Interestingly, the blue LEDs, which used an indium top contact, had an incubation time (burn-in time) of tens of seconds, which may indicate that a reaction took place between the indium metal and the oxide layer, perhaps leading to the formation on a transparent and conducting ITO layer. Progress in making green and blue LEDs continues, for example through the deposition of selected metals in the structure (Steiner, Kozlowski, and Lang, 1995) or the

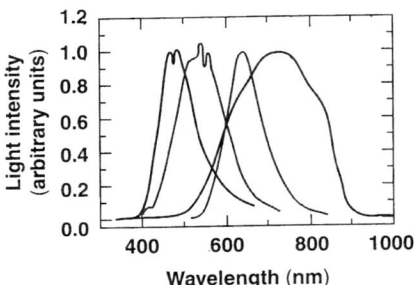

FIG. 30. Electroluminescence spectra of four Au/π-Si/c-Si LEDs prepared under different conditions. (From Steiner *et al.*, 1995.)

use of other porous semiconductors such as SiC (Mimura, Matsumoto, and Kanemitsu, 1995). Thus, π-Si LEDs cover the three primary colors and are good candidates for color display applications.

Other LEDs with a much broader EL spectrum than PL spectrum have also been fabricated. These devices may have a nearly flat EL spectrum from ~ 1 μm to 2.5 μm (Penczek *et al.*, 1995), while the PL spectrum of the same material peaks at 800 nm. The mechanism for bright, white-light EL may be thermal emission ("black-body radiation") or may be related to the presence of very hot carriers, consistent with the very large biases ($V > 10$ V) and large current densities ($J > 1$ A/cm^2) required. Note that c-Si p-n junctions (Newman, 1955) and MOSFETs (Toriumi *et al.*, 1987) can produce broadband EL, whose origin is still under debate (Deboy and Kolzer, 1994). Whether the mechanisms for broadband emission in c-Si and in π-Si devices are similar is uncertain at this time.

LEDs operating in the infrared are also attractive. For example, 1.5 μm radiation is eye-safe and compatible with fiber-communication systems. In addition, it is not absorbed by the c-Si wafer, which may be an advantage for the integration of π-Si LEDs in close proximity to microelectonic circuits. We have very recently been able to produce EL at 1.54 μm (using Er-doped oxidized π-Si; see Section II.4) and at 1.1 μm (using recrystallized oxidized π-Si; see Section II.3). (Tsybeskov *et al.*, 1996d; 1997b.)

6. Compatibility with Microelectronics

Ultimately, the success or failure of π-Si based technology rests on the compatibility of π-Si LEDs with standard microelectronic processing. The LEDs of Tsybeskov *et al.* (1996a; 1996b) are compatible with many aspects of Si microelectronic processing. For example, the annealing step used to cover the nanocrystals with a thin oxide layer is performed at $\geqslant 900°C$ in an inert atmosphere (N_2) or in diluted oxygen (10% O_2 in N_2). Once the devices have undergone this step, the poly-Si film that separates the porous layers from the top metal contact can be grown by low pressure chemical vapor deposition (LPCVD) at 650°C and annealed at 900°C after implantation with phosphorus ions, without affecting the luminescence. Preliminary investigations of the influence of other processing steps such as ion implantation, reactive ion etching, and chemical vapor deposition on π-Si (Peng et al., 1993; 1994; Duttagupta et al., 1995d) have been promising.

For device applications, uniform porous layers are highly desirable. Uniformity is required in depth and laterally. Many samples are not uniform as a function of depth. Spatially-resolved PL and Raman measurements show that the first few microns often have a different nanostructure from that of deeper layers (Ettedgui et al., 1993; Kozlowski and Lang, 1992). As a result, the PL spectrum and intensity change in depth. The lateral uniformity is another issue. One distinguishes three length scales: (1) Uniformity over large areas ($\gg 1$ cm^2) can be maintained if the electrochemical cell is designed carefully. (2) Uniformity on a scale of $0.1-100$ μm is more difficult to maintain. Very high porosity layers tend to crack when they are removed from the solution and dried (except when supercritical drying is used (Canham et al., 1994)). This leads to a surface with a morphology remininiscent of a dry lake bed (Ettedgui et al., 1993). Spatially resolved photoluminescence (SRPL) maps of such surfaces taken with 1 μm spatial resolution show large differences in the local PL intensity, with regions that appear dark (no PL) and bright (visible PL). Even layers that show no evidence of cracking are often not homogeneous on a micron scale; SRPL maps have shown that the surface of apparently uniform samples may in fact be made of a very large number of small (~ 1 μm) regions that emit very bright PL spaced by $\geqslant 1$ μm (Ettedgui et al., 1993). The production of π-Si layers that appear homogeneous on a 1 μm scale require a proper choice of the anodization parameters (Ettedgui et al., 1993). (3) Uniformity on a scale of 10 nm is impossible to achieve since the crystallite size is in the nanometer range.

If π-Si LEDs are to be integrated with very large scale integration (VLSI) circuits, they must be very small and anodization should not affect the neighboring Si electronic circuitry. Several techniques that allow the manufacture of miniature π-Si areas (down to 1 μm^2 and below) and protect the c-Si regions immediately adjacent to them have been developed (Duttagupta et al., 1995a Duttagupta et al., 1995b; Kozlowski and Lang, 1992). They include localized amorphization by ion implantation to prevent anodization, localized low-energy ion bombardment to seed the pore

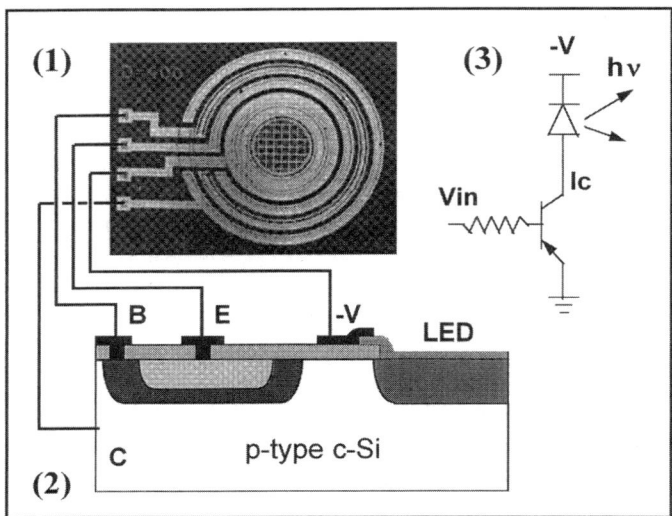

FIG. 31. An integrated π-Si LED/bipolar transistor circuit. (1) a top-view photograph, (2) a schematic cross section, and (3) is the equivalent circuit. (From Hirschman et al., submitted.)

formation, and protection of selected areas by a combination of Si nitride and photoresists.

By using a combination of the techniques discussed in Section VI, we have very recently built an integrated bipolar transistor/π-Si LED structure (Fauchet et al., 1997; Hirschman et al., 1996). The complete structure, shown in Fig. 31, was fabricated using accepted Si microelectronic fabrication procedures. It was possible to turn the LED on and off by applying a small current pulse to the base of the bipolar transistor. Arrays of such integrated structures can easily be fabricated. This represents a critical first step toward the integration of Si-based LEDs into VLSI circuits.

VII. Conclusions and Outlook

After the 1990 report by Canham of strong, visible PL from π-Si, the understanding of the unusual optical properties of this form of nanoscale Si has improved to the point where most of the initial disagreements have been resolved. A universally accepted model for the luminescence is not yet available, but the number of realistic models has been reduced to two.

Opportunities for significant contributions continue to exist in the areas of preparation and processing of π-Si, measurement of nonoptical properties, development of nanoscale characterization tools to perform spectroscopic studies on isolated nanostructures, and theoretical modeling. Light-emitting devices made of π-Si have finally achieved significant milestones in terms of efficiency, stability, spectral coverage and compatibility with microelectronics. However, the devices demonstrated to date are far from optimized. Improvements can be expected to result from better material control, device design, and integration.

Acknowledgments

This work was supported by grants from Rochester Gas & Electric, the New York State Energy Research & Development Authority, the Army Research Office, and the NSF Science & Technology Center for Photoinduced Charge Transfer, with additional support from Xerox and the sponsors of the Laboratory for Laser Energetics. The results from the Rochester group were obtained together with my immediate collaborators, Dr. L. Tsybeskov, Dr. C. Peng, Dr. J. Rehm, Mr. S. Duttagupta, Mr. K. Hirschman and Mr. J. von Behren, and with contributions from Prof. Y. Gao, Prof. G. McLendon, and Prof. S. Kurinec. I am also grateful to Dr. L. Brus, Dr. L. Canham, Prof. R. Collins, Prof. F. Koch, and Dr. M. Hybertsen for stimulating discussions and to many other colleagues and friends for their contributions. The interpretations and opinions presented in this chapter are those of the author.

References

Allan, G., Delerue, C., and Lannoo, M. (1996). *Phys. Rev. Lett.* **76**, 2961.
Anedda, A., Bongiovanni, G., Cannas, M., Congiu, F., Mura, A., and Martini, M. (1993). *J. Appl. Phys.* **74**, 6993.
Bambha, N. K., Nighan, Jr., W. L., Campbell, I. H., Fauchet, P. M., and Johnson, N. M. (1988). *J. Appl. Phys.* **63**, 2316.
Bomchil, G. and co-workers (1988). *Microelectron. Eng.* **8**, 293.
Bradfield, P. L., Brown, T. G., and Hall, D. G. (1989). *Appl. Phys. Lett.* **55**, 100.
Bassous, E., Freeman, M., Halbout, J.-M., Iyer, S. S., Kesan, V. P., Munguia, P., Pesarcik, S. F., and Williams, B. L. (1992). *Mat. Res. Soc. Symp. Proc.* **256**, 23.
Brandt, M. S., Fuchs, H. D., Stutzmann, M., Weber, J., and Cardona, M. (1992). *Solid State Commun.* **81**, 307.
Bsiesy, A., Muller, F., Ligeon, M., Gaspard, F., Herino, R., Romestain, R., and Vial, J. C. (1993). *Phys. Rev. Lett.* **71**, 637.
Bustarret, E., Ligeon, M., and Rosenbauer, M. (1995). *Phys. Stat. Sol. (b)* **190**, 111.

Calcott, P. D. J., Nash, K. J., Canham, L. T., Kane, M. J., and Brumhead, D. (1993). *J. Phys: Condensed Matter* **5**, L91; (1993) *Mat. Res. Soc. Symp. Proc.* **283**, 143; (1993) *J. Lumines.* **57**, 257.

Calcott, P. D. J., Nash, K. J., Canham, L. T., and Kane, M. J. (1995). *Mat. Res. Soc. Symp. Proc.* **358**, 465.

Campbell, I. H., and Fauchet, P. M. (1986). *Solid State Commun.* **58**, 739.

Canham, L. T. (1990). *Appl. Phys. Lett.* **57**, 1046.

Canham, L. T., Cullis, A. G., Pickering, C., Dosser, O. D., Cox, T. I., and Lynch, T. P. (1994). *Nature* **368**, 133.

Canham, L. T. (1995). *Phys. Stat. Sol. (b)* **190**, 9.

Collins, R. T., Tischler, M. A., and Stathis, J. H. (1992). *Appl. Phys. Lett.* **61**, 1649.

Cullis, A. G., and Canham, L. T. (1991). *Nature* **353**, 335.

Cullis, A. G. (1993). Photograph shown in *Scientific American*, December, p. 22.

Davies, G. (1989). *Physics Reports* **176**, 83.

Deboy, G., and Kolzer, J. (1994). *Semicond. Sci. Technol.* **9**, 1017.

Delerue, C., Allan, G., and Lannoo, M. (1993). *Phys. Rev. B* **48**, 11024.

Dorofeev, A. M., and co-workers (1995). *J. Appl. Phys.* **77**, 2679.

Duttagupta, S. P., Fauchet, P. M., Peng, C., Kurinec, S. K., Hirschman, K., and Blanton, T. N. (1995a). *Mat. Res. Soc. Symp. Proc.* **358**, 647.

Duttagupta, S. P., Peng, C., Fauchet, P. M., Kurinec, S. K., and Blanton, T. N. (1995b). *J. Vac. Sci. Technol. B* **13**, 1230.

Duttagupta, S. P., Peng, C., Tsybeskov, L., and Fauchet, P. M. (1995c). *Mat. Res. Soc. Symp. Proc.* **380**, 73.

Duttagupta, S. P., Tsybeskov, L., Fauchet, P. M., Ettedgui, E., and Gao, Y. (1995d). *Mat. Res. Soc. Symp. Proc.* **358**, 381.

Duttagupta, S. P., Chen, X. L., Jenekhe, S. A., and Fauchet, P. M. *Solid State Commun.* **101**, 33.

Duttagupta, S. P., and co-workers. Submitted for publication, 1997.

Ettedgui, E., Peng, C., Tsybeskov, L., Gao, Y., Fauchet, P. M., Mizes, H. A., and Carver, G. A. (1993). *Mat. Res. Soc. Symp. Proc.* **283**, 173.

Fauchet, P. M., and Nighan, Jr., W. L. (1986). *Appl. Phys. Lett.* **48**, 721.

Fauchet, P. M., and Campbell, I. H. (1988). *Crit. Rev. Solid State and Mater. Sci.* **14**, S79.

Fauchet, P. M., Ettedgui, E., Raisanen, A., Brillson, L. J., Seiferth, S., Kurinec, S. K., Gao, Y., Peng, C., and Tsybeskov, L. (1993). *Mat. Res. Soc. Symp. Proc.* **298**, 271.

Fauchet, P. M., Hulin, D., Vanderhaghen, R., Mourchid, A., and Nighan, W. L. Jr., J. (1992). *J. Non-Cryst. Solids* **141**, 76.

Fauchet, P. M. (1994). In *Porous Silicon* (eds. Z. C. Feng and R. Tsu), World Scientific, Singapore, pp. 429–465.

Fauchet, P. M., and co-workers (1994a). In *Semiconductor Silicon 1994: Proceedings of the 7th International Symposium on Silicon Materials Science and Technology* (eds. H. R. Huff, W. Bergholz, and K. Sumino), The Electrochemical Society, Pennington, pp. 499–510.

Fauchet, P. M., and co-workers (1994b). In *Advanced Photonics Materials for Information Technology* (ed. S. Etemad), SPIE, Bellingham, WA, Vol. 2144, pp. 34–50.

Fauchet, P. M., Tsybeskov, L., Peng, C., Duttagupta, S. P., von Behren, J., Kostoulas, Y., Vandyshev, J. V., and Hirschman, K. D. (1995). *J. Selected Topics Quantum Electron.* **1**, 1126.

Fauchet, P. M., Tsybeskov, L., Duttagupta, S. P., and Fauchet, P. M. (1997). *Thin Solid Films* **297**, 254.

Favennec, P. N., L'Harldon, H., Moutonnet, D., Salvi, M., and Gauneau, M. (1993). *Mat. Res. Soc. Symp. Proc.* **301**, 181.

Fischer, P. B., Dai, K., Chen, E., and Chou, S. Y. (1993). *Vac. Sci. Technol.* **B11**, 2524.

George, T., Anderson, M. S., Pike, W. T., Lin, T. L., and Fathauer, R. W. (1992). *Appl. Phys. Lett.* **60**, 2359.
Goldstein, A. N., Echer, C. M., and Alivisatos, A. P. (1992). *Science* **256**, 1425.
Goldstein, A. N. (1996). *Appl. Phys. A* **62**, 33.
Halimaoui, A., Oules, C., Bomchil, G., Bsiesy, A., Gaspard, F., Herino, R., Ligeon, M., and Muller, F. (1991). *Appl. Phys. Lett.* **59**, 304.
Harvey, J. F., Shen, H., Lux, R. A., Dutta, M., Pamulapati, J., and Tsu, R. (1992). *Mat. Res. Soc. Symp. Proc.* **256**, 175.
Herino, R., and co-workers (1987). *J. Electrochem. Soc.* **134**, 1994.
Hill, N. A., and Whaley, K. B. (1996). *J. Electron. Mater.* **25**, 269.
Hirschman, K. D., Tsybeskov, L., Duttagupta, S. P., and Fauchet, P. M. (1996). *Nature* **384**, 338.
Hummel, R., Morrone, A., Ludwig, M., and Chang, S. S. (1993). *Appl. Phys. Lett.* **63**, 271.
Hummel, R. E., Ludwig, M. H., Chang, S.-S., Fauchet, P. M., Vandyshev, Ju. V., and Tsybeskov, L. (1995). *Solid State Commun.* **95**, 553.
Hybertsen, M. S. (1994). *Phys. Rev. Lett.* **72**, 1514.
Iyer, S. S., and Xie, Y.-H. (1993). *Science* **260**, 40.
Kalkhoran, N. M., Namavar, F., and Maruska, H. P. (1992). *Mat. Res. Soc. Symp. Proc.* **256**, 84.
Kanemitsu, Y. (1993). *Phys. Rev. B* **48**, 12,357.
Kanemitsu, Y., Uto, H., Masumoto, Y., Matsumoto, T., Futagi, T., and Mimura, H. (1993). *Phys. Rev. B* **48**, 2827.
Kimura, T., Yokoi, A., Horiguchi, H., Saito, R., Ikoma, T., and Sato, A. (1994). *Appl. Phys. Lett.* **65**, 983.
Koch, F. (1993). *Mat. Res. Soc. Symp. Proc.* **298**, 319.
Koch, F., Petrova-Koch, V., Muschik, T., Nikolov, A., and Gavrilenko, V. (1993). *Mat. Res. Soc. Symp. Proc.* **283**, 197.
Koshida, N., and Koyama, H. (1992). *Appl. Phys. Lett.* **60**, 347.
Koshida, N., Koyama, H., Yamamoto, Y., and Collins, G. J. (1993). *Appl. Phys. Lett.* **63**, 2655.
Koyama, H., and Koshida, N. (1993). *J. Appl. Phys.* **74**, 6365.
Koyama, H., and Koshida, N. (1994). *Jpn. J. Appl. Phys.* **33**, L1737.
Kozlowski, F., and Lang, W. (1992). *J. Appl. Phys.* **72**, 5400.
Kux, A., and Ben-Chorin, M. (1995). *Phys. Rev. B* **51**, 17535.
Kux, A., Kovalev, D., and Koch, F. (1995). *Appl. Phys. Lett.* **66**, 49.
Lang, W., Steiner, P., and Kozlowski, F. (1993). *J. Lumines.* **57**, 341.
Lauerhaas, J. M., Credo, G. M., Heinrich, J. L., and Sailor, M. J. (1992). *J. Am. Chem. Soc.* **114**, 1911.
Lazarouk, S., Jaguiro, P., Katsouba, S., Masini, G., La Monica, S., Maiello, G., and Ferrari, A. (1996). *Appl. Phys. Lett.* **68**, 1646.
Lee, E. J., Ha, J. S., and Sailor, M. J. (1995). *Mat. Res. Soc. Symp. Proc.* **358**, 387.
Lehmann, V., and Gösele, U. (1991). *Appl. Phys. Lett.* **58**, 856.
Lehmann, V., and co-workers (1993). *Jpn. J. Appl. Phys.* **32**, 2095.
Linnros, J., and Lalic, N. (1995). *Appl. Phys. Lett.* **66**, 3048.
Littau, K. A., Szajowski, P. J., Miller, A. J., Kortan, A. R., and Brus, L. E. (1993). *J. Phys. Chem.* **97**, 1224.
Lockwood, D. J., and Wang, A. G. (1995). *Solid State Commun.* **94**, 905.
Loni, A., Simons, A. J., Calcott, P. D. J., and Canham, L. T. (1995a). *J. Appl. Phys.* **77**, 3557.
Loni, A., Simons, A. J., Cox, T. I., Calcott, P. D. J., and Canham, L. T. (1995). *Electron. Lett.* **31**, 1288.
Lu, Z. H., Lockwood, D. J., and Baribeau, J.-M. (1995). *Nature* **378**, 258.
Mauckner, G., Hamann, J., Rebitzer, W., Baier, T., Thonke, K., and Sauer, R. (1995). *Mat. Res. Soc. Symp. Proc.* **358**, 489.

McColm, I. J. (1990). *Ceramic Hardness*, Plenum Press, New York.
Michel, J., Benton, J. L., Ferrante, R. F., Jacobson, D. C., Eaglesham, D. J., Fitzgerald, E. A. Xie, Y.-H., Poate, J. M., and Kimerling, L. C. (1991). *J. Appl. Phys.* **70**, 2672.
Mimura, H., Matsumoto, T., and Kanemitsu, Y. (1995). *Mat. Res. Soc. Symp. Proc.* **358**, 635.
Namavar, F., Lu, F., Perry, C. H., Cremins, A., Kalkhoran, N. M., and Soref, R. A. (1995). *J. Appl. Phys.* **77**, 4813.
Nassiopoulos, A. G., Grigoropoulos, S., Papadimitrou, D., and Gogolides, E. (1995). *Phys. Stat. Sol. (b)* **190**, 91.
Newman, R. (1955). *Phys. Rev.* **100**, 700.
Ohno, T., Shiraishi, K., and Ogawa, T. (1992). *Phys. Rev. Lett.* **69**, 2400.
Penczek, J., Knoesen, A., Lee, H. W. H., and Smith, R. L. (1995). *Mat. Res. Soc. Symp. Proc.* **358**, 641.
Peng, C., Tsybeskov, L., and Fauchet, P. M. (1993). *Mat. Res. Soc. Symp. Proc.* **283**, 121.
Peng, C., Tsybeskov, L., Fauchet, P. M., Seiferth, F., Kurinec, S. K., Rehm, J. M., and McLendon, G. L. (1993). *Mat. Res. Soc. Symp. Proc.* **298**, 179.
Peng, C., Fauchet, P. M., Rehm, J. M., McLendon, G. L., Seiferth, F., and Kurinec, S. K. (1994). *Appl. Phys. Lett.* **64**, 1259.
Peng, C., and Fauchet, P. M. (1995). *Appl. Phys. Lett.* **67**, 2515.
Peng, C., Fauchet, P. M., Hirschman, K. D., and Kurinec, S. K. (1995). *Mat. Res. Soc. Symp. Proc.* **358**, 689.
Peng, C., Hirschman, K. D., and Fauchet, P. M. (1996). *J. Appl. Phys.* **80**, 295.
Perry, C. H., Lu, F., Namavar, F., Kalkhoran, N. M., and Soref, R. A. (1992). *Mat. Res. Soc. Symp. Proc.* **256**, 153.
Petrova-Koch, V., Muschik, T., Kux, A., Meyer, B. K., Koch, F., and Lehmann, V. (1992). *Appl. Phys. Lett.* **61**, 943.
Petrova-Koch, V., Muschik, T., Kovalev, D. I., Koch, F., and Lehmann, V. (1993). *Mat. Res. Soc. Symp. Proc.* **283**, 179.
Petrova-Koch, V., Muschik, T., Polisski, G., and Kovalev, D. (1995). *Mat. Res. Soc. Symp. Proc.* **358**, 483.
Pickering, C., Beale, M. I. J., Robbins, D. J., Pearson, P. J., and Greef, R. (1984). *J. Phys. C: Solid State Phys.* **17**, 6535.
Prokes, S. (1993). *Appl. Phys. Lett.* **62**, 3244.
Proot, J. P., Delerue, C., and Allan, G. (1992). *Appl. Phys. Lett.* **61**, 1948.
Rehm, J. M., McLendon, G. L., Tsybeskov, L., and Fauchet, P. M. (1995). *Appl. Phys. Lett.* **66**, 3669.
Rehm, J. M., McLendon, G. L., and Fauchet, P. M. (1996a). *J. Am. Chem. Soc.* **118**, 4490.
Rehm, J. M., McLendon, G. L., and Fauchet, P. M. (1996b). In *Advanced Luminescent Materials* (eds. D. J. Lockwood, P. M. Fauchet, N. Koshida, and S. R. J. Brueck), The Electrochemical Society, Pennington, NJ, pp. 212–221.
Ren, S. Y., and Dow, J. D. (1992). *Phys. Rev. B* **45**, 6492.
Richter, A., Steiner, P., Kozlowski, F., and Lang, W. (1991). *IEEE Elec. Dev. Lett.* **12**, 691.
Robinson, M. B., Dillon, A. C., Haynes, D. R., and George, S. M. (1992). *Appl. Phys. Lett.* **61**, 1414.
Sakamoto, T., Tokioka, H., Takanabe, S., Niwano, Y., Goto, Y., Namizaki, H., Wada, O., and Kurokawa, H. (1995). *Mat. Res. Soc. Symp. Proc.* **358**, 933.
Sanders, G. D., and Chang, Y.-C. (1992). *Phys. Rev. B* **45**, 9202.
Saunders, W. A., Atwater, H. A., Vahala, K. J., Flagan, R. C., and Sercel, P. C. (1993). *Mat. Res. Soc. Symp. Proc.* **283**, 77.
Schuppler, S., Friedman, S. L., Marcus, M. A., Adler, D. L., Xie, Y.-H., Ross, F. M., Harris, T. D., Brown, W. L., Chabal, Y. J., Brus, L. E., and Citrin, P. H. (1994). *Phys. Rev. Lett.* **72**, 2648.

Schuppler, S., Friedman, S. L., Marcus, M. A., Adler, D. L., Xie, Y. H., Ross, F. M., Chabal, Y. J., Harris, T. D., Brus, L. E., Brown, W. L., Chaban, E. E., Szajowski, P. F., Christman, S. B., and Citrin, P. H. (1995). *Phys. Rev. B* **52**, 4910.

Simons, A. J., Cox, T. I., Loni, A., Canham, L. T., Uren, M. J., Reeves, C., Cullis, A. G., Calcott, P. D. J., Houghton, M. R., and Newey, J. P. (1996). In *Advanced Luminescent Materials* (eds. D. J. Lockwood, P. M. Fauchet, N. Koshida, and S. R. J. Bruek), The Electrochemical Society, Pennington, NJ, pp. 73–86.

Smith, R. L., and Collins, S. D. (1992). *J. Appl. Phys.* **71**, R1.

Steiner, P., Kozlowski, F., Sandmaier, H., and Lang, W. (1993). *Mat. Res. Soc. Symp. Proc.* **283**, 343.

Steiner, P., Kozlowski, F., and Lang, W. (1995). *Mat. Res. Soc. Symp. Proc.* **358**, 665.

Sturm, J. C., and co-workers (1993). *Mat. Res. Soc. Symp. Proc.* **298**, 69.

Takagahara, T., and Takeda, K. (1992). *Phys. Rev. B* **46**, 15578.

Takagi, H., Ogawa, H., Yamazaki, Y., Ishizaki, A., and Nakagiri, T. (1990). *Appl. Phys. Lett.* **56**, 2379.

Taskin, T., Gardelis, S., Evans, J. H., Hamilton, B., and Peaker, A. R. (1995). *Electron. Lett.* **31**, 2132.

Tamura, H., Ruckschloss, M., Wirschem, T., and Veprek, S. (1994). *Appl. Phys. Lett.* **65**, 1537.

Toriumi, A., Yoshimi, M., Iwase, M., Akiyama, Y., and Taniguchi, K. (1987). *IEEE Trans. Elec. Dev.* **ED-34**, 1501.

Tsybeskov, L., Peng, C., Duttagupta, S. P., Ettedgui, E., Gao, Y., Fauchet, P. M., and Carver, G. E. (1993). *Mat. Res. Soc. Symp. Proc.* **298**, 307.

Tsybeskov, L., and Fauchet, P. M. (1994). *Appl. Phys. Lett.* **64**, 1983.

Tsybeskov, L., Vandyshev, J. V., and Fauchet, P. M. (1994). *Phys. Rev. B* **49**, 7821.

Tsybeskov, L., Duttagupta, S. P., and Fauchet, P. M. (1995). *Solid State Commun.*, **95**, 429.

Tsybeskov, L., Duttagupta, S. P., Hirschman, K. D., and Fauchet, P. M. (1996a). In *Advanced Luminescent Materials* (eds. D. J. Lockwood, P. M. Fauchet, N. Koshida, and S. R. J. Bruek), The Electrochemical Society, Pennington, NJ, pp. 34–47.

Tsybeskov, L., Duttagupta, S. P., Hirschman, K. D., and Fauchet, P. M. (1996b). *Appl. Phys. Lett.* **68**, 2058.

Tsybeskov, L., Moore, K. L., Hall, D. G., and Fauchet, P. M. (1996c). *Phys. Rev. B* **54**, R8361.

Tsybeskov, L., and co-wokers (1996d). *Appl. Phys. Lett.* **69**, 3411.

Tsybeskov, L., Moore, K. L., Fauchet, P. M., and Hall, D. G. (1997a). *Mat. Res. Soc. Symp. Proc.* **452**, 523.

Tsybeskov, L., and co-workers (1997b). *Appl. Phys. Lett.* **70**, 1790.

Uhlir, Jr., A. (1956). *Bell. Syst. Tech. J.* **35**, 333.

van Buuren, T., Tiedje, T., Dahn, J. R., and May, B. M. (1993). *Appl. Phys. Lett.* **63**, 2911.

van Buuren, T., Eisebitt, S., Patitsas, S., Ritchie, S., Tiedje, T., Young, J. F., and Gao, Y. (1995). *Mat. Res. Soc. Symp. Proc.* **358**, 441.

Vervoort, L., Bassani, F., Mihalcescu, I., Vial, J. C., and Arnaud d'Avitaya, F. (1995). *Phys. Stat. Sol. (b)* **190**, 123.

Vezin, V., and co-workers (1992). *Appl. Phys. Lett.* **60**, 2625.

Vial, J. C., Bsiesy, A., Fishman, G., Gaspard, F., Herino, R., Ligeon, M., Muller, F., Romestain, R., and MacFarlane, R. M. (1993). *Mat. Res. Soc. Symp. Proc.* **283**, 241.

Vingsbo, O., and co-workers (1986). In *Microindentation Techniques in Materials Science and Engineering, ASTM STP 889* (eds. P. J. Blau and B. R. Lawn), American Society for Testing and Materials, Philadelphia, p. 257.

von Behren, J., Ucer, K. B., Tsybeskov, L., Vandyshev, Ju. V., and Fauchet, P. M. (1995a). *J. Vac. Sci. Technol. B* **13**, 1225.

von Behren, J., Tsybeskov, L., and Fauchet, P. M. (1995b). *Appl. Phys. Lett.* **66**, 1662.

von Behren, J., Kostoulas, Y., Ucer, K. B., and Fauchet, P. M. (1996). *J. Non-Cryst. Solids* **198–200**, 957.
Wadayama, T., Yamamoto, S., and Hatta, A. (1994). *Appl. Phys. Lett.* **65**, 1653.
Wang, L. W., and Zunger, A. (1994). *J. Chem. Phys.* **100**, 2394.
Wang, J., Zhang, F.-L, Wang, W.-C., Zheng, J.-B., Hou, X.-Y., and Wang, X. (1994). *J. Appl. Phys.* **75**, 1070.
Wilson, W. L., Szajowski, P. J., and Brus, L. E. (1993). *Science* **262**, 1242.
Wolford, D. J., Scott, B. A., Reimer, J. A., and Bradley, J. A. (1983). *Physica* **117B&118B**, 920.
Xie, Y. H., Hybertsen, M. S., Wilson, W. L., Ipri, S. A., Carver, G. E., Brown, W. L., Dons, E., Weir, B. E., Kortan, A. R., Watson, G. P., and Liddle, A. J. (1994). *Phys. Rev. B* **49**, 5386.
Young, J. F., Gong, T., Fauchet, P. M., and Kelly, P. J. (1994). *Phys. Rev. B* **50**, 2208.
Zaidi, S. H., Chu, A.-S., and Brueck, S. R. J. (1995). *Mat. Res. Soc. Symp. Proc.* **358**, 957.
Zhang, L., Coffer, J. L., Gnade, B. E., Xu, D., and Pinizzotto, R. F. (1995). *Mat. Res. Soc. Symp. Proc.* **358**, 671.
Zheng, B., Michel, J., Ren, F. Y. G., Kimerling, L. C., Jacobson, D. C., and Poate, J. M. (1994). *Appl. Phys. Lett.* **64**, 2842.

CHAPTER 7

Theory of Radiative and Nonradiative Processes in Silicon Nanocrystallites

C. Delerue, G. Allan, and M. Lannoo

INSTITUT D'ELECTRONIQUE ET DE MICROÉLECTRONIQUE DU NORD
DÉPARTEMENT ISEN—B.P.69
59652 VILLENEUVE D'ASCQ CEDEX, FRANCE

I. INTRODUCTION	253
II. ELECTRONIC PROPERTIES	254
III. OPTICAL TRANSITIONS AND RADIATIVE LIFETIME	258
IV. EXCHANGE SPLITTING AND SYMMETRY OF THE CRYSTALLITES	262
V. ATOMIC RELAXATION, STOKES SHIFT, AND SELF-TRAPPED EXCITON	269
1. *Stokes Shift for the Delocalized States*	270
2. *The Existence of Self-trapped Excitons*	271
VI. NONRADIATIVE RECOMBINATION	279
1. *Recombination on Surface Dangling Bonds*	279
2. *Nonradiative Auger Recombination*	286
VII. SCREENING IN NANOCRYSTALLITES AND COULOMB CHARGING EFFECTS	292
1. *Hydrogenic Impurities*	292
2. *Coulomb Effects and Effective Dielectric Constant*	295
VIII. CONCLUSION	298
References	299

I. Introduction

Near threshold optical transitions in an indirect bandgap material like silicon (Si) have very small oscillator strength preventing its use in optoelectronics. An interesting field of research is thus to devise Si based materials where this selection rule is broken, leading to efficient luminescence. A major step in this direction has been the observation of intense photoluminescence (PL) from porous silicon (π-Si) (Canham, 1990). This has stimulated considerable theoretical activity on the optical properties of Si nanocrystallites and quantum wires. The aim of this chapter is to review what has been achieved in this field over recent years and to discuss the basic problems that have not yet been cleared up.

This chapter is organized as follows. We first review the predictions concerning the electronic structure of crystallites and quantum wires and discuss their range of accuracy. We then consider their optical properties: optical absorption, luminescence and compare calculated radiative recombination rates with observed values. In the third part, we analyze in detail the two level model due to exchange splitting invoked to explain some experimental data on π-Si. This is followed by aspects related to atomic relaxation, for example, the Stokes shift and a recent calculation supporting the existence of self-trapped excitons. We then describe theoretical work on nonradiative recombination: (i) caused by surface defects like dangling bonds and (ii) the intrinsic Auger process, showing that they give a coherent account of the experimental data. The final step is to investigate screening and Coulomb charging effects in crystallites with application to donor and acceptor impurities as well as carrier injection.

II. Electronic Properties

Quantum confinement is characterized by several consequences: (i) the fundamental gap exhibits a blue shift, (ii) in nanocrystallites the filled and the empty states become quantized, and (iii) optical dipole transitions across the fundamental gap become allowed. All three effects increase with decreasing size of the crystallites. A number of calculations of these effects have been performed over the last few years, not only for Si nanocrystals (Ren and Dow, 1992; Proot, Delerue, and Allan, 1992; Delerue, Allan, and Lannoo 1993; Takagahara and Takeda, 1992; Huaxiang, Ling, and Xide, 1993; Fishman, Mihalcescu, and Romestain, 1993a; Delley and Steigmeier, 1993; Wang and Zunger, 1994a; Hirao, Uda, and Murayama, 1993; Hirao, 1995; Gavartin, Matthai, and Morrison, 1995; Hill and Whaley, 1995) but also for quantum wires (Ohno, Shiraishi, and Ogawa, 1992; Hybertsen, 1992; Read *et al.*, 1993; Voos *et al.*, 1992; Sanders and Chang, 1992; Buda, Kohanoff, and Parrinello, 1992; Mintmire, 1993; Hybertsen and Needels, 1993; Shik, 1993; Polatoglou, 1993; Yeh, Zhang, and Zunger, 1994a; Yeh, Zhang, and Zunger, 1994b; Lee *et al.*, 1995) since both possibilities have been invoked for π-Si. They essentially belong to four classes: effective mass approximation (EMA), empirical tight-binding (ETB), empirical pseudopotential (EPS), and finally *ab initio* local density approximation (LDA). We do not here recall the basic principles of these methods but make a comparative presentation of their predictions together with a discussion of their expected accuracy.

7 Radiative and Nonradiative Processes in Silicon Nanocrystallites

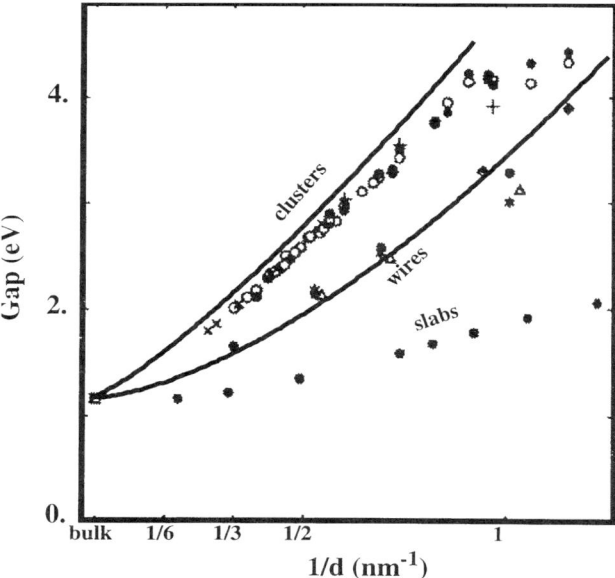

FIG. 1. Energy gap versus confinement parameter $1/d$ for hydrogen terminated Si clusters, wires and slabs ($d = aN/8$ for slabs, $d = a(0.5N/\pi)^{1/2}$ for wires, $d = a(0.75N/\pi)^{1/3}$ for clusters where $a = 0.54$ nm and N is the number of Si atoms in the unit cell). LDA: (●, ○) Delley and Steigmeier (1995), (+) Hirao et al. (1993), (◆) Mintmire (1993), (△) Read et al. (1993), (∗) Buda et al. (1992). EPS: (×) Wang and Zunger (1994a). ETB: (straight lines) present authors (Proot et al., 1992) for clusters; same method used for wires with a [100] axis and a square basis. (□) is the bulk bandgap.

In Fig. 1 we give the predicted blue shifts versus size as obtained from LDA calculations compiled in Delley and Steigmeier (1995) for hydrogen terminated clusters, wires and slabs. They are compared to the results obtained in our group (Proot, Delerue, and Allan, 1992; Delerue, Allan, and Lannoo, 1993) using a nonorthogonal ETB method with parameters providing an extremely good fit to the bulk band structure. EPS result of Wang and Zunger (1994a) are also plotted for comparison. First of all one can notice that the blue shift decreases when going from crystallites to wires and slabs which is natural since the confinement effect occurs in a smaller number of dimensions. A second point is the strikingly good agreement between the ETB, EPS, and LDA predictions, which gives some confidence about the reliability of these theoretical values. At this stage it is important to notice that the LDA gap values include a rigid shift of 0.6 eV since it is known that LDA underestimates the bulk bandgap by this amount.

Another point of interest concerns the accuracy of the EMA for this

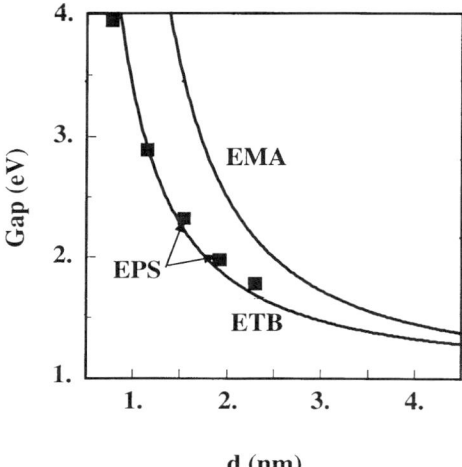

FIG. 2. Comparison of EMA (Voos et al., 1992), ETB (present authors (Proot et al., 1992)), and EPS (■) (Yeh et al., 1994a; Yeh et al., 1994b) calculations for wires with a [100] axis and a square basis.

problem. In Fig. 2 we compare the EMA predictions for wires (Voos et al. 1992) to the same ETB values that are plotted in Fig. 1 and to the EPS results of Yeh, Zhang, and Zunger (1994a; 1994b). We find that the EMA calculations grossly overestimate the blue shift. They can thus be discarded, since the EMA can be considered as an approximation to the best ETB or EPS description, which match the effective masses.

One can, however, wonder why semi-empirical techniques should provide quantitative estimates of the one-electron gap. This question is of primary importance since there exist other ETB calculations (Ren and Dow, 1992; Hill and Whaley, 1995) that provide significantly lower values for the gap. The basic point is that semi-empirical calculations are based on the postulate of "transferability" of the parameters from the known bulk band structure (to which they are fitted) to the unknown crystallite case. If this is accepted, then an essential criterion by which a particular semi-empirical model can be judged is how well it describes the bulk band structure. In this regard the ETB and EPS models of Proot, Delerue, and Allan (1992); Delerue, Allan, and Lannoo (1993); Wang and Zunger (1994a) both give extremely good fits to the Si bulk band structure over a large energy interval (18 eV for Proot, Delerue, and Allan (1992); Delerue, Allan, and Lannoo (1993) with a root mean square error of 0.1 eV (Mattheiss and Patel (1981); Allen, Broughton, and McMahan (1986)) containing the valence band as well as the conduction band. It is thus not surprising that they produce identical results for the crystallites. On the contrary, the sp^{3}s* model of Ren and Dow (1992); Hill and Whaley (1995) gives in comparison a very poor description of the conduction band, which is much too flat, and consequently must underestimate the bandgap, as it indeed does. This can be judged by the conduction band effective masses, practically exact for Proot, Delerue, and Allan (1992); Delerue, Allan, and Lannoo (1993) ($m_l = 1.09$, $m_t = 0.2$) but much too large for Ren and Dow (1992); Hill and Whaley (1995) ($m_l = 1.62$, $m_t = 0.74$).

The previous criterion for transferability only strictly applies to clusters that are large enough for the Hamiltonian in the interior to become identical to the bulk one (i.e., roughly speaking, with diameter larger than 1 nm). Then the solution of the problem at energy E can be obtained as a combination of the bulk solution at this energy plus the appropriate boundary conditions. This means that a second criterion must be that these are correctly simulated in the model. In Proot, Delerue and Allan (1992); Delerue, Allan, and Lannoo (1993), we have used Si–H terminations with coupling parameters large enough to avoid Si–H states in the gap in order to simulate an interface with a material with a large bandgap (like SiO_2). This approach is justified by the agreement with pseudopotential (Wang and Zunger, 1994a) and local density (Delley and Steigmeier, 1995) calculations treating real Si–H bonds. This agreement with the corrected LDA calculations in Fig. 1 is also explained by the same arguments, since the LDA Hamiltonian plus a rigid shift of 0.6 eV of the conduction states with respect to the valence states (so-called "scissor operator") gives a quite accurate representation of the bulk Hamiltonian and also of the Si–H terminations.

Another piece of evidence comes from the comparison between the predictions of the different ETB models presented in Fig. 3. Curves (a) and (b) are obtained from fits of practically equivalent quality to both bulk-Si valence and conduction bands. This is not true of (c) and (d) where the fit to the conduction band becomes poorer and poorer. For these, the crystallite gap prediction is too small, as could be anticipated from the fact that the conduction band is too narrow, that is, corresponds to effective masses which are much too high.

All the previous calculations share in common the fact that they are one-particle theories. One must therefore correct the predicted one-particle gap by the exciton binding energy, coming from the electron–hole attraction, eventually corrected by exchange terms. As the crystallites sizes of interest are smaller than the bulk exciton Bohr radius, the calculation reduces to the average value of this interaction $\langle e^2/r_{eh} \rangle$. As shown in Section VI the resulting correction to the one-particle gap remains small, of the order of 0.2 eV for the usual sizes.

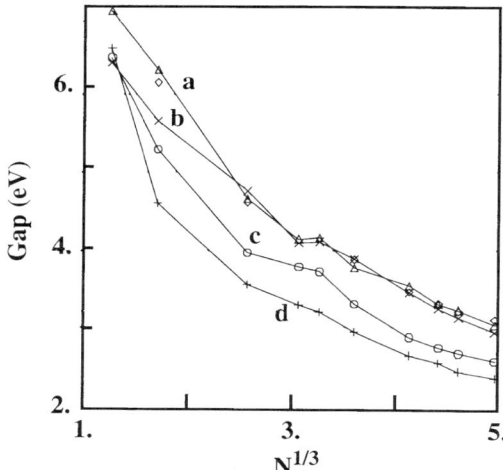

FIG. 3. Comparison between different ETB models for Si clusters passivated with hydrogen. (a) Nonorthogonal tight binding parameters (Mattheiss et al., 1981). (b) Third nearest neighbors parameters (Tserbak et al., 1993). (c) Second nearest neighbors parameters (Kauffer et al., 1976). (d) First nearest neighbors sp^{3}s* parameters (Vogl et al., 1983).

The main conclusion that can be drawn from this section is that empirical ETB or EPS as well as corrected LDA techniques are likely to give reliable predictions for crystallites, but with the condition of providing an extremely good description of the bulk band structure. This is the case of the values presented in Fig. 1.

III. Optical Transitions and Radiative Lifetime

We consider here hydrogen terminated clusters with a practically spherical shape and calculate the optical transitions between the quantum confined states of Section II (see, Proot, Delerue and Allan (1992); Delerue, Allan, and Lannoo (1993)).

The radiative recombination time τ is defined by the Fermi golden rule and given by Dexter (1958)

$$\frac{1}{\tau} = \frac{16\pi^2}{3} n \frac{e^2}{h^2 m^2 c^3} E_0 |\langle i_{BC}|p|f_{BV}\rangle|^2 \qquad (1)$$

7 RADIATIVE AND NONRADIATIVE PROCESSES IN SILICON NANOCRYSTALLITES

where $|i_{BC}\rangle$ is the initial state of the electron in the conduction band, $|f_{BV}\rangle$ is the final state in the valence band (i.e., the hole state), E_0 is the energy of the transition, and n is the refractive index of π-Si for which we take a value of 1.33 corresponding to a layer porosity of 74% (Sagnes et al., 1993). The momentum matrix element $|\langle i_{BC}|p|f_{BV}\rangle|^2$ is developed in the tight binding basis and the calculations are done by replacing the atomic orbitals by gaussians (Petit, Allan, and Lannoo, 1986a). This procedure has been used to predict the optical cross section of the isolated dangling bond in Si (Petit, Lannoo, and Allan, 1986b).

The results are given in Fig. 4 where we have plotted the Boltzmann average of Eq. (1) at temperature $T = 300$ K versus the crystallite gap. For crystallites with an optical gap larger than 2.3 eV (sizes lower than 2.5 nm), the radiative recombination is quite efficient with a characteristic time lower than 10 μs. The mixing of different wave-vector states is important and the clusters acquire optical properties intermediate between those of an indirect

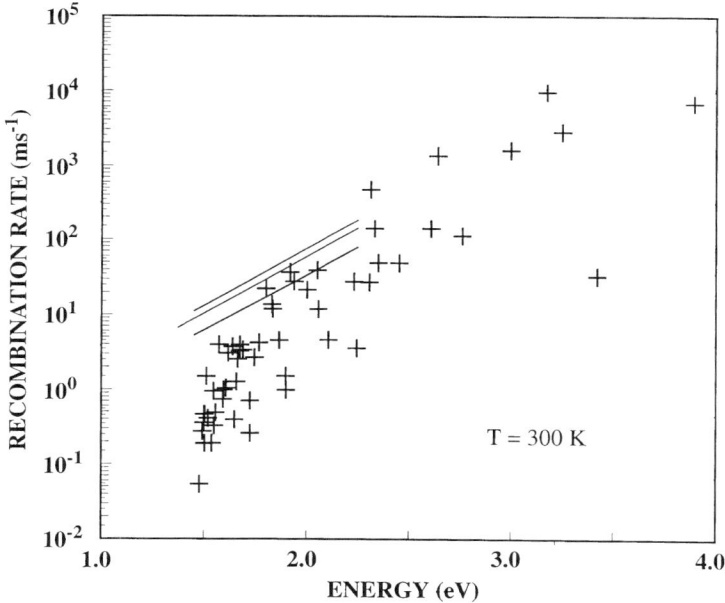

FIG. 4. Calculated recombination rate (ms^{-1}) of an excited electron–hole pair in silicon crystallites (+) with respect to the photon energy at 300 K. The spin degeneracy is not included: its inclusion would divide the calculated recombination rates by a factor 2. Continuous lines plot the experimental dependence (Vial et al., 1992) of decay rates on photon energy for three 65% porosity layers that differ by oxidation level.

gap and a direct gap material (Sanders and Chang, 1992; Steigmeier, Delley, and Auderset, 1992). Another feature of Fig. 4 is the strong decrease of the radiative recombination rate for lower photon energy. This is obviously due to the indirect nature of the Si bandgap, which gives a radiative recombination rate equal to zero in the limit of bulk Si (in a first order dipole theory). The decrease is very abrupt for energies lower than 2 eV and then the radiative recombination becomes quite slow. For comparison the dependence of PL decay rates given in Vial et al. (1992) is also plotted. We see that on average the predicted radiative recombination rates are about one order of magnitude lower than the experimental decay rate. This is particularly true for photon energy lower than 2.0 eV. This means that, even if the confinement brings some momentum mixing to allow first order optical transitions, the radiative recombination due to dipolar transitions is not efficient enough to account for the observed values. However, for large clusters, one must take into account the contribution of phonon assisted transitions, which is the dominant mechanism in the bulk situation. This possibility has been investigated in Hybertsen (1994) in the framework of EMA. The results demonstrate that the phonon assisted transitions dominate as long as the blue shift is smaller than 1.5 eV. This is completely coherent with the difference between experiment and theory evident in Fig. 4. This shows that the quantum confinement effect can account for the observed values of the radiative recombination time (also taking into account that in the range 300–500 K, the dominant mechanism for relaxation is probably nonradiative (Vial et al., 1992)).

Let us now consider optical absorption, which is expected to be *a priori* much less sensitive to surface properties (Sagnes et al., 1993). From comparison between excitation spectra of π-Si and absorption spectra of Si molecules, it has been recently deduced that the luminescence of π-Si could come from Si molecules (Kanemitsu et al., 1992). One argument was that the sharp absorption edge near 3.2 eV in the excitation spectrum is much higher than the predicted bandgap of crystallites of ~ 3 nm diameter (~ 1.5 eV). However, the threshold of the optical absorption has been measured below 2.0 eV and also shows a blue shift compared to bulk Si (Sagnes et al., 1993). These shifts can be reasonably related to quantum size effects (Sagnes et al., 1993). Therefore we have performed a theoretical calculation of the optical absorption of Si nanostructures in order to see if it is compatible with the experimental results. The optical absorption coefficient $\alpha(hv)$ for a photon energy hv is given by

$$\alpha(hv) \sim \frac{1}{hv} \sum_{n,n'} |\langle n|p|n'\rangle|^2 \delta(E_{n,n'} - hv) \qquad (2)$$

with notations as defined above.

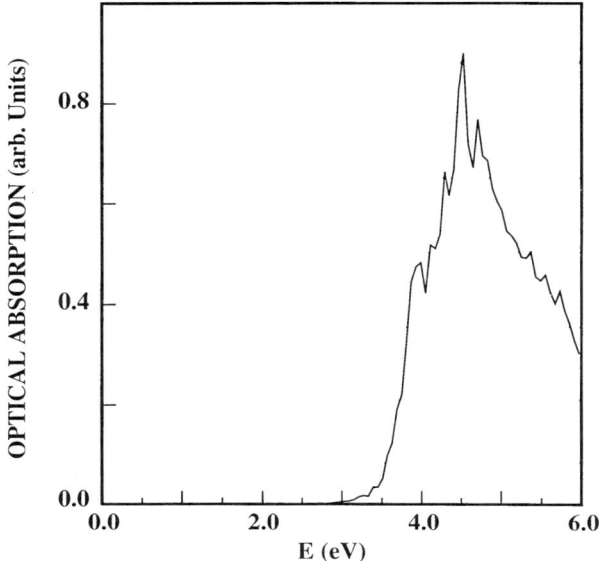

FIG. 5. Optical absorption coefficient α with respect to the photon energy $E = h\nu$ calculated for a silicon crystallite with diameter of 3.86 nm. The bandgap is calculated at 1.67 eV (without exciton binding energy).

In Fig. 5 we present the absorption coefficient for a crystallite with a bandgap of 1.67 eV. The absorption edge is shifted to near 3.5 eV, which corresponds to the direct-gap absorption of bulk Si. Nothing is visible in the figure between 1.67 and ~3.0 eV on this scale. The reason is that the optical matrix element for transitions with energy between 1.67 and 3.0 eV is several orders of magnitude lower than for transitions above 3.0 eV. The optical adsorption becomes very close to that of bulk Si (Madelung, 1991) but with a blue shift of the absorption edge, that is, it is very close to the absorption of an indirect semiconductor. Therefore the excitation spectrum of π-Si reported in Kanemitsu et al. (1992) with an excitation edge above 3.0 eV is compatible with the hypothesis of quantum confinement. In Fig. 6, we plot the absorption coefficient of the same crystalline near the band edge (1.67 eV), but on a magnified scale, showing that the absorption threshold is at the bandgap energy (1.67 eV) because the transition is dipole allowed even if this is only with fairly weak oscillator strength, compatible with the small radiative lifetime in luminescence. Therefore the optical threshold is also subject to a blue shift depending on the size of the crystallites in agreement with experiments (Sagnes et al., 1993).

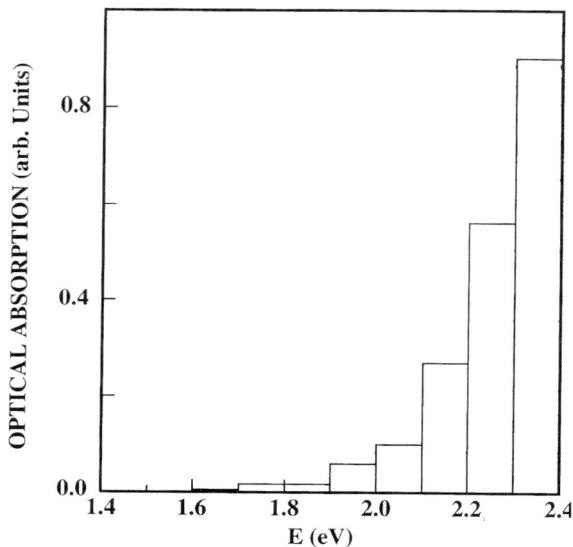

FIG. 6. Same as Fig. 5, but only the energy region near the bandgap is plotted. The amplitudes of the bars represent the integrated absorption coefficient over the width of the bar.

IV. Exchange Splitting and Symmetry of the Crystallites

Although π-Si is a very heterogeneous material on a microscopic scale, some fine structures clearly appear in the excitation spectrum of the visible luminescence at 2 K (Calcott et al., 1993a; Suemoto et al., 1993) of some π-Si samples. In particular, an onset of a few millielectron volts and, at higher energy, structures associated with phonon-assisted transitions (Calcott et al., 1993a; 1993b) are observed. In addition, the lifetime of the visible luminescence decreases going from low (4 K) to higher temperatures (~ 100–200 K) in parallel with an increase of the luminescence intensity (Calcott et al., 1993a; Zheng, Wang, and Chen, 1992; Vial et al., 1993). Both effects have been interpreted on the basis of a "two-level model" resulting from the electron–hole exchange splitting, which although being smaller than 0.15 meV in bulk Si, could reach a few millielectron volts in π-Si because of the strong confinement (Calcott et al., 1993a; Zheng, Wang, and Chen, 1992; Fishman, Romestain, and Vial, 1993b).

Although the two-level model of excitons in nanocrystallites is appealing, it is by nature oversimplified since the exciton states are derived from degenerate valence and conduction bands. We have thus performed detailed calculations for the excitons, taking into account the manifold of hole and electron states, and their couplings due to Coulomb, exchange, and spin–orbit interactions (see, Martin et al., 1994).

As the first step, we need the one-electron states of the confined Si crystallites. This is done here using a tight-binding framework with the parameters of Tserbak, Polatoglou, and Theodorou (1993). Spin–orbit coupling is not yet included in this calculation. The dangling bonds at the surface are saturated by hydrogen atoms to avoid spurious localized states in the bandgap. The Hamiltonian matrix is diagonalized using an inverted Lanczos iteration procedure, which allows the treatment of large Si crystallites (up to ~ 2000 Si atoms) with arbitrary shape. The calculated energies are close to those discussed in Section II of Proot, Delerue, and Allan (1992); Delerue, Allan, and Lannoo (1993), which includes the overlaps between atomic orbitals. We do not use here the same method, because the treatment of the overlaps in the problems of excitons is computationally tricky and the tight-binding parameters of Tserbak, Polatoglou, and Theodorou (1993) give a band structure of reasonable quality. Optical matrix elements are obtained as in Proot, Delerue, and Allan (1992); Delerue, Allan, and Lannoo (1993), that is, without the assistance of phonons. This point will be discussed later.

As the second step, we calculate the excitonic spectrum. We write the exciton wavefunction as a linear combination of Slater determinants built from the one-electron states. The total Hamiltonian matrix is then expressed on the basis of all these Slater determinants and diagonalized. Actually, this is impossible even with the best available computers, because the dimension of the basis is too large. To reduce considerably the number of basis states we only take into account the Slater determinants corresponding to single electron–hole excitations. If $|\psi_i^c\rangle$ and $|\psi_j^v\rangle$ are, respectively, conduction and valence one-electron states of energy E_i^c and E_j^v, the exciton wave function Ψ_{exc} is written as

$$\Psi_{\text{exc}} = \sum_{i,j} a_{ij} |\psi_j^v \to \psi_i^c\rangle \qquad (3)$$

where $|\psi_j^v \to \psi_i^c\rangle$ is the Slater determinant corresponding to the excitation of one electron from the state $|\psi_j^v\rangle$ to the state $|\psi_i^c\rangle$ and a_{ij} are the variational parameters. Then we write the matrix elements of the total Hamiltonian H between the determinants. The Hamiltonian matrix can be obtained following simple rules:

1. The diagonal terms contain the one-particle excitation energies $E_i^c - E_j^v$, where E_i^c and E_j^v are the one-electron eigenvalues corresponding to $|\psi_i^c\rangle$ and $|\psi_j^v\rangle$.

FIG. 7. Two-level model for the recombination of excitons in π-Si (Calcott et al., 1993a; Calcott et al., 1993b). The lowest excitonic state is split due to the exchange interaction between the electron and the hole. The upper level has a much smaller lifetime than the lower one. Thermal equilibrium between the two levels could explain the temperature dependence of the radiative lifetime (Calcott et al., 1993b; Vial et al., 1993).

2. The diagonal and nondiagonal terms have two electron repulsion terms $e^2/(\varepsilon r_{ll'})$ (between electrons l and l') screened by the appropriate dielectric constant ε. Equivalently, they can be expressed as $e^2/(\varepsilon r_{eh})$ in a two-particle electron–hole formalism. This screening partially takes into account higher order excitations. We use $\varepsilon = 1 + 10.4/(1+(0.92/R)^{1.18})$ with R the average crystallite radius[1] in nanometer units, which accounts for the reduction of the dielectric constant due to the confinement (see Section VI).

Finally, the matrix of the total Hamiltonian is diagonalized by standard methods. We also take advantage of the strong confinement in crystallites. The splitting between the levels can be of the order of several tenths of an electron volt, which is large compared to the other couplings. Therefore, the expansion in Eq. (3) can be limited to a reasonable number of Slater determinants. In practice, we have found that the Slater determinants built from the 12 lowest spin states of the conduction band and the 12 highest spin states of the valence band are sufficient. Once the excitonic wave functions are known, we calculate the radiative lifetimes as in Proot, Delerue, and Allan (1992); Delerue, Allan, and Lannoo (1993).

Let us now discuss the validity of the two-level model (Fig. 7). In this model, the lowest exciton level is split into two levels as a result of the exchange interaction between the electron and the hole. The triplet state is the lowest one and the splitting is thought to be in the range of 10 meV because of confinement. At low temperature ($\lesssim 20$ K), only the triplet state is populated and the lifetime is long, in the millisecond range (it is not infinite because the spin–orbit coupling slightly mixes the $S = 0$ and $S = 1$ states). At higher temperatures, the singlet state becomes populated and the

---

[1] The effective dielectric constant being mainly related to the bandgap of the cluster (see Section VI), we use for an asymmetric crystallite of given bandgap the effective dielectric constant corresponding to the spherical cluster with the same bandgap.

radiative recombination is enhanced. The onset energy in selectively excited PL at 2 K has also been explained by the exchange splitting (Calcott et al., 1993a). At 2 K, the luminescence comes from the triplet state but excitons are predominantly photocreated in the singlet state, because the oscillator strength is inversely proportional to the radiative lifetime.

Wires with varying diameter were also suggested as the luminescent structures in π-Si on the basis of structural analysis. We have thus studied the excitonic spectrum of crystallites with spherical, ellipsoidal, and undulating ellipsoidal shapes. The ellipsoids are defined by $(x/a)^2 + (y/b)^2 + (z/b)^2 = 1$ where a is larger than b. This can be seen as a deformation of a sphere in one direction x. For the undulating ellipsoids, we introduce an angular fluctuation on the surface of the ellipsoid. This is done using a combination of spherical harmonic functions with arbitrary axes and arbitrary amplitudes. These complex shapes should simulate properly the undulating wires provided the undulations are large enough to localize the excitons.

In Fig. 8 we plot a typical low-energy excitonic spectrum calculated for a spherical crystallite including spin–orbit coupling. It is quite complex, with many levels, because the electron and hole states are derived from degenerate bands. It was shown in previous works that there are interactions between the conduction states of different minima and that the induced splitting (valley-orbit splittings) are of the order of 1–10 meV (Proot,

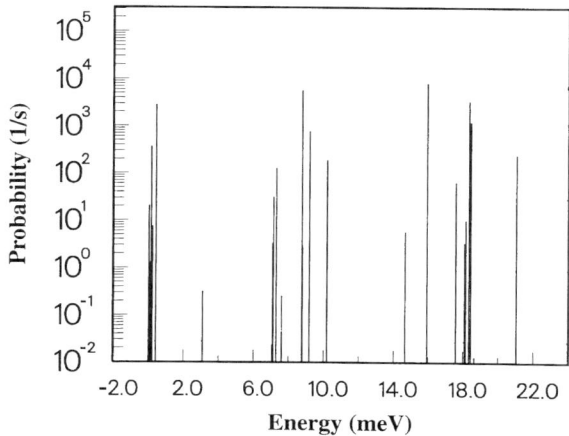

FIG. 8. Calculated excitonic structure of a spherical silicon crystallite of diameter 3.86 nm. The levels are indicated by vertical bars. The zero of energy corresponds to the lowest exciton level. The height of the bars represents the calculated radiative recombination rate (inverse of the radiative lifetime).

Delerue, and Allan, 1992; Delerue, Allan, and Lannoo, 1993; Hybertsen and Needels, 1993). The exciton wavefunctions built from these one-electron wavefunctions are also mutually coupled by the Coulomb interaction. Furthermore, we see that the recombination rate of the lowest states (<1 meV in Fig. 8) is not two or three orders of magnitude smaller than those of higher excitonic states, so the radiative lifetime only slightly depends on temperature. This point is verified on all the spherical crystallites even if the excitonic spectrum strongly depends on the crystallite size, because of the large variation in the ordering of the valley-orbit split levels. The main explanation for this comes from the spin–orbit coupling. If this is neglected, some levels have a finite lifetime and the others have an infinite one since the total spin ($S = 0$ and $S = 1$) is a good quantum number. The spacing of the levels is between 1 and 10 meV, which corresponds to the average valley-orbit splitting and the average exchange integral. On including the spin–orbit coupling, the singlet and the triplet state are completely mixed together, because this one is of the same amplitude as the spacing between the levels ($\lambda = 15$ meV) (Tserbak, Polatoglou, and Theodorou, 1993). As a consequence, all the levels have, on the average, similar lifetimes and the two-level model is not valid. This would be the same for cubic crystallites or any kind of crystallites in which the x, y, z axes are equivalent by symmetry. This result can be of interest from the perspective of making artificial Si nanostructures with well-controlled shapes.

For the ellipsoids, the degeneracy of the highest state in the valence band is lifted by the anisotropy. The same is obtained for wires using effective-mass calculations (Voos et al., 1992). The highest valence state behaves like p_x for a crystallite in the direction x. If the splitting between the p_x-like state and the others (p_y-like and p_z-like) is large compared to the spin–orbit coupling ($\lambda = 15$ meV (Tserbak, Polatoglou, and Theodorou, 1993), then the latter is quenched and the total spin S approximately remains a good quantum number. Because of the strong confinement, already for $a = b\sqrt{2}$ the anisotropy leads to a large splitting (>50 meV) between the p_x-like state and the p_y-p_z-like states. However, the excitonic spectrum remains quite complex, with several low-lying states having strongly varying lifetimes, because of the remaining degeneracy of the conduction states.

For the undulating ellipsoids, we fix the maximum amplitude of the fluctuation at 25% of the average radius of the ellipsoid. Note that the choice of another amplitude (e.g., 35%) gives similar results. We find that all the orbital degeneracies in the valence and conduction bands are lifted. The excitonic spectrum becomes much simpler with the lowest state having systematically a much longer lifetime than the first higher state (see e.g., Fig. 9). The two-level model then becomes valid to describe such anisotropic crystallites.

7 RADIATIVE AND NONRADIATIVE PROCESSES IN SILICON NANOCRYSTALLITES   267

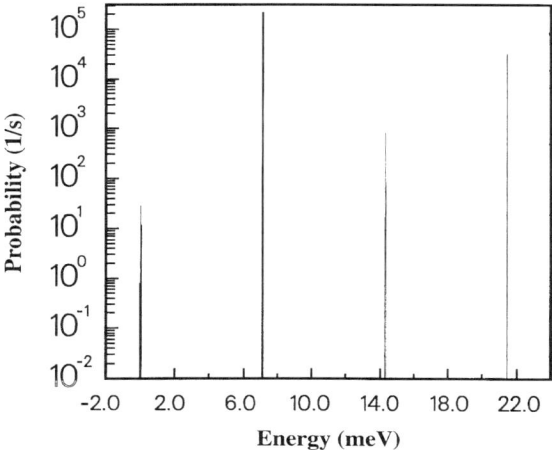

FIG. 9. Calculated excitonic structure of a silicon crystallite with complex shape built from an ellipsoid with a long axis of 2.4 nm, a short axis of 1.8 nm, and 25% of surface undulations. The levels are indicated by vertical bars. The height of the bars represents the calculated radiative recombination rate (inverse of the radiative lifetime).

In Fig. 10 we plot the calculated splitting Δ between the two lowest excitonic levels for undulating ellipsoids. All the crystallites are characterized by $a = b\sqrt{2}$ and a surface undulation of 25%. The trends in the results do not depend on the ratio between a and b, provided a is significantly larger than b. The calculation is done for different orientations of the long axis x of the ellipsoids: [100], [110], [111], and an average over all directions of space. We confirm that the coinfinement enhances the exchange splitting by two orders of magnitude. The [100] direction gives the smallest splitting and the [111] direction the largest one. All other directions of space are intermediate between these two curves. The explanation of this strong dependence on the direction of the main axis of the crystallite is detailed in Appendix B of Martin et al. (1994).

In Fig. 10 we compare the calculated splittings with the onset in selectively excited PL measured in Calcott et al. (1993a). If the experimental onset corresponds to the smallest exchange splitting among all the crystallites luminescing at the same energy, then we should compare the experimental data with the calculated splittings for the [100]-oriented crystallites. We see that there is a factor 3–5 of discrepancy. The agreement is slightly better for the [111]-oriented crystallites, but to our knowledge there is no experimental indication that crystallites are predominantly oriented in the

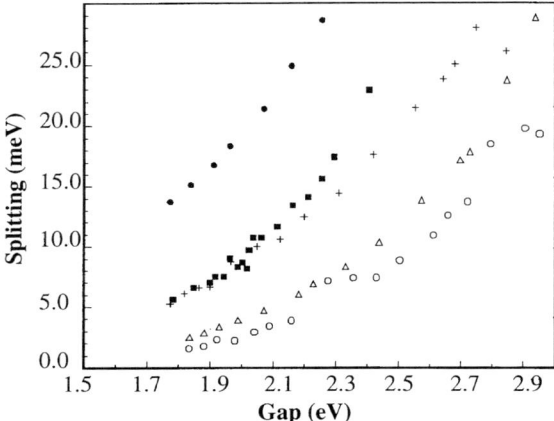

FIG. 10. Splitting between the two lowest calculated excitonic levels in asymmetrical silicon crystallites with respect to their excitonic gap. Crystallites have undulating ellipsoidal shapes with a longer axis in the [100] direction (○), [110] direction (△), and [111] direction (+). (×) correspond to the average of the splitting over all the orientations of the longer axis of the crystallite. (■) are the first onsets measured by selectively excited PL and (●) are the energy splittings derived from the fit of the temperature dependence of the luminescence lifetime (Calcott et al., 1993b). (●) at (1.8 eV, 10 meV) is from Vial et al. (1993).

⟨111⟩ directions. The difference with the splittings deduced from the fit of the temperature dependence of the lifetime is even more important. We can note that there is already a large difference between the values at ∼1.8 eV measured in Calcott et al. (1993a); Vial et al. (1993). We conclude that the onset of the selectively excited PL cannot be explained by the exchange splitting alone, although we have confirmed that this effect may be present in π-Si.

Before discussing the large discrepancy between theory and experiments, let us compare our calculation with other theoretical work. Published calculations (Calcott et al., 1993a; Vial et al. (1993) are based on the EMA in which the exchange splitting is written as $\Delta = J \int |\varphi(r, r)|^{2'} d^3 r$, where $\varphi(r_e, r_h)$ is the envelope function for the exciton and J is defined as twice the exchange integral per unit inverse volume for the conduction-band-minimum and valence-band-maximum states in bulk Si. This simple approach is sufficient to show that the exchange interaction increases with the confinement. However, the determination of J is difficult. In Calcott et al. (1993a), J is indirectly obtained from the exchange splitting of the Ga-2 isoelectronic donor in bulk Si; in Vial et al. (1993); Fishman, Romestain, and Vial (1993b), it is derived from the experimental knowledge of the higher limit for the exchange splitting of the free exciton in bulk silicon (<0.15 meV)

(Merle et al., 1978). In addition, these experimental values are probably influenced by the effect of the spin–orbit coupling in the same manner as discussed above and, therefore, cannot be easily and directly obtained from experiments.

The calculated lifetimes associated with the two lower levels of the undulating ellipsoids exhibit an important scattering as explained in Martin et al. (1994), which is consistent with the highly nonexponential decay of the luminescence (Vial et al., 1992) (this nonexponential character can also come from other factors). The difference of two or three orders of magnitude between the lifetimes of the upper and lower levels is in good agreement with experiments (Calcott et al., 1993a). However, our calculated lifetimes are between one and three orders of magnitude larger than the experimental ones, confirming our previous conclusions. The difference is particularly important for the smaller bandgaps. Of course, nonradiative recombination may be important, even if it seems to occur mainly at high temperature (>100 K) (Vial, et al., 1992; Vial et al., 1993). Radiative recombination also occurs with the assistance of phonons like in bulk Si, as discussed in Section III (Hybertsen, 1994), in agreement with experiments. In that case, the influence of the exchange splitting remains because it does not depend on the mechanism of the radiative recombination.

In Xie et al. (1994) substantial differences in the low temperature decays of the luminescence were reported between the π-Si samples and those of Calcott et al. (1993a) (this was also reported for dispersed Si nanocrystallites (Wilson, Szajowski, and Brus, 1993). In symmetrical crystallites, we predict that the effect of the exchange splitting vanishes, because of the mixing introduced by the spin–orbit coupling. It would be interesting to look at the differences in morphology between these samples as suggested in Xie et al. (1994). In the same spirit, anisotropic elastic strains that are present in π-Si can also lift the degeneracies, playing the same role as the anisotropy of the confinement.

Finally, one must mention recent calculations (Nash et al., to be published) showing that the influence of the spin–orbit coupling in asymmetrical crystallites (or wires) is weak but not negligible, as discussed above. In particular, because of this small influence, the spin–spin splittings observed by optically detected magnetic resonance (ODMR) (Brandt and Stutzmann, 1995) are found compatible with the confinement model in contrast to the conclusions of Brandt and Stutzmann (1995).

V. Atomic Relaxation, Stokes Shift, and Self-trapped Exciton

We want to discuss here the difference in transition energies between absorption and luminescence. One likely origin is the electron–phonon interaction which leads to lattice reorganization in the excited state. We shall consider two situations: (1) transitions between delocalized states and (2) the existence of self-trapped excitons.

1. STOKES SHIFT FOR THE DELOCALIZED STATES

In the following, we show that the lattice relaxation in the excitonic state must be substantial. To understand its influence on selectively excited PL, let us suppose that the electronic system couples to one lattice displacement coordinate Q and the equilibrium in the ground state corresponds to $Q = 0$. The situation is summarized in the configuration coordinate diagram (see Fig. 11), which represents the total energy of a particular crystallite with respect to Q. Since the selectively excited luminescence is measured at low temperature ($<2\,\text{K}$), only the lowest vibronic state is populated in the ground state. Under laser excitation, the maximum of absorption occurs for the vertical transitions (see arrows (i) in Fig. 11). Then the system relaxes to its lowest vibronic state, which is centered at $Q = Q_0$. Finally, the radiative

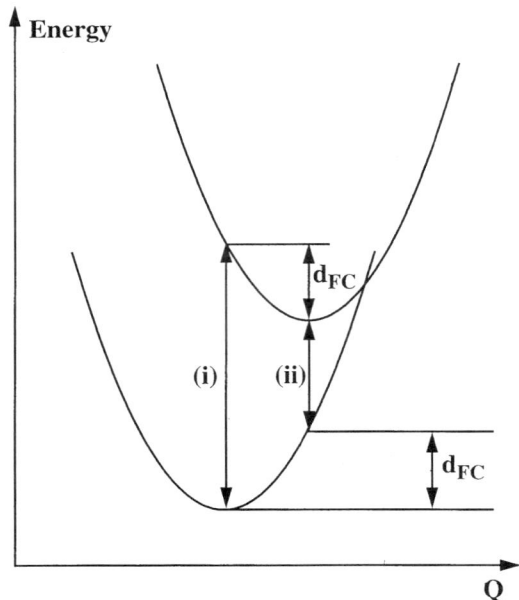

FIG. 11. Configuration coordinate diagram representing the variation of the total energy versus the lattice-displacement coordinate Q. The lowest curve corresponds to the system in its ground state, the upper curve in the excitonic state.

recombination appears with a maximum probability at the vertical transition (see arrows (ii) in Fig. 11). In this simplified view, if the luminescence energy is equal to E_l, the excitation energy is $E_{abs} = E_l + 2d_{FC}$ where d_{FC} is the Franck–Condon shift equal to the energy gain due to lattice relaxation after capture. Therefore, the onset in the selectively excited luminescence (the Stokes shift) would be equal to $2d_{FC}$. Of course, the situation is actually more complex, with transitions (in excitation or luminescence) to the difference vibronic states of the final states. In the case of a strong lattice coupling, this leads to a gaussian broadening of the lines centered at vertical transitions.

The lattice relaxation occurs in the excitonic state because an electron is transferred from a bonding-like (valence) to an antibonding-like (conduction) state, which tends to weaken the bonds. The local amplitude of the distortion is directly connected to the electron–hole density. The confinement increases this density and, therefore, the Stokes shift is enhanced by the confinement. Variation of the excitonic energy gap with respect to the lattice deformations is described by the deformation potentials.

We have used a simple model described in Martin *et al.* (1994) using these deformation potentials and based on standard elasticity theory to show that d_{FC} could be substantial in nanostructures. The atomic displacements are obtained by minimization of the total energy. The exciton energy depends on the local elastic deformations through the deformation potentials, which are taken from measurements on π-Si. The Franck–Condon shift d_{FC} is calculated as the difference in energy between the stable situation of the ground and excited (excitonic) states of the crystallite. Results for d_{FC} are somewhat uncertain due to the scatter in the experimental data for the deformation potentials. Anyway, we find small but substantial values for d_{FC}, for example, of order 25 meV for a gap of 2.6 eV. Comparison with Fig. 10 shows that the Franck–Condon shift could explain at least partially the difference between the onset observed in the selectively excited PL and the calculated exchange splitting.

2. The Existence of Self-trapped Excitons

In spite of the numerous calculations performed so far, the actual gap of Si crystallites is not yet known with certainty. This is illustrated in Fig. 12, which presents a compilation of values due to Lockwood (1994) comparing optical absorption and luminescence.

The large difference between the two curves would correspond to a huge Stokes shift (~ 1 eV for a crystalline diameter $d \sim 1.5$ nm), much larger than the predicted values discussed in Section V.1 (< 100 meV). In fact, as shown

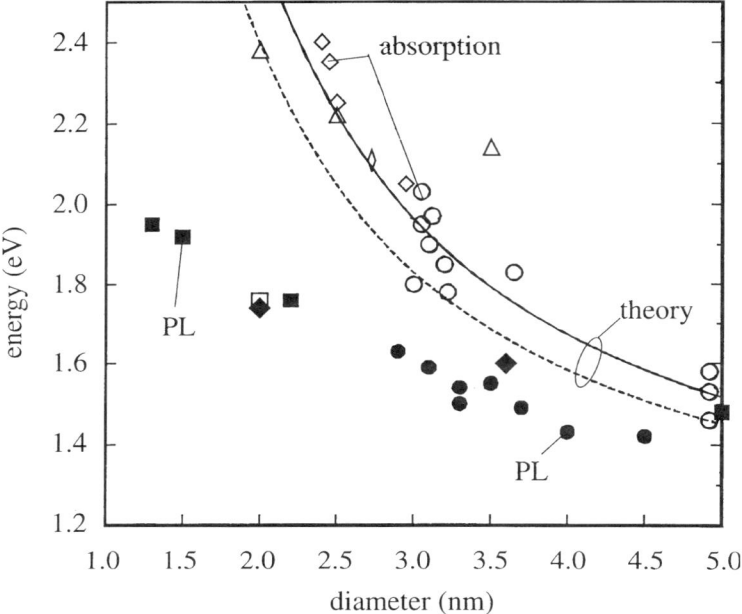

FIG. 12. Compilation (Lockwood, 1994) of optical bandgaps of silicon crystallites and π-Si samples obtained from optical absorption (empty symbols) and PL (full symbols). Straight line: one-electron bandgap calculated for spherical silicon crystallites (Proot et al., 1993). Dashed line: the same but including the excitonic binding energy of Eq. (17).

in Fig. 12, optical absorption energy gaps are in excellent agreement with calculated values for crystallites. Only the luminescence energies differ greatly and, for small crystallites, are practically independent of the size, while predicted values behave as $d^{-1.45}$ (Proot, Delerue, and Allan, 1992; Delerue, Allan, and Lannoo, 1993). Such a behavior is more consistent with the eventual existence of deep luminescence centers such as the "surface" states postulated by Koch et al., 1993) and Kanemitsu (1994). The problem with such possible surface states is that nothing is presently known regarding their nature and origin. The only states that have been really identified are "dangling bond" states at the surface which will be discussed in the next section.

We shall thus investigate here the possibility of the existence of intrinsic surface states that might behave as luminescent systems. We shall see, both from empirical tight binding and first principle local density calculations, that such states indeed exist in the form of "self-trapped excitons". This possibility is not restricted to the case of Si crystallites but, from general considerations discussed in the following, is likely to be valid for all types of semiconductor crystallites (see Allan, Delerue, and Lannoo, 1996).

To illustrate the physical basis of such self-trapped excitons let us consider an isolated single covalent bond. This bond is characterized by a σ bonding state filled with two electrons and an empty σ^* antibonding state (Fig. 13(a)). The origin of the binding is the gain in energy resulting from having the two electrons in the lower bonding state. Optical absorption in this system leads to the situation in Fig. 13(b), with essentially no binding. In such a case the repulsive force between the atoms dominates, so that the molecule eventually dissociates. If, on the other hand, the molecule is embedded in an elastic medium, then it cannot dissociate, but one ends up with a large distance between the constituent atoms and a much reduced separation between the σ and σ^* states (Fig. 13(c)). The resulting luminescence energy is thus much smaller than the optical absorption energy, corresponding to a Stokes shift of the order of the binding energy, that is, ~ 1 eV.

The applicability of this model to a Si crystallite essentially depends on the possibility of localizing the electron–hole excitation on a particular covalent bond, that is, of creating a self-trapped exciton. For this, one must be able to draw a configuration coordinate diagram like the one shown in Fig. 14 where the configuration coordinate Q corresponds to the stretching of the covalent bond. For small Q the ground and first excited states are delocalized over the crystallite and show a normal parabolic behavior. However, for Q larger than a critical value Q_c the system localizes the electron–hole pair on one particular single bond, leading to a larger bond length Q_e and a smaller luminescence energy. This self-trapped state can be stable or metastable. An interesting point is that it may exist only for small

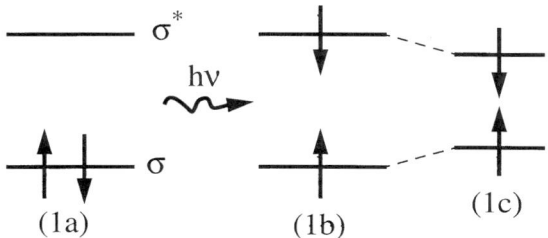

FIG. 13. Bonding (σ), antibonding (σ^*) splitting: (a) in the normal state, (b) in the "normal" excited state with unchanged bondlength, and (c) in the relaxed excited state.

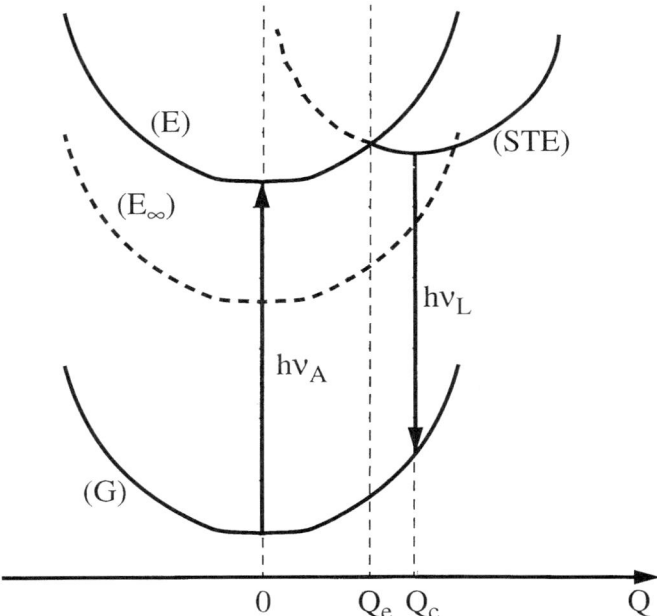

FIG. 14. Schematic configuration coordinate diagram showing the energies of the ground state (G), the normal excitonic state (E), and the self-trapped exciton state (STE). The curve (E_∞) corresponds to a very large crystallite with no blue shift, showing that the STE state might not exist for large crystallites.

enough crystallites, in view of the important blue shift, as shown in Fig. 14. Such a self-trapped exciton is likely to be favored at surfaces of crystallites where the elastic response of the environment is likely to be weaker than in the bulk. In the following we show that self-trapped excitons provide a sound physical basis for a new origin of the luminescence of Si nanocrystallites.

For the first calculation, we use a total energy semi-empirical tight binding (TB) technique which allows the treatment of quite large crystallites (~180 Si atoms). The Hamiltonian includes interactions up to the second neighbors and the total energy is the sum of one-electron energies plus repulsion terms between first and second neighbors. The parameters (Vinchon, Spanjaard, and Desjonqueres, 1992) of the system are fitted on the band structure, the lattice parameter, the elastic constants and the cohesive energy of bulk Si. The calculation reproduces the reconstruction of the (2 × 1) 100 Si surface. Details can be found in Vinchon, Spanjaard, and

7 RADIATIVE AND NONRADIATIVE PROCESSES IN SILICON NANOCRYSTALLITES

Desjonqueres (1992). The second calculation technique is based on an *ab initio* local density calculation using the DMol code (DMol). Because of computation limits, clusters studied by LDA are restricted to a maximum of ~30 Si atoms. This is not a severe restriction since we are interested here in localized surface states. With the two techniques, the total energy is minimized to get stable atomic configuration for the ground and first excited states. For the first excited state, the occupation of one-electron states (TB or LDA eigenstates) is chosen to correspond to a single particle excitation.

The first interesting situation is obtained when stretching the Si–Si bond of a surface dimer whose dangling bonds are saturated by hydrogen atoms (Fig. 15). Then the stable situation for the excited state corresponds to the surface Si atoms being almost returned to their original lattice sites (Fig. 15). The electron and the hole are localized on the weakly interacting Si dangling bonds (second nearest neighbors) which form bonding and antibonding states whose gap is equal to 0.52 eV in the TB calculation for a 1.67 nm crystallite (123 Si atoms). As expected for a localized state, this bandgap only slightly depends on the crystallite size: we obtain values between 0.5 and 1.0 eV depending on the local configuration of the dimer. These results are totally confirmed by the LDA calculation on a small cluster with 12 Si atoms, which has a somewhat larger bonding–antibonding gap of 0.99 eV. The physics underlying the trapping mechanism is the same as discussed above. The calculated configuration coordinate diagrams given in Figs. 16 and 17 fully correspond to the general schematic picture of Fig. 14. It is interesting to note that TB and LDA (Fig. 16) calculations give fairly similar

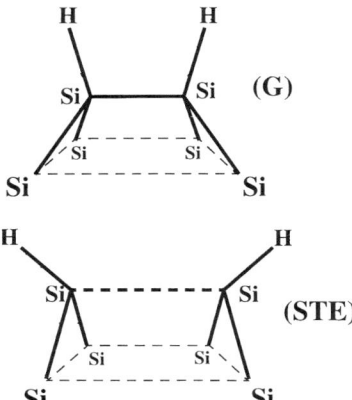

FIG. 15. Schematic side views of the cluster surface dimer in the ground state (G) and in the self-trapped state (STE). The distance d between the dimer atoms is close to the bulk nearest neighbor interatomic distance (G) and the next nearest neighbor one (STE).

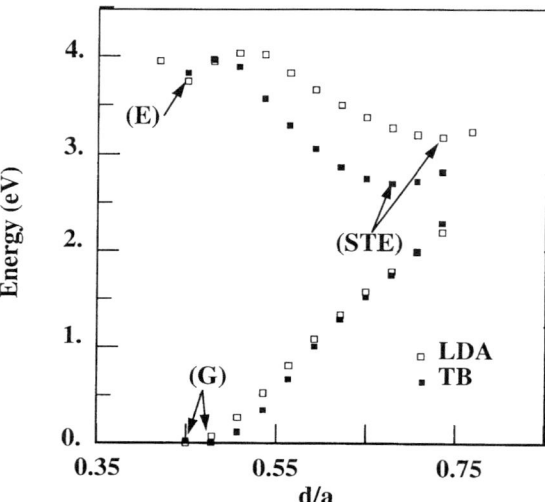

FIG. 16. Empty symbols: energy of a crystallite with 12 silicon atoms in the ground state and in the excitonic state as a function of the dimer interatomic distance d ($a = 0.54$ nm) obtained from the local density calculation. Full symbols: the energies are calculated with the TB technique for the cluster with the same atomic positions for comparison with the LDA results.

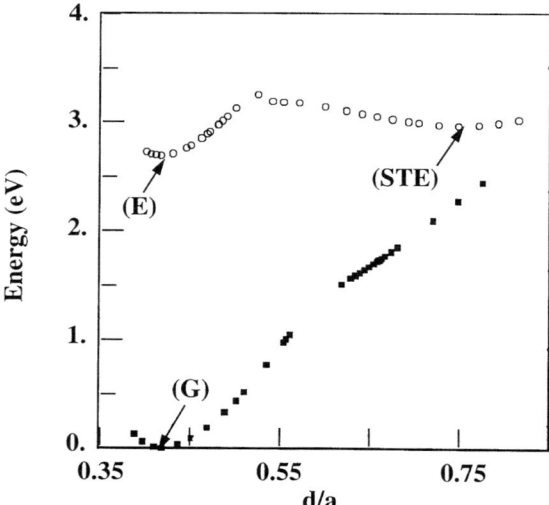

FIG. 17. Total energy (TB technique) of a spherical crystallite with 123 silicon atoms in the ground state and in the excitonic state as a function of the dimer interatomic distance d ($a = 0.54$ nm). The arrows indicate the energy minima.

curves for this diagram, especially as regards the flatness of the energy surface around the self-trapped exciton minimum. Note that this self-trapped state can even be a stable situation for the excited state of very small crystallites.

Dimers are also expected at the Si–SiO$_2$ interface. We have studied using LDA (reliable TB parameters are not available for Si–SiO$_2$) the cluster of Fig. 18, which corresponds to an atomic configuration expected at the Si(001)–SiO$_2$ interface (Pasquarello, Hybertsen, and Car, 1995). To account for the rigidity of the remaining medium, only the two Si atoms of the dimer and their oxygen neighbors are allowed to relax. Again, we find that the exciton can be trapped on the dimer. Because of the constraints due to the oxide chain, the dimer cannot relax as easily as the hydrogenated surface: we calculate the Si–Si dimer distance to be 3.24 Å in the self-trapped state and an optical gap of 1.51 eV. The exciton recombination would then probably be radiative in the near visible.

A third interesting situation is obtained when considering a Si–Si bond inside a crystallite (Fig. 19). For small crystallites, we obtain stable situations for the excited state corresponding to stretched Si–Si bonds (Fig. 19). Then three backbonds of each Si atom are nearly in the same plane. With TB (LDA), we obtain a typical Si–Si distance of 3.2 Å (3.0 Å) in the self-trapped state and an optical gap of 1.3 eV (2.2 eV). From the calculated configuration coordinate diagram of Fig. 20, we see that such a self-trapped state is only stable in very small crystallites with a bandgap larger than ~2.6 eV.

One can make some general statements about the conditions favoring the existence of such self-trapped states for a given bond. Among these we can list:

1. The elastic response of the environment should be as weak as possible. This is best realized near surfaces, for example, for Si–H where

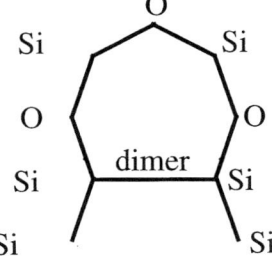

FIG. 18. Schematic view of part of the cluster used for the calculations at the crystallite-SiO$_2$ interface (see text).

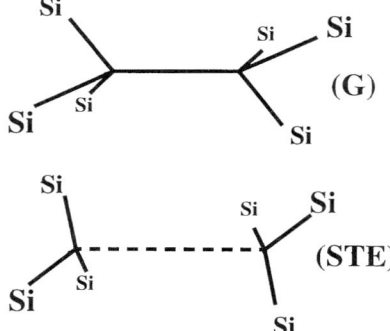

FIG. 19. Schematic views of a cluster Si–Si bond in the ground state (G) and in the self-trapped state (STE).

displacing the hydrogen does not cost any elastic energy. For obvious reasons, this is more difficult for "normal" Si–Si bonds and, in that case, self-trapped states are only possible in small crystallites.
2. The size of the crystallite should be as small as possible, favoring a large blue shift and the stabilization of locally distorted excited states.
3. The capture of the exciton is easier when it allows the release of local stresses. This is the case of the Si–Si dimer where the stresses correspond to the bending of the backbonds in the normal state.

Such self-trapped states are likely to be metastable in most cases. The question then arises as to if and how they can be excited. One answer is provided by the well documented example of the EL2 defect in GaAlAs, which exhibits a similar configuration coordinate diagram and can be optically excited with a long lifetime (Vincent and Bois, 1978). Here, extrapolation of the energy curves of the self-trapped states in Figs. 17 and 20 suggest that a vertical transition energy of ~ 3.5 to ~ 5.0 eV would be necessary to excite them directly from the ground state. Once in the metastable state there is also the possibility that the system returns to the normal excited state by thermal excitation over the barrier. This process would thus be in competition with direct radiative recombination from the self-trapped state.

In conclusion, the total energy calculation in this section shows the existence of self-trapped excitons in Si crystallites. These give a luminescence energy almost independent of size and explain the huge Stokes shift observed for small crystallites. They give sound support to the model of intrinsic surface states and reconcile it with the effect of quantum confinement necessary for a description of the optical absorption data. In fact the

7 RADIATIVE AND NONRADIATIVE PROCESSES IN SILICON NANOCRYSTALLITES

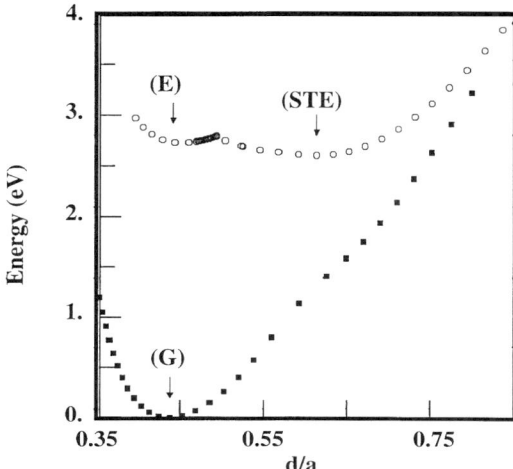

FIG. 20. Energy of the crystallite in the ground state and in the excitonic state as a function of the bond interatomic distance d ($a = 0.54$ nm) obtained with the TB calculation (cluster with 123 silicon atoms).

luminescence could originate both from self-trapped states and from the direct transition between the conduction and valence states. This could explain the observation of the phonon structures in the excitation spectrum of the PL under resonant conditions (Calcott, et al., 1993a; Rosenbauer et al., 1995) in which it is not possible to excite directly metastable self-trapped states. Finally, such self-trapped excitons are not specific to Si crystallites and should also manifest themselves in crystallites obtained from other semiconductors.

VI. Nonradiative Recombination

We consider here two possible channels for nonradiative recombination: surface defects like dangling bonds and the intrinsic Auger recombination. We show that both effects allow for detailed interpretation of some experimental data.

1. RECOMBINATION ON SURFACE DANGLING BONDS

The degradation of the luminescence is correlated with an increase in the density of dangling bonds (Prokes, Carlos, and Bermudez, 1992; Brandt and

Stutzmann, 1992) due, for example, to hydrogen desorption. The adsorption of oxygen induced by light is also a cause for light emission degradation (Brandt and Stutzmann, 1992; Robinson et al., 1992; Xu, Gal, and Gross, 1992), but this might be due to formation of a thin layer of oxide. Indeed, this process is known to introduce many dangling bonds, which in this case are not as easily passivated by hydrogen as during anodic oxidation (Poindexter and Caplan, 1983). Dangling bonds are expected to be efficient nonradiative recombination centers, but a quantitative estimation of their influence is needed. From previous work on the $Si-SiO_2$ interface the electron–hole recombination on Si dangling bonds is known to be due to the multiphonon capture of the electron and hole (Goguenheim and Lannoo, 1990). We discuss here a calculation of the corresponding capture rate based on what is known both theoretically and experimentally for dangling bonds at the $Si-SiO_2$ interface (see, Proot, Delerue, and Allan, 1992; Delerue, Allan, and Lannoo, 1993)).

Dangling bonds correspond to coordination defects in which the Si atom has only three equivalent covalent bonds. The best known case is the P_b center at $Si(111)-SiO_2$ interface. Such defects are also expected to occur at the surface of crytallites in π-Si as was demonstrated by electron paramagnetic resonance (Brandt and Stutzmann, 1992; Von Bardeleben et al., 1993). The P_b center can exist in three charge states $+, 0, -$ with ionization levels $\varepsilon(-, 0) = \varepsilon_c - 0.3$ eV and $\varepsilon(0, +) = \varepsilon_v + 0.3$ eV, where ε_c and ε_v are the bulk Si band edges (Poindexter and Caplan, 1983; Goguenheim and Lannoo, 1991). To transpose these properties to the case of Si crystallites one can apply the following arguments: (i) Deep gap states are fairly localized in real space and only experience the local situation. Energy should not be shifted by the quantum size effect. (ii) Valence and conduction states of the crystallite exhibit a blue shift $\Delta E_{c,v}$ due to confinement.

The situation we want to discuss now corresponds to a crystallite with an electron–hole pair and a neutral dangling bond at the surface. The electron–hole recombination on the dangling bond can be seen as a two step process: first a carrier is captured by the neutral dangling bond, then a second carrier is captured by the charged dangling bond. Cross sections corresponding to the capture of an electron or a hole by a neutral Si dangling bond at the $Si-SiO_2$ interface are measured in the $10^{-14}-10^{-15}$ cm^2 range at 170 K (Goguenheim and Lannoo, 1991). Cross sections for the capture by a charged dangling bond are not known experimentally. Therefore we first concentrate on the capture of a carrier on a neutral dangling bond. The capture cross section has a thermally activated behavior which is usually approximated by (Henry and Lang, 1977)

$$\sigma \sim \exp\left(-\frac{E_b}{kT}\right) \quad (4)$$

with E_b equal to the barrier height

$$E_b = \frac{(E_0 - d_{FC})^2}{4 d_{FC}} \quad (5)$$

where E_0 is the ionization energy of the defect and d_{FC} is the Franck–Condon shift equal to the energy gain due to lattice relaxation after capture (Goguenheim and Lannoo, 1990). d_{FC} is related to the phonon energy $h\nu$ by $d_{FC} = Sh\nu$ where S is the so-called Huang–Rhys factor. It can be shown that Eq. (4) is valid only under restrictive conditions that are fulfilled in the case of the dangling bond in bulk Si (Goguenheim and Lannoo, 1990): strong electron–phonon coupling ($S \gg 1$), high temperature, and $E_0 \sim d_{FC}$. The fact that the Franck–Condon shift is close to the ionization energy in bulk Si ($E_0 \sim d_{FC}$) explains why the cross section is weakly thermally activated (Goguenheim and Lannoo, 1991). This situation is summarized on the configuration coordinate diagram of Fig. 21, which is valid both for the capture of a hole or an electron by a dangling bond (energies are similar for the two processes (Goguenheim and Lannoo, 1991).

Because of quantum confinement, ionization energy E_0 in crystallites is different frrom its bulk Si value. The dangling bond state is fairly localized and its energy will remain constant on an absolute energy scale when the confinement is varied. Thus the change ΔE in the ionization energy E_0 is due to the shift of the band edges. We have seen above that the shift is not the same for the hole and for the electron but the difference remains small. For example it is close to 0.3–0.4 eV (for the hole and electron) for a blue shift of 0.7 eV. On the contrary, the Franck–Condon shift is unaffected by the confinement because it only depends on the local atomic relaxation (Fig. 21).

From Eq. (4) we expect a strong decrease of the cross section with ionization energy and, therefore, with confinement. However to estimate this change, Eq. (4) is no longer valid because condition $E_0 \sim d_{FC}$ is not valid any more when ΔE is large. We are also interested in the dependence of cross section over a wide range of temperatures for which Eq. (4) is not accurate enough. To improve on this we have made use in Proot, Delerue, and Allan (1992); Delerue, Allan, and Lannoo (1993) of a recently proposed analytic expression of the capture coefficient which remains valid over the whole temperature range, for any ionization energy, and for any strength of the coupling between lattice and defect (Goguenheim and Lannoo, 1990). The capture cross section σ is written as $\sigma_0 R$ where σ_0 is a coefficient whose dependence on T and E_0 is weak and R is a dimensionless function which

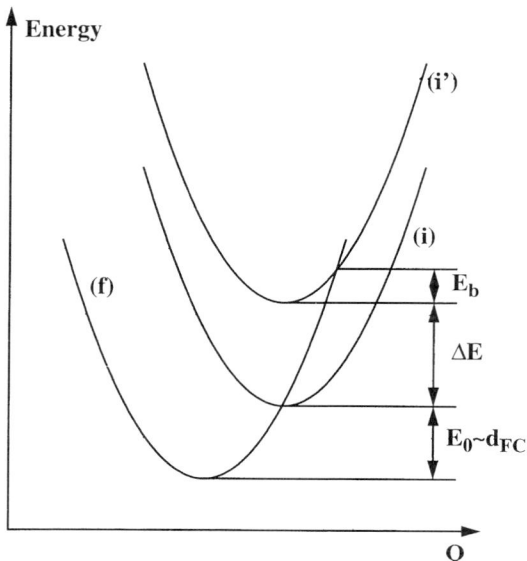

FIG. 21. Configuration coordinate diagram representing the variation of the total energy versus the atomic displacement for two charge states of the defect (initial (i) and final (f)). Two initial states are indicated; one (i) in bulk silicon (ionization energy E_0) and the other (i') in a silicon crystallite (ionization energy $E_0 + \Delta E$). The situation in bulk silicon corresponds to a negligible barrier for capture. In silicon crystallites, the increase in ionization energy creates a barrier E_b for the recombination (in a classical picture).

generalizes Eq. (4) in terms of the same parameters. R is given by[2] (Goguenheim and Lannoo, 1990)

$$R = \frac{1}{\sqrt{2\pi}}\left(\left(\frac{E_0}{h\nu}\right)^2 + z^2\right)^{-1/4} \exp\left\{-S\coth\left(\frac{h\nu}{2kT}\right) + \frac{E_0}{2kT} + \left(\left(\frac{E_0}{h\nu}\right)^2 + z^2\right)^{1/2}\right.$$

$$\left. - \frac{E_0}{h\nu} a\sinh\left(\frac{E_0}{h\nu z}\right)\right\} \quad (6)$$

with

$$z = \frac{S}{\sinh\left(\frac{h\nu}{2kT}\right)}.$$

[2]There is a misprint in the last term of Eq. (29) of Goguenheim and Lannoo (1990). The correct equation is Eq. (6) of the present paper.

We evaluate $\sigma(T)$ from values known for the P_b center and from the calculated blue shift, to obtain E_0. We take the cross section equal to 10^{-15} cm^2 at 170 K which is a lower limit of the measured values (Goguenheim and Lannoo, 1991). This is valid for capture of electrons and holes, because the measured values are quite close for the two processes at the neutral center. We also take $hv = 20$ meV and $S = 15$ (Goguenheim and Lannoo, 1991). The dependence of σ on the ionization energy is very strong, over several decades. In the approximate classical terms of Eq. (4), this is due to increase of energy barrier for carrier capture with ionization energy. Dependence of σ on temperature is weak when the shift in ionization energy is small, because energy barrier is negligible. But when ionization energy increases, dependence becomes important.

We now compare the nonradiative capture rate W due to a single neutral dangling bond in a crystallite with intrinsic radiative recombination. In Fig. 22 we plot W with respect to the energy gap in the crystallites at

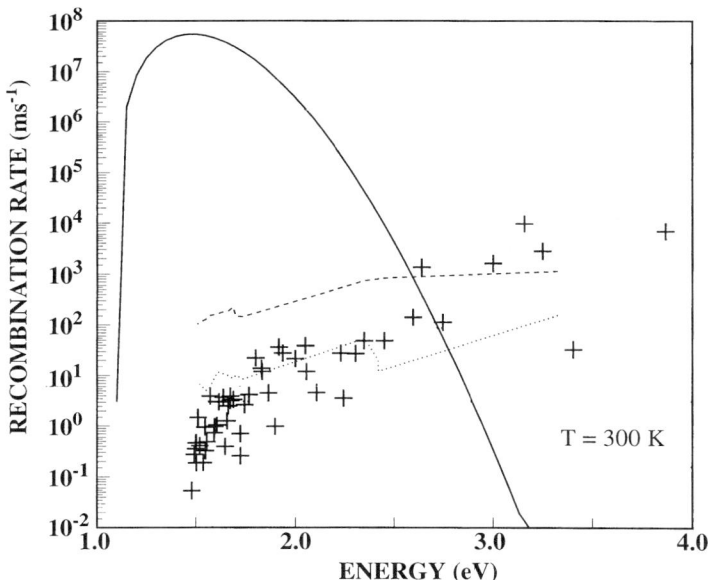

FIG. 22. Capture rates ($T = 300$ K) of an electron or a hole in silicon crystallites due to a nonradiative capture on a single neutral silicon dangling bond plotted with respect to the excitonic bandgap energy of the crystallites (continuous line). Crosses give the radiative recombination rates of the electron–hole pairs in the same crystallites. The other curves are the radiative capture rates of carriers on a neutral dangling bond (hole capture: dashed line and electron capture: dotted line).

$T = 300$ K. For comparison we also show calculated radiative recombination rates (see Fig. 4). W decreases at high energy because of the increase in the barrier. At energies close to the bulk bandgap, W also decreases very quickly because the volume of the corresponding crystallite tends to infinity and the probability to be captured by a single dangling bond vanishes. For photon energies in the range of interest for the visible luminescence of π-Si (1.4 eV–2.2 eV), the nonradiative capture by neutral dangling bonds is much faster than the radiative recombination, particularly at $T = 300$ K. We deduce that the presence of one neutral Si dangling bond at the surface of a crystallite in π-Si kills its luminescence above 1.1 eV, in agreement with experiments (Meyer et al., 1993).

We have also calculated the radiative capture rate of an electron and a hole by a single neutral dangling bond in the Si crystallites previously investigated. We first calculate the electronic structure of the crystallites with one Si dangling bond at the surface. This is simply done by removing one hydrogen atom of the crystallite. The radiative capture rates are calculated using Eq. (1). Results are plotted in Fig. 22 (hole capture: dashed line, electron capture: dotted line). We see that on average the radiative capture time is between 1 and 10 μs for the capture of a hole and between 10 and 100 μs for the capture of an electron. Therefore for crystallites with an optical bandgap lower than 2.2 eV, the nonradiative capture is much faster than the radiative capture. For crystallites with higher optical bandgap, the situation can be inverted depending on the temperature. In that case we see in Fig. 22 that the intrinsic radiative recombination and the radiative capture have comparable efficiency.

From the above discussion, we conclude that, for crystalites with optical bandgap lower than 2.2 eV, the presence of a neutral dangling bond leads to the nonradiative capture of the electron or the hole (more probably the electron because the confinement energy is slightly lower for the electron than the hole and therefore the capture barrier is smaller). This means that the other carrier remains in the conduction band or the valence band. We now discuss its capture by the charged dangling bond. The situation is summarized in the configuration coordinate diagram of Fig. 23. The lower energy curve corresponds to the ground state of the crystallite with one neutral dangling bond at the surface. The higher curve describes the system after optical excitation of an electron in the conduction band. The intermediate curve is the total energy after capture of the first carrier on the dangling bond. These two higher curves are equivalent to those labelled (i') and (f) in Fig. 21. The vertical shift ΔE is the confinement energy. It is equal to the confinement energy ΔE_c of the conduction band with respect to the bulk Si conduction band in the case where the first particle to be captured is an electron. It is equal to ΔE_v in the case of the capture of the hole. E_{b1}

7 RADIATIVE AND NONRADIATIVE PROCESSES IN SILICON NANOCRYSTALLITES

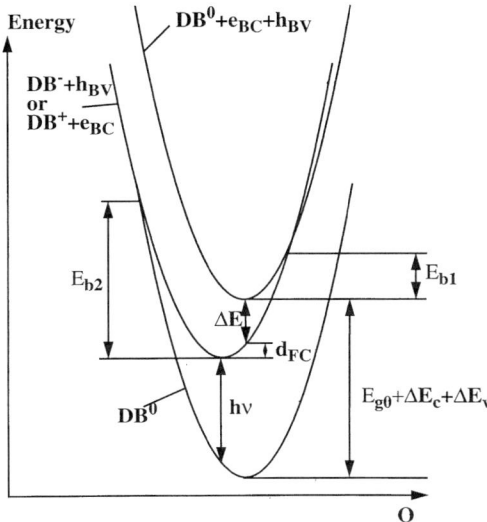

FIG. 23. Configuration coordinate diagram representing the variation of the total energy of a crystallite with one dangling bond at the surface. The ground state (lower curve) corresponds to filled valence states, empty conduction states and the dangling bond in the neutral charge state (DB^0). The higher curve (equivalent to curve (i') in Fig. 21) represents the same system after excitation of an electron in the conduction band leaving a hole in the valence band ($DB^0 + e_{BC} + h_{BV}$). The intermediate curve (equivalent to curve (f) in Fig. 21; the curve (i) in Fig. 21 is not reproduced here) describes the system after capture of a carrier by the dangling bond. Two situations are possible: the capture of the electron, the hole remaining in the valence band ($DB^- + h_{BV}$) or the capture of the hole, the electron remaining in the conduction band ($DB^+ + e_{BC}$). ΔE is the energy shift of the conduction or the valence band due to confinement compared to the bulk silicon band structure (conduction band for the capture of the electron, valence band for the capture of the hole).

is the energy barrier (from a classical point of view) for the capture of the first carrier, which has been analyzed above. E_{b2} is the energy barrier for the multiphonon capture of the second carrier, which we want to discuss now. Compared to the first capture, the second one involves much larger energies (the sum of the thermal ionization energies is equal to the bandgap energy). From Fig. 23 and Eq. (5), we can calculate the energy barrier E_{b2}. We obtain

$$E_{b2} = \frac{(E_{g0} + \Delta E - 2d_{FC})^2}{4d_{FC}} \quad (7)$$

where E_{g0} is the bulk Si bandgap energy and ΔE is the confinement energy of the band corresponding to the carrier involved in the second capture (ΔE_c for the electron, ΔE_v for the hole). For a confinement energy ΔE of 0.3 eV, E_{b2} is equal to 0.53 eV which is a very large barrier for a multiphonon capture. Therefore, the capture cross section for the second carrier should be strongly reduced compared to the first one (inserting the appropriate values in Eq. (6) gives a reduction factor of 3.10^{-7} at $T = 300$ K and 5.10^{-11} at $T = 10$ K). This means that, due to the large barrier E_{b2}, the second capture may become a radiative process. The average energy of the emitted photon ($h\nu$ in Fig. 23) should be equal to $E_{g0} + \Delta E - 2d_{FC}$, which is about 0.8 eV for a ΔE of 0.3 eV. This is just the range of values for the infrared emission from π-Si, which has been reported recently and interpreted as due to the radiative recombination on Si dangling bonds (Meyer *et al.*, 1993). Our study concludes that this interpretation is coherent with the hypothesis of quantum confinement and that the photon emission would correspond to the capture of the second carrier, more probably the hole, by the corresponding charged dangling bond.

From the previous discussions we can thus draw the following conclusions for the crystallites of interest: (i) nonradiative capture by neutral dangling bonds is large enough to kill the luminescence; (ii) nonradiative capture by charged dangling bonds is negligible and is dominated by radiative capture which falls in the energy range of observed infrared luminescence.

2. NONRADIATIVE AUGER RECOMBINATION

We now discuss the role of the Auger effect, which is one of the most efficient channels for electron–hole recombination in bulk Si. First we present results of tight binding calculations showing that it remains just as efficient even in very confined systems with a lifetime in the nanosecond range. We then show that Auger recombination gives a coherent interpretation of several recent experiments on π-Si. A detailed account of this work is given in Delerue *et al.* (1995) and Mihalescu *et al.* (1995).

We again consider spherical Si crystallites of radius R whose surface is passivated by hydrogen atoms. Let us assume that there is one electron–hole pair in a crystallite plus one additional carrier (electron or hole) introduced by injection or another excitation. In the Auger process, the electron–hole pair recombines nonradiatively transferring its energy to the third carrier. In bulk Si, the Auger probability per unit time and volume is equal to $Apn^2 + Bnp^2$ where p and n are respectively the hole and electron concentrations. A and B are not known accurately, values being reported between 10^{-30} and 10^{-32} cm^6/s (Landsberg, 1991; *Properties of Silicon*,

7 RADIATIVE AND NONRADIATIVE PROCESSES IN SILICON NANOCRYSTALLITES 287

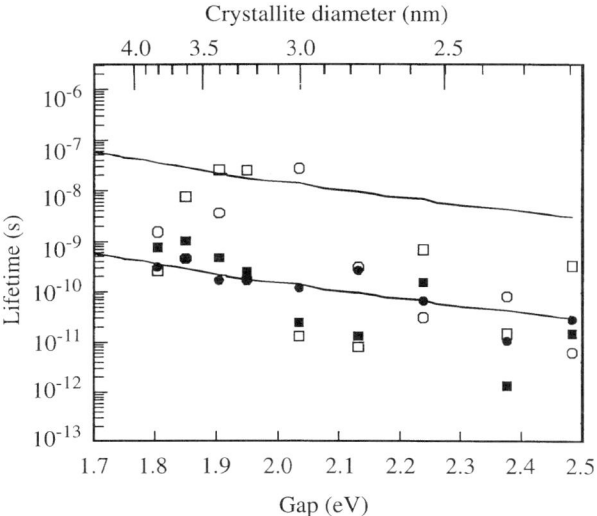

FIG. 24. Auger recombination time versus energy gap for spherical silicon crystallites. The squares (circles) represent the results from the direct calculation for the Auger process with two electrons and one hole (two holes and one electron). Empty symbols correspond to a level broadening induced by the electron-lattice coupling and full symbols to a level broadening of $\sim 0.1\,\text{eV}$ (see text). The two continuous curves represent upper and lower bounds of an extrapolation from the bulk values.

1988). Let us first extrapolate these data to nanocrystallites taking the concentrations n and p to correspond to one carrier confined in a spherical volume $4\pi R^3/3$. We see in Fig. 24 that such extrapolated Auger lifetimes τ lie between 0.1 and 100 nsec. for crystallite radius $R < 2.5$ nm which is several orders of magnitude faster than the radiative lifetime. The recombination should thus occur by the Auger process instead of by light emission.

However, the transfer of the Auger coefficients A and B from bulk to nanocrystallite semiconductors is questionable. In crystallites, the energy quantization (level spacing ~ 10 meV) makes it impossible to find a state for the third carrier with an excitation energy exactly matching the exciton energy. In consequence, the Auger effect is only possible if one takes into account the broadening of the electronic levels, which can come for example from the coupling with neighboring crystallites through a Si oxide barrier or a Si bridge ("undulating wires" (Cullis and Canham, 1991)) or from the electron lattice coupling. We have taken this effect into account by replacing the Auger final state by a gaussian density of states. For undulating wires where the barrier between crystallites is small, broadening is important and we have adjusted the width of the gaussian to obtain a continuous density of excited states (~ 0.1 eV is enough). For pure electron-lattice coupling we have used previously calculated values of the Franck–Condon shift d_{FC}, that is, the relaxation energy of the crystallite following electron–hole excitation is estimated to be between 14 and 1 meV for diameters between 2 and 4 nm (Martin et al., 1994). In that case the use of a gaussian is completely justified (Bourgoin and Lannoo, 1983) and its width is directly proportional to d_{FC}. The probability of Auger recombination ($=1/\tau$) is calculated with the Fermi rule (Landsberg, 1991)

$$\frac{1}{\tau} = \frac{2\pi}{h} \sum_f \left| \left\langle \Psi_i \left| \frac{e^2}{\varepsilon r} \right| \Psi_f \right\rangle \right|^2 \rho(E_f - E_i) \tag{8}$$

where Ψ_i and Ψ_f are respectively the initial and final states of energy E_i and E_f, and ρ is the density of final states discussed above. For the initial state Ψ_i, we take the Slater determinant corresponding to two electrons in the lowest conduction state and one hole in the highest valence state (eeh process). For the final states, we explore all the possible Slater determinants corresponding to filled valence states and one excited electron in the conduction band. The use of Slater determinants is justified because in such confined systems the electron and hole wave functions remain almost uncorrelated (Martin et al., 1994). For the same reason, one can show that the results can be easily extrapolated to the case of higher excitation or injection conditions just by rescaling ($1/\tau \approx pn^2$ or np^2). In the case of a hhe Auger process, the situation is formally symmetric. The Slater determinants are built from the one-electron spin–orbitals obtained from a tight binding calculation (Martin et al., 1994) and the matrix elements of the screened Coulomb potential in Eq. (8) are calculated as in Martin et al. (1994). Other details of the calculation are given elsewhere (Mihalescu et al., 1995). The calculated eeh Auger lifetimes plotted in Fig. 24 are mostly in the range 0.1 to 1 nsec with similar values for the hhe process. The scattering is relatively large because the electron–hole energy can be more or less close to the possible excitation energies of the third carrier. These results are close to the previous simple estimation (Fig. 24) because broadening induced by the electron–phonon coupling is sufficiently large to smooth the effect of quantization. We have verified that additional sources of broadening do not change drastically the Auger lifetime. The good agreement with extrapolated bulk values for small gaps gives confidence in our results. We recall that Fig. 24 gives an upper limit for the lifetime, which must strongly decrease at higher excitation with more than three carriers in a crystallite ($1/\tau \approx pn^2$ or np^2).

We now discuss the experimental implications of the Auger effect, starting with time-resolved PL of π-Si, combined with photoacoustic measurements to estimate the absorbed light in the samples (Monchalin et al., 1984). Time evolutions of the PL typically show two components (Calcott et al., 1993a): (1) a fast one with a lifetime smaller than 20 nsec. associated with a defect either in the oxide or at the interface with Si (Tsybeskov, Vandyshev, and Fauchet, 1994), and (2) a slow one which is often attributed to recombination in confined Si structures (Vial et al., 1992; Calcott et al., 1993a). Typical results are plotted in Fig. 25. The intensity of the fast component and the photoacoustic signal vary linearly with the excitation intensity showing that the absorbed light is proportional to the incident one. In contrast, the intensity of the slow component saturates at high flux.

The Auger effect gives a simple and natural explanation of this saturation. Let us assume that the "slow band" comes from radiative recombination in luminescent crystallites, that is, those without efficient nonradiative recombination centers. As long as the excitation intensity remains weak, there is only one electron–hole pair per luminescent crystallite and luminescence follows linearly the excitation power. At higher flux, when two electron–hole pairs are created in the same crystallite, the first one quickly recombines nonradiatively by the Auger effect. The absorption coefficient at 337 nm being between 10^4 and 9×10^4 cm^{-1} (Koch, Petrova-Koch, and Muschick, 1993), one can suppose that the excitation light is entirely

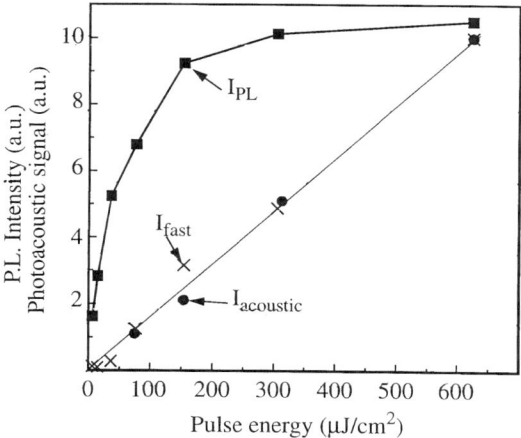

FIG. 25. Comparison of the PL intensity (for the slow and fast components) and the photoacoustic signal as a function of the excitation intensity for a 1 μm thick, 65% porosity anodic oxidized sample.

absorbed in a thickness of 1 μm. With about 10^{18} crystallites/cm³, the saturation should occur when the number of absorbed photons per crystallite is of the order of one, strongly supporting our interpretation. As for bulk Si, the experimental determination of the Auger coefficients is not straightforward, but the estimations are coherent with our calculated values. Limited by time resolution of the experimental system, one can just deduce that experimental Auger lifetime is faster than ~ 1 nsec. This is confirmed by recent ultrafast PL experiments showing that the Auger lifetimes certainly are in the subnanosecond range (Fauchet, 1995) (the number of photons per crystallite is not estimated).

In the same spirit, let us consider the voltage quenching of the PL (Bsiesy et al., 1993b). In these experiments, the PL is observed on n-type π-Si samples cathodically polarized in an aqueous solution of sulfuric acid. The measured spectra are very close to those taken "in air" for a polarization between 0 and -1 V, but they are narrowed and blue shifted at increasing potential leading to a complete quenching of the emitted light at -1.5 eV (Bsiesy et al., 1993b). Actually the blue shift is only due to the fact that the red part (low energy) of the spectrum is quenched first and the energy cut-off depends linearly on the potential. Let us consider crystallites of radius R characterized by a blue shift of the conduction band $\Delta E_c(R)$. We can define the ionization level $\varepsilon_R(1/0)$ for the filling of the lowest conduction state by one electron (Bourgoin and Lannoo, 1983) and an electron injection level $\varepsilon_F(V)$ (the effective Fermi level) in π-Si that depends linearly on the applied voltage. All the crystallites with $\varepsilon_R(1/0)$ lower than $\varepsilon_F(V)$ have at least one electron injected in the conduction band. When excited, they are non-luminescent because of fast Auger recombination mediated by the injected electron. The situation corresponds to the larger crystallites characterized by a small $\Delta E_c(R)$ (Fig. 26). Therefore the PL from the crystallites with a small bandgap ("red part") is quenched by the Auger effect.

The Auger effect, in addition to Coulomb charging effects (see Section VII), also explains the peculiar spectral width (~ 0.25 eV) of the electroluminescence (EL) of n-type π-Si cathodically polarized in a persulfate aqueous solution (Bsiesy et al., 1993b). In this problem, only the selectivity in the electron injection has to be considered, because the holes provided by the solvated species have a broad energy distribution (~ 1 eV) far inside the valence band (Bsiesy et al., 1993a). This is confirmed by the fact that the same EL features can be obtained when the hole source is changed (e.g., H_2O_2) (Bsiesy et al., to be published). Under injection, because of the charging effects, energy of the lowest conduction band level depends on its population (Lannoo, Delerue, and Allan, 1995) (see Section VII). If we define $\varepsilon_R(2/1)$ as the ionization level for filling by a second electron, the difference $U(R) = \varepsilon_R(2/1) - \varepsilon_R(1/0)$ is the average electron–electron interac-

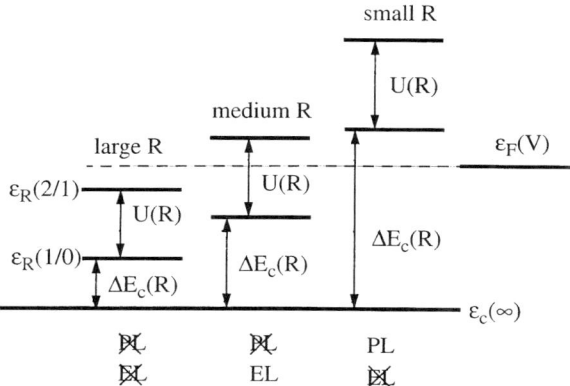

FIG. 26. Ionization levels in crystallites of different sizes R corresponding to the filling of the lowest conduction state by one electron $\varepsilon_R(1/0)$ or by two electrons $\varepsilon_R(2/1)$. By definition, the difference between $\varepsilon_R(1/0)$ and the minimum of the conduction band of the bulk crystal $\varepsilon_c(\infty)$ is $\Delta E_c(R)$, which is due to the confinement. In each case, the possibility of PL and EL is indicated.

tion energy. For a crystallite in an aqueous medium characterized by a large dielectric constant ($\varepsilon_{out} \approx 80$), $U(R)$ is in the 0.1 eV range ($U(R) = 0.11$ eV for $R = 1.5$ nm (Lannoo, Delerue, and Allan, 1995). Due to the Auger process, only crystallites with one electron can give luminescence when a hole is injected, meaning that for a given polarization V (i.e., $\varepsilon_F(V)$), EL is only possible in crystallites for which $\varepsilon_F(V)$ lies between $\varepsilon_R(1/0)$ and $\varepsilon_R(2/1)$. In that case, we see in Fig. 26 that $\Delta E_c(R)$ must be restricted to an energy window defined by $U(R)$. As from Proot, Delerue, and Allan (1992); Delerue, Allan, and Lannoo (1993) we know that $\Delta E_g(R) \approx 2.5 \Delta E_c(R)$, we find that the width of the EL spectrum is approximately equal to $2.5 \times U(R) \approx 0.27$ eV in excellent agreement with experiments (Bsiesy et al., 1993a).

An interesting consequence of the interpretation based on the combination of a selective injection and the Auger effect is that the same ionization level $\varepsilon_R(1/0)$ plays a role in both PL and EL experiments. At a given polarization, we see in Fig. 26 that a crystallite giving EL must be inefficient in PL and vice versa. This conclusion has been confirmed by simultaneous PL and EL experiments on π-Si samples polarized in an aqueous solution showing that for any polarization the high energy edge of the EL spectrum always corresponds exactly to the low energy edge of PL (Horry, 1995). This is a clear demonstration that the injection of one electron necessary for EL leads to a quenching of PL in a crystallite.

The Auger recombination thus gives a unified and simple explanation of three distinct experiments mixing electrical and optical properties. These conclusions can be generalized to nanocrystallites of other semiconductors with an indirect bandgap: (i) radiative lifetimes are several orders of magnitude larger than Auger lifetimes, which means that the Auger effect kills luminescence at high optical excitation or at high injection conditions; (ii) the Auger effect can give promising fast nonlinear optical properties; and (iii) Coulomb charging effects control the injection of carriers.

VII. Screening in Nanocrystallites and Coulomb Charging Effects

In this section we discuss different aspects of the screening in crystallites: hydrogenic impurities, Coulomb charging effects and the effective dielectric constant (see Lannoo, Delerue, and Allan, 1995; Allan et al., 1996).

1. Hydrogenic Impurities

We start with one single donor within a spherical crystallite of radius $R = 3a_0^3 N/32\pi$, where a_0 is the bulk lattice constant and N is the number of Si atoms. The bare potential energy of the donor electron at a distance r from the nucleus of charge $+e$ is $V_b(r) = -e^2/r$. This will polarize the electron gas of the crystallite and, within linear screening theory, will result in a self-consistently screened potential $V(r)$, which can be written as

$$V(r) = \int \varepsilon^{-1}(r, r') V_b(r') dr' \tag{9}$$

The problem is to know if there is a local relationship between V and V_b and, if so, whether or not the ratio V/V_b is related to the bulk dielectric constant. A complete answer to the question necessitates a full first-order self-consistent calculation. At present this cannot be achieved by *ab initio* methods for the clusters of interest with size in the 2 nm range. We have thus used a semi-empirical linear combination of atomic orbitals (LCAO) technique as described in Allen, Broughton, and McMahan (1986). Following Lannoo (1974); Lannoo and Bourgoin (1981), this can be made self-consistent by adding to the original matrix elements charge dependent Coulomb terms. It is reduced to a calculation of potentials due to point atomic charges, which allows the final result to be cast under a form similar to Eq. (9), but in a matrix formulation whose size is equal to the number of atoms.

We then apply it to spherical crystallites containing one donor. These crystallites are saturated by hydrogen atoms to avoid dangling bonds. Typical results for the impurity at the center are given in Fig. 27, where the ratio V_j/V_{bj} is plotted versus the distance from the center. Results for the Si atoms within the cluster are seen to scatter around a straight line ending on the point $V/V_b = 1$ when $r = R$. This result can be given a straightforward interpretation in the classical picture where the crystallite is considered as a continuous medium of dielectric constant ε_{in} embedded in another medium of dielectric constant ε_{out}. Then, taking the center of the clusters as the origin, the potential energy of an electron at point **r** due to the charge $+e$ at point **r**' can be obtained from simple electrostatics as the sum of the direct interaction $-e^2/(\varepsilon_{in}|r - r'|)$ and a corrective term (Böttcher, 1973; Brus, 1983, 1984):

$$V_{in}(\mathbf{r}, \mathbf{r}') = -e^2 \sum_{n=0}^{\infty} \frac{(\varepsilon_{in} - \varepsilon_{out})(n + 1)\mathbf{r}^n \mathbf{r}'^n P_n(\cos \theta)}{\varepsilon_{in}[\varepsilon_{out} + n(\varepsilon_{in} + \varepsilon_{out})]R^{2n+1}} \tag{10}$$

where θ is the angle between **r** and **r**'. $V_{in}(\mathbf{r}, \mathbf{r}')$ is due to the surface polarization charge density, which we shall call the "image" charge density by analogy with planar situations. The calculation shown in Fig. 27 corresponds to the centered impurity, where the potential energy simplifies to

$$V(r) = -e^2 \left[\frac{1}{\varepsilon_{in} r} + \frac{1}{R}\left(\frac{1}{\varepsilon_{out}} - \frac{1}{\varepsilon_{in}}\right) \right] \tag{11}$$

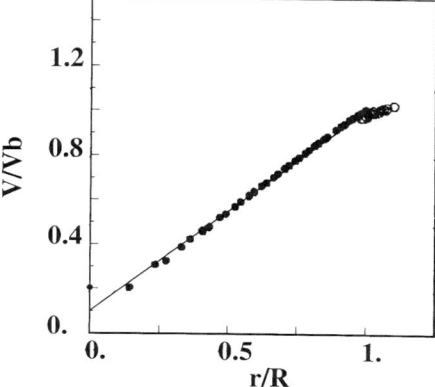

FIG. 27. Ratio of the self-consistent values of the potential to the bare one as a function of the position in a cluster with 705 silicon atoms. The open circles for $r > R$ correspond to hydrogen atoms. The straight line corresponds to the classical expression (see Eq. 11).

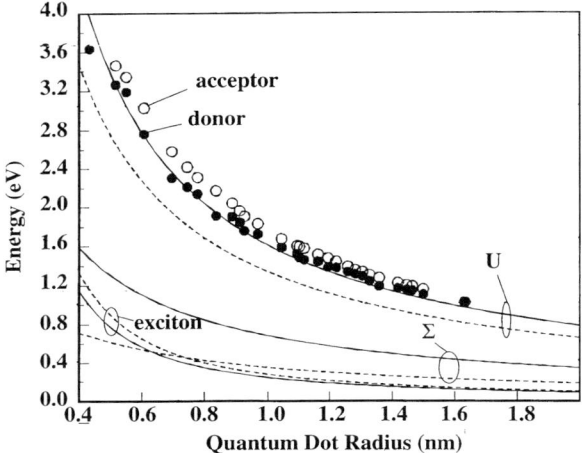

FIG. 28. Energy levels of hydrogenic impurities (● donor, ○ acceptor) obtained with the self-consistent calculation ($\varepsilon_{out} = 1$), self-energy $\Sigma(R)$, Coulomb energy $U(R)$ and exciton binding energy $E_{BX}(R)$ as a function of the particle radius R (continuous lines for $\varepsilon_{out} = 1$, dashed lines for $\varepsilon_{out} = 1.77$ (π-Si (Sagnes *et al.*, 1993)).

When $\varepsilon_{out} = 1$ we get $V/V_b = 1$ at the surface, as found in the numerical calculation. The dispersion of the calculated points around the straight line (Eq. 11) in Fig. 27 shows that the effective dielectric constant ε_{in}: (i) can only be defined as an average property and (ii) cannot be determined precisely since the intercept of the straight line at $r = 0$ is subject to some uncertainty.

We now consider the donor binding energy E_B, which we define as the difference between the lowest conduction states of the same crystallite with one excess electron, with or without the donor impurity. The computed values are plotted in Fig. 28, which shows that E_B can take fairly large values, in agreement with the conclusions of Tsu and Babic (1993, 1994). A fairly accurate expression is found in first order perturbation theory by calculating the average potential from Eq. (11) with the wavefunction $\sin kr/r$ of effective mass theory. This gives

$$E_B(R) = \left(\frac{1}{\varepsilon_{out}} + \frac{1.44}{\varepsilon_{in}}\right)\frac{e^2}{R} \qquad (12)$$

In the bulk, the binding energy is equal to the ionization energy since the conduction band states form a continuum. This is no more true in crystallites since these states now form a discrete spectrum. In a perfect crystallite ionization will occur via the continuum of states above the potential barrier, which exists at the surface, with an ionization energy $I_0(R)$. For the doped crystallite the ionization energy simply becomes $I_0(R) + E_B(R)$. It is thus strongly enhanced for small crystallites so that the donor electron should remain trapped. For π-Si this is in contradiction with electron paramagnetic resonance measurements concluding that hydrogenic impurities are always ionized, but also that the corresponding carriers are not found as free carriers (Von Bardeleben et al., 1993). To provide an explanation for this discrepancy we consider the following reaction: the initial situation consists of a neutral donor and a neutral dangling bond in two different crystallites, the final one resulting from electron transfer between the donor and dangling bond defect. Taking the bottom of the bulk conduction band as the origin of energies, the energy of the extra electron in the initial situation is $\Delta E_c - E_B(R)$ where ΔE_c is the blue shift of the crystallite containing the donor. In the final situation it becomes $\varepsilon_{db} - e^2/(\varepsilon_{out}d)$ where ε_{db} is the binding energy of the dangling bond state and $-e^2/(\varepsilon_{out}d)$ comes from the electrostatic attraction by the ionized donor in the other crystallite at distance d. The total difference in energy between the final and initial state becomes

$$\Delta E = -(\Delta E_c + \varepsilon_{db}) + E_B(R) - e^2/(\varepsilon_{out}d) \tag{13}$$

One can show that this equation still holds true within the same crystallite with $d = R$. With ε_{out} equal to 1.77 measured for a porous layer with a porosity of 74% (Sagnes et al., 1993), typical values obtained taking $R = 1.5$ nm are $\Delta E_c = 0.33$ eV (Proot, Delerue, and Allan, 1992; Delerue, Allan, and Lannoo, 1993), $\varepsilon_{db} = 0.3$ eV and $E_B(R) = 0.72$ eV (Eq. 12) so that ΔE is negative when $d < 9.3$ nm. Because of the high density of dangling bonds in π-Si (Von Bardeleben et al., 1993), this means that donor states should remain ionized, their electron being trapped at defects like dangling bonds since the previous reasoning remains valid for all deep defects.

2. Coulomb Effects and Effective Dielectric Constant

The following interesting question concerns the self-energy of particles and Coulomb charging effects. Up to now the level structure of crystallites has been obtained from semi-empirical calculations, which implicitly contain the same self-energy corrections (due to exchange and correlation) as in the bulk. In crystallites we thus need to determine corrections introduced by the finite size of the system. We do this within an electrostatic formulation, which includes the change in self-energy due to the "image" charge distribution on the surface. Let us then put an extra electron in the lowest conduction state ψ_c. The image contribution Σ to its self-energy is $1/2 \langle \psi_c | V_{in}(\mathbf{r}, \mathbf{r}) | \psi_c \rangle$ with V_{in} given by Eq. (10) with opposite sign. Taking again $\psi_c \propto \sin kr/r$, which provides an excellent approximation, we find

$$\Sigma = \frac{1}{2}\left(\frac{1}{\varepsilon_{out}} - \frac{1}{\varepsilon_{in}}\right)\frac{e^2}{R} + \delta\Sigma \qquad (14)$$

where the first term comes from $n = 0$ in Eq. (10) and $\delta\Sigma$ is the corrective term due to the remaining sum. This last term simplifies greatly when $\varepsilon_{in} + \varepsilon_{out} \gg \varepsilon_{out}$ in which case one obtains

$$\delta\Sigma = 0.47 \frac{e^2}{\varepsilon_{in} R}\left(\frac{\varepsilon_{in} - \varepsilon_{out}}{\varepsilon_{out} + \varepsilon_{in}}\right) \qquad (15)$$

which is usually small but not negligible. This gives the shift in energy of the extra electron in the lowest conduction band state, which represents the experimentally important quantity (Fig. 29). The injection of a second electron leads to an additional upwards shift given by the average repulsion with the other electron and its "image charge" (Fig. 29). With an electron distribution close to $\sin kr/r$ this shift is given by

$$U(R) = \left(\frac{1}{\varepsilon_{out}} + \frac{0.79}{\varepsilon_{in}}\right)\frac{e^2}{R} \qquad (16)$$

Under injection of n electrons (n small) the lowest filled conduction states will thus exhibit a shift approximately equal to $\Sigma + (n-1)U$ (Fig. 29). Finally, the situation for holes is completely symmetrical.

We now consider the same corrections applied to exciton states. To get

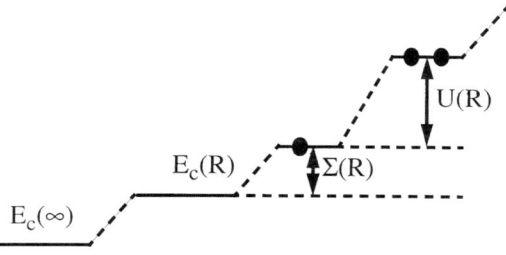

FIG. 29. Shift of the lowest conduction level due to the injection of one electron (Σ) or two electrons ($\Sigma + U$). The situation for holes is symmetrical.

the proper value for the exciton binding energy E_{BX} we must add the contribution of direct electron–hole interaction $-e^2/\varepsilon_{in} r_{eh}$ and different image terms: self-energies of the electron and of the hole (both given by Eq. 14) plus the interaction of one particle with the image charge distribution of the other. This last one is equal to $(1/\varepsilon_{in} - 1/\varepsilon_{out})e^2/R$ and obviously almost exactly compensates the sum of the two self energies, so that the image contribution to the exciton binding energy reduces to the constant term $-2\delta\Sigma$. With a carrier distribution close to $\sin kr/r$ we have

$$E_{BX}(R) = 1.79 e^2/\varepsilon_{in} R - 2\delta\Sigma \tag{17}$$

To estimate $\Sigma(R)$, $U(R)$, $E_{BX}(R)$ numerically, the problem is still to know what value of ε_{in} should be used. Wang and Zunger (1994b) treated this problem by computing the quantity $1 + 4\pi\chi$ where χ is the quantum polarizability of the whole sphere. In the bulk this procedure is exact provided one neglects local field effects, which introduces an error of the order of 10–20%. Assuming this to hold true in crystallites, the procedure of calculating $1 + 4\pi\chi$ might be justified if the following conditions are met: (i) one can define macroscopic quantities as averages over unit cells and treat them as continuous variables and (ii) the local ratio between the macroscopic polarization and field is constant and equal to χ. We have seen in Fig. 27 that this seems practically verified. From that point of view $1 + 4\pi\chi$ represents one particular way of calculating an average ε_{in}. To get some feeling about the accuracy of such an averaging procedure we have estimated ε_{in} using three different approaches: (1) a fit of the self-consistent donor potential of Fig. 27 by Eq. (11), (2) the classical donor potential Eq. (11) is used in the LCAO calculation and ε_{in} is adjusted to fit the self-consistent donor binding energy (Fig. 28), and (3) a least square minimization of $V - V_b/\varepsilon_{in}$ in a case corresponding to a bound exciton with the electron confined on the central atom and the hole in the highest valence state. Figure 30 shows that there are substantial differences between the three results, mostly for small crystallite radii, due to the different ways of computing the spatial average of $\varepsilon^{-1}(\mathbf{r}, \mathbf{r}')$. Our results lie between those of Tsu and Babic (1993, 1994); Wang and Zunger (1994b) and their average value can be well approximated by $\bar{\varepsilon}_{in} - 1 = (11.4 - 1)/(1 + (0.92/R)^{1.18})$ with R in nanometer units. Figure 28 gives $\Sigma(R)$, $U(R)$, $E_{BX}(R)$ computed with $\bar{\varepsilon}_{in}$ for $\varepsilon_{out} = 1$ and $\varepsilon_{out} = 1.77$ (π-Si). The values of U are so large that the injection of more than one electron- or hole-into Si nanocrystallites in π-Si must be very difficult if not impossible. However, when $\varepsilon_{out} \to \infty$ corresponding to the experimental case where π-Si samples are in aqueous electrolyte ($\varepsilon_{out} \approx 80$), U reduces to ~ 0.15 eV, so that the injection of two carriers becomes easier (Chazalviel, Ozanam, and Dubin, 1994). This could

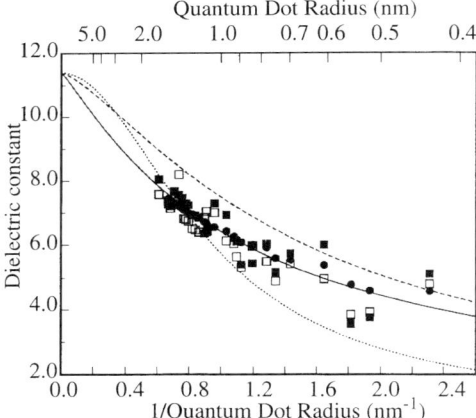

FIG. 30. Plot of the calculated ε_{in} obtained by a fit of the donor potential (■) and of the self-consistent donor binding energy (□) with classical laws and by a least square minimization of $V - V_b/\varepsilon_{in}$ in a case corresponding to a bound exciton (●). The continuous line is a fit of these values ($\tilde{\varepsilon}_{in} = 1 + (11.4 - 1)/(1 + (0.92/R)^{1.18})$). The dashed curve corresponds to $\tilde{\varepsilon}_s$ of Wang and Zunger (1994b) and the dotted line to the generalized Penn model of Tsu and Babic (1993, 1994).

explain the different transport properties of π-Si in "air" and in electrolyte.

To summarize this section, we have discussed the binding energies of impurities, showing why they should remain in ionized form. We have also shown that Coulomb charging effects, such as those occurring under carrier injection conditions, are quite substantial. Finally, we have justified the approximate use of an effective dielectric constant, which, however, turns out to be about half the bulk value for most crystallites of interest.

VIII. Conclusion

In this chapter we have discussed in detail the theoretical calculations that have been performed so far on various aspects concerning Si nanocrystallites. One of the main conclusions is that theory has proved capable of predictions of a quantitative nature for well defined physical situations. In the future there will thus be need to check such predictions in experimental situations where one deals with controlled size and shape distributions. Up to now this has not yet been the case, since most of the experimental information has been obtained from π-Si, which is a much more complex material.

REFERENCES

Allan, G., Delerue, C., and Lannoo, M. (1996). *Phys. Rev. Lett.* **76**, 2961.
Allan, G., Delerue, C., Lannoo, M., and Martin, E. (1995). *Phys. Rev. B* **52**, 11982.
Allen, P. B., Broughton, J. Q., and McMahan, A. K. (1986). *Phys. Rev. B* **34**, 859.
Böttcher, C. J. F. (1973). *Theory of Electric Polarization.* Elsevier, Amsterdam, Vol. I, 2nd edn.
Bourgoin, J., and Lannoo, M. (1983). In *Point Defects in Semiconductors II* (ed. M. Cardona), Springer Verlag, New York, p. 103.
Brandt, M. S., and Stutzmann, M. (1992). *Appl. Phys. Lett.* **61**, 2569.
Brandt, M. S., and Stutzmann, M. (1995). *Solid State Commun.* **93**, 473.
Brus, L. E. (1983). *J. Chem. Phys.* **79**, 5566; *ibid.* (1984), 4403.
Bsiesy, A., Muller, F., Ligeon, M., Gaspard, F., Hérino, R., Romestain, R., and Vial, J. C. (1993a). *Phys. Rev. Lett.* **71**, 637.
Bsiesy, A., Muller, F., Mihalcescu, I., Ligeon, M., Gaspard, F., Hérino, R., Romestain, R., and Vial, J. C. (1993b). In "Light Emission from Silicon" (eds. J. C. Vial, L. T. Canham, and W. Lang), *J. Lumines.* **57**. Elsevier, North-Holland, p. 29.
Bsiesy, A., and co-workers (to be published).
Buda, F., Kohanoff, J., and Parrinello, M. (1992). *Phys. Rev. Lett.* **69**, 1272.
Calcott, P. D. J., Nash, K. J., Canham, L. T., Kane, M. J., and Brumhead, D. (1993a). *J. Phys. Condens. Mater.* **5**, L91.
Calcott, P. D. J., Nash, K. J., Canham, L. T., Kane, M. J., and Brumhead, D. (1993b). In *Microcrystalline Semiconducors-Materials Science & Devices* (eds. P. M. Fauchet, C. C. Tsai, L. T. Canham, I. Shimizu, and Y. Aoyagi), *MRS Symposia Proceedings* **283**. Materials Research Society, Pittsburgh.
Canham, L. T. (1990). *Appl. Phys. Lett.* **57**, 1046.
Chazalviel, J.-N., Ozanam, F., and Dubin, V. M. (1994). *J. Phys. I France* **4**, 1325.
Cullis, A. G., and Canham, L. T. (1991). *Nature (London)* **353**, 335.
Delerue, C., Allan, G., and Lannoo, M. (1993). *Phys. Rev. B* **48**, 11024.
Delerue, C., Lannoo, M., Allan, G., Martin, E., Mihalcescu, I., Vial, J. C., Romestain, R., Müller, F., and Bsiesy, A. (1995). *Phys. Rev. Lett.* **75**, 2228.
Delley, B., and Steigmeier, E. F. (1993). *Phys. Rev. B* **47**, 1397.
Delley, B., and Steigmeier, E. F. (1995). *Appl. Phys. Lett.* **67**, 2370.
Dexter, D. L. (1958). In *Solid State Physics* (eds. F. Seitz and D. Turnbull), Academic Press, New York, Vol. 6, p. 30.
DMol User Guide, version 2.3.5. San Diego: Biosym. Technologies, 1993.
Fauchet, P. M. (1995). Private communication.
Fishman, G., Mihalcescu, I., and Romestain, R. (1993). *Phys. Rev. B* **48**, 1464.
Fishman, G., Romestain, R., and Vial, J. C. (1993). *J. Phys.* **IV3**, 355.
Gavartin, J. L., Matthai, C. C., and Morrison, I. (1995). *Thin Solid Films* **255**, 39.
Goguenheim, D., and Lannoo, M. (1900). *J. Appl. Phys.* **68**, 1059.
Goguenheim, D., and Lannoo, M. (1991). *Phys. Rev. B* **44**, 1724.
Henry, C. H., and Lang, D. V. (1977). *Phys. Rev. B* **15**, 989.
Hill, N. A., and Whaley, K. B. (1995). *Phys. Rev. Lett.* **75**, 1130.
Hirao, M., Uda, T., and Murayama, Y. (1993). *Mater. Res. Soc. Symp. Proc.* **283**, 425.
Hirao, M. (1995). In "Microcrystalline and Nanocrystalline Semiconductors" (eds. R. W. Collins, C. C. Tsai, M. Hirose, F. Koch, and L. Brus), *Mat. Res. Soc. Symp. Proc.* **358**. Materials Research Society, Pittsburgh, p. 3.
Horry, M. A. (1995). Ph.D. thesis (University of Grenoble) and presentation at the EMRS Spring Meeting in Strasbourg. Symposium on porous silicon: material, technology and devices.

Huaxiang, Fu, Ling, Ye, and Xide, Xie (1993). *Phys. Rev. B* **48**, 10978.
Hybertsen, M. S. (1992). In *Light Emission from Silicon* (eds. S. S. Iyer, L. T. Canham, and R. T. Collins), Materials Research Society, Pittsburgh, p. 179.
Hybertsen, M. S., and Needels, M. (1993). *Phys. Rev. B* **48**, 4608.
Hybertsen, M. S. (1994). *Phys Rev. Lett.* **72**, 1514.
Kanemitsu, Y., Suzuki, K., Uto, H., Masumoto, Y., Masumoto, T., Kyushin, S., Higushi, K., and Matsumoto, H. (1992). *Appl. Phys. Lett.* **61**, 2446.
Kanemitsu, Y. (1994). *Phys. Rev. B* **49**, 16845.
Kauffer, E., Pecheur, P., and Gerl, M. (1976). *J. Phys. C* **9**, 2319.
Koch, F., Petrova-Koch, V., and Muschick, T. (1993). In "Light Emission from Silicon" (eds. J. C. Vial, L. T. Canham, and W. Lang), *J. Lumines.* **57**. Elsevier, North-Holland, p. 271.
Landsberg, Peter T. (1991). In *Recombination in Semiconductors*, Cambridge University Press.
Lannoo, M. (1974). *Phys. Rev. B* **10**, 2544.
Lannoo, M., and Bourgoin, J. (1981). In *Point Defects in Semiconductors I* (ed. M. Cardona), Springer Verlag, New York.
Lannoo, M., Delerue, C., and Allan, G. (1995). *Phys. Rev. Lett.* **74**, 3415.
Lee, S.-G., Cheong, B.-H., Lee, H.-K., and Chang, K. J. (1995). *Phys. Rev. B* **51**, 1762.
Lockwood, D. J. (1994). *Solid State Commun.* **92**, 101.
Madelung, O. (Ed.) (1991). *Semiconductors. Group IV and II–V Compounds*. Data in Science and Technology, Springer Verlag.
Martin, E., Delerue, C., Allan, G., and Lannoo, M. (1994). *Phys. Rev. B* **50**, 18258.
Mattheiss, L. F., and Patel, J. R. (1981). *Phys. Rev. B* **23**, 5384.
Merle, J. C., Capizzi, M., Firioni, P., and Frova, A. (1978). *Phys. Rev. B* **17**, 4821.
Meyer, B. K., Hofmann, D. M., Stadler, W., Petrova-Koch, V., Koch, F., Omling, P., and Emanuelsson, P. (1993). *Appl. Phys. Lett.* **63**, 2120.
Mihalescu, I., Vial, J. C., Bsiesy, A., Müller, F., Romestain, R., Martin, E., Delerue, C., Lannoo, M., and Allan, G. (1995). *Phys. Rev. B* **51**, 17605.
Mintmire, J. W. (1993). *J. Vac. Sci. Technol. A* **11**, 1733.
Monchalin, J. P., Bertrand, L., Rousset, G., and Lepoutre, F. (1984). *J. Appl. Phys.* **56**, 190.
Nash, K. J., Calcott, P. D. J., Canham, L. T., and Needs, R. J. (1995). *Phys. Rev. B* **51**, 17698.
Ohno, T., Shiraishi, K., and Ogawa, T. (1992). *Phys. Rev. Lett.* **69**, 2400.
Pasquarello, A., and Hybertsen, M. S., Car, R. (1995). *Phys. Rev. Lett.* **74**, 1024.
Petit, J., Allan, G., and Lannoo, M. (1986a). *Phys. Rev. B* **33**, 8595.
Petit, M., Lannoo, M., and Allan, G. (1986b). *Solid State Commun.* **60**, 861.
Poindexter, I. H., and Caplan, P. J. (1983). *Prog. Surf. Sci.* **14**, 201.
Polatoglou, H. M. (1993). *J. Lumines.* **57**, 117.
Prokes, S. M., Carlos, W. E., and Bermudez, V. M. (1992). *Appl. Phys. Lett.* **61**, 1447.
Proot, J. P., Delerue, C., and Allan, G. (1992). *Appl. Phys. Lett.* **61**, 1948.
Properties of Silicon, published by INSPEC, The Institution of Electrical Engineers, EMIS Datareviews Series No. 4, 1988.
Read, A. J., Needs, R. J., Nash, K. J., Canham, L. T., Calcott, P. D. J., and Qteish, A. (1992). *Phys. Rev. Lett.* **69**, 1232; *ibid.* (1993). *Phys. Rev. Lett.* **70**, 2050.
Ren, S. Y., and Dow, J. D. (1992). *Phys. Rev. B* **45**, 6492.
Robinson, M. B., Dillon, A. C., Haynes, D. R., and George, S. M. (1992). *Appl. Phys. Lett.* **61**, 1414.
Rosenbauer, M., Finkbeiner, S., Bustarret, E., Weber, J., and Stutzmann, M. (1995). *Phys. Rev. B* **51**, 10539.
Sagnes, I., Halimaoui, A., Vincent, G., and Badoz, P. A. (1993). *Appl. Phys. Lett.* **62**, 1155.
Sanders, G. D., and Chang, Yia-Chung (1992). *Phys. Rev. B* **45**, 9202.
Shik, A. (1993). *J. Appl. Phys.* **74**, 2951.

Steigmeier, E. F., Delley, B., and Auderset, H. (1992). *Physica Scripta* **T45**, 305.
Suemoto, T., Tanaka, K., Nakajima, A., and Itakura, T. (1993). *Phys. Rev. B* **70**, 3659.
Takagahara, T., and Takeda, K. (1992). *Phys. Rev. B* **46**, 15578.
Tserbak, C., Polatoglou, H. M., and Theodorou, G. (1993). *Phys. Rev. B* **47**, 7104.
Tsu, R., and Babic, D. (1993). In *Optical Properties of Low Dimensional Silicon Structures* (eds. D. C. Bensahel, L. T. Canham, and S. Ossicini), NATO ASI Series, Kluwer, Academic Publishers, Dordrecht, p. 203.
Tsu, R., and Babic, D. (1994). *Appl. Phys. Lett.* **64**, 1806.
Tsybeskov, L., Vandyshev, J. V., and Fauchet, P. (1994). *Phys. Rev. B* **49**, 7821.
Vial, J. C., Bsiesy, A., Gaspard, F., Hérino, R., Ligeon, M., Muller, F., Romestain, R., and Macfarlane, R. M. (1992). *Phys. Rev. B* **45**, 14171.
Vial, J. C., Bsiesy, A., Fishman, G., Gaspard, F., Hérino, R., Ligeon, M., Muller, F., Romestain, R., and Macfarlane, R. M. (1993). In *Microcrystalline Semiconductors-Materials Science & Devices*, p. 241.
Vincent, G., and Bois, D. (1978). *Solid State Commun.* **27**, 431.
Vinchon, T., Spanjaard, D., and Desjonqueres, M. C. (1992). *J. Phys. C: Con. Mat.* **4**, 5061.
Vogl, P., Hjalmarson, H. P., and Dow, J. (1983). *J. Phys. Chem. Solids* **44**, 365.
Von Bardeleben, H. J., Stievenard, D., Grosman, A., Ortega, C., and Siejka, J. (1993). *Phys. Rev. B* **47**, 10899.
Voos, M., Uzan, Ph., Delalande, C., Bastard, G., and Halimaoui, A. (1992). *Appl. Phys. Lett.* **61**, 1213.
Wang, L. W., and Zunger, A. (1994a). *J. Phys. Chem.* **98**, 2158.
Wang, L. W., and Zunger, A. (1994b). *Phys. Rev. B* **73**, 1039.
Wilson, W. L., Szajowski, P. F., and Brus, L. E. (1993). *Science* **262**, 1242.
Xie, Y. H., Hybertsen, M. S., Wilson, W. L., Ipri, S. A., Carver, G. E. Brown, W. L., Dons, E., Weir, B. E., Kortan, A. R., Watson, G. P., and Liddle, A. J. (1994). *Phys. Rev. B* **50**, 8138.
Xu, Z. Y., Gal, M., and Gross, M. (1992). *Appl. Phys. Lett.* **60**, 1375.
Yeh, Chin-Yu, Zhang, S. B., and Zunger, A. (1994a). *Appl. Phys. Lett.* **64**, 3545.
Yeh, Chin-Yu, Zhang, S. B., and Zunger, A. (1994b). *Phys. Rev. B* **50**, 14405.
Zheng, X. L., Wang, W., and Chen, H. C. (1992). *Appl. Phys. Lett.* **60**, 986.

CHAPTER 8

Silicon Polymers and Nanocrystals

Louis Brus

CHEMISTRY DEPARTMENT
COLUMBIA UNIVERSITY
NEW YORK, NY, 10027

I. INTRODUCTION 303
II. SILICON POLYMERS IN ONE, TWO, AND THREE DIMENSIONS 304
   1. *Band Structure* 304
   2. *Luminescence* 307
III. PASSIVATED SILICON NANOCRYSTALS 308
   1. *Theory of Optical Properties* 308
   2. *Nanocrystal Synthesis, Characterization, and Luminescence* 309
   3. *Comparison Between Nanocrystals and Macroscopic Crystalline Si* . 319
   4. *Physical Size Regimes for Individual Nanocrystals* 321
IV. ELECTRON TRANSPORT IN POROUS NANOCRYSTAL MATERIALS 322
References 326

I. Introduction

Electronic computing and long distance optical communication both depend upon silicon (Si) based materials: in the first case, elemental Si crystalline wafers, and in the second case, silica optical fibers. These two technologies are extraordinarily powerful and constantly growing in importance. Nevertheless, at present the interconnection between these two technologies is rather awkward and inefficient. Optical signals cannot be directly generated in crystalline Si, which is an indirect gap material. The bandgap luminescence is electric dipole forbidden. The natural radiative rate Γ_r of a Si electron–hole singlet excitonic pair, due to vibronic interaction with transverse optic (TO) phonons, is quite slow $-2 \times 10^4 \, \text{s}^{-1}$ when the pair is weakly bound near 4 K by the Coulomb attraction (Haynes *et al.*, 1961; Cuthbert, 1970). By contrast, Γ_r for a bound pair in a direct gap semiconductor, or in a molecule with a strongly allowed transition, is on the order of $10^9 \, \text{s}^{-1}$.

The indirect gap nature of crystalline Si (c-Si) is a consequence of translational symmetry. In principle, any effect which breaks translational symmetry makes the transition formally allowed. This chapter considers Si materials as function of dimensionality, surface chemistry, and finite nanocrystal size, in order to gain insight into the nature of the luminescence process. Interest in Si luminescence has been stimulated by discovery (Canham, 1990; Koyama et al., 1991; Lehmann and Gösele, 1991; Petrova-Koch et al., 1992;) of the efficient luminescence and quantum nature of porous Si (π-Si), and by the need for optical interconnects in future generations of ultra-large scale integration (ULSI) Si microcircuits. The final section discusses electron transport through assemblies of Si nanocrystals, in order to understand the remarkable difference between wet and dry π-Si electronic properties. This chapter describes the understanding my colleagues and I have developed from work in AT&T Bell Laboratories, now part of Lucent Technologies, on various Si materials. The chemistry and physics of direct gap nanocrystals (Brus, 1991; Wang and Herron, 1991; Weller, 1993; Banyai and Koch, 1993), and a chemical perspective on these same Si optical properties (Brus, 1994), have been given elsewhere.

II. Silicon Polymers in One, Two, and Three Dimensions

1. BAND STRUCTURE

All Si polymers show the sp^3 tetrahedral local bonding that is characteristic of diamond and zinc-blende lattices, as well as silane chemical compounds such as Si_2H_6. Consider three prototypical polymers: linear trans polysilane $(SiH_2)_n$, puckered sheet polysilyne $(SiH)_n$, and normal diamond lattice crystalline Si_n. These are termed 1D-Si, 2D-Si, and 3D-Si. In the first polymer, each Si atom is bonded to two other Si atoms and to two H "capping" atoms, creating a zig-zag chain of capped Si atoms. In 2D-Si, each Si is bonded to three other Si and one H atom, creating a capped puckered sheet. In all three cases the valence band is strongly delocalized and is formed by Si–Si bonding orbitals with a totally symmetric maximum at the Γ point, as shown in Figs. 1 and 2. At the valence band maximum, the spatial degeneracy is equal to the Si–Si bonding dimensionality.

While the valence bands are similar, the conduction bands are quite different. 1D-Si shows a 3.89 eV direct gap (Takeka and Shiraishi, 1989; Takeka, Shiraishi, and Matsumoto, 1990). In 2D-Si the conduction band has two local minima, with the direct 2.61 eV gap slightly higher than the indirect 2.48 eV gap at the * point (see Fig. 1) near the Brillouin zone

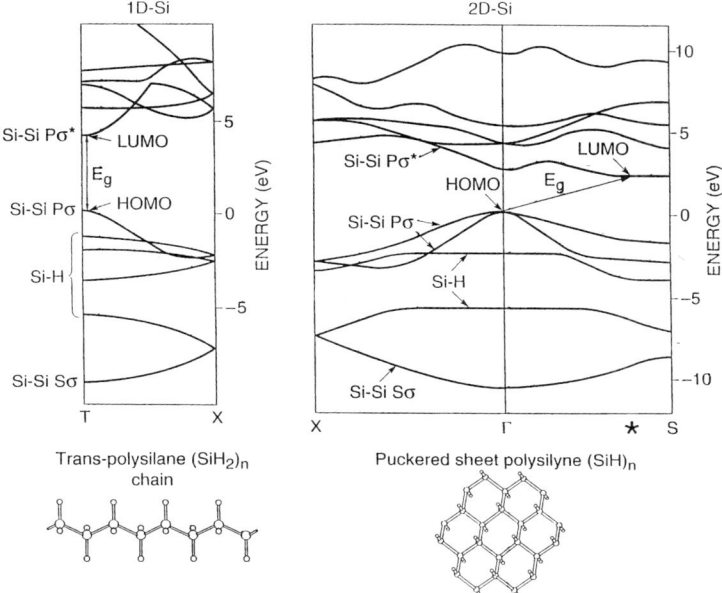

FIG. 1. Band structures of 1D-Si and 2D-Si. (From Takeda et al., 1989.)

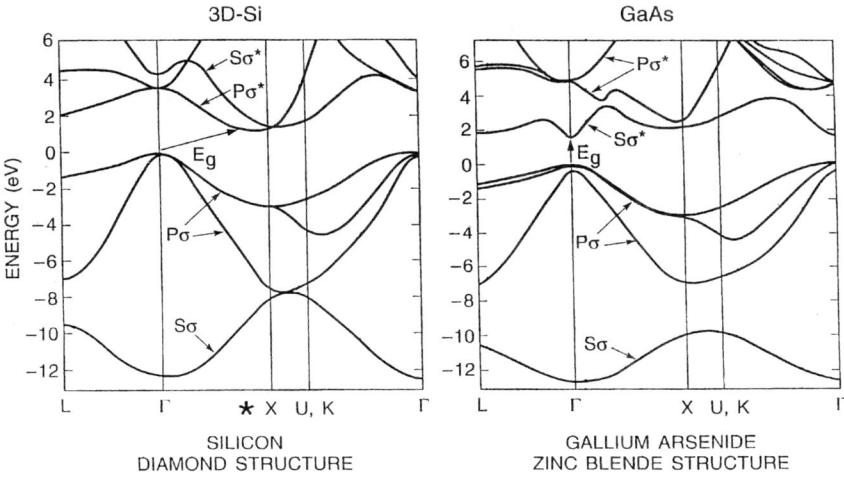

FIG. 2. Band structures of Si and GaAs. (From Cohen and Chelikowsky, 1988.)

boundary (Takeka and Shiraishi, 1989; Takeda, Shiraishi, and Matsumoto, 1990; Van de Walle and Northrup, 1993). In 3D-Si the direct gap is a 3.4 eV saddle point, while the indirect gap is now 2.3 eV lower at 1.1 eV (Cohen and Chelikowsky, 1988). This systematic progression reflects the relative energies of two types of antibonding Si–Si states. Silicon p antibonding states come down in energy with respect to Si s antibonding states as dimensionality increases. Also, s-p hybridization inside the unit cell is a strong function of electron wavevector q, and increases across the Brillouin zone in each material. Near the zone boundary, the functions are strongly mixed hybrid s-p in character. It is the hybridization of the functions inside the unit cell that determines their relative stability, and whether the polymeric material is direct or indirect.

In 2-D Si the conduction band local minimum at Γ is an in plane s-p hybrid on the opposite side of the Si atom from the Si–H bond (Van de Walle and Northrup, 1993). The doubly degenerate antibonding, out-of-plane p state is about 1.5 eV higher at Γ. However, as q increases, these bands mix so that the (lowest) conduction band state near the zone edge is nearly a sp^3 hybrid. The mixing decreases the energy of the lower component, and creates the double minimum conduction band. In Fig. 1 the two bands show the general appearance of an avoided crossing as a function of q. In 3D-Si the situation is inverted, in that near Γ the antibonding p state is lowest, at 3.4 eV, while the antibonding s state is about 1 eV higher. As q increases these states mix strongly for propagation along (100) to give a sp hybrid LUMO at 1.1 eV near the X point.

3D-GaAs is quite similar to 3D-Si, in that the valence bands have very similar dispersion. sp^3 delocalization is so strong that the electronegativity difference between Ga and As makes little difference. However, in the conduction bands at Γ in Fig. 2 the s and p type states reverse with respect to their ordering in Si. This occurs because the s band is stabilized by becoming partially charge transfer in character, with the promoted electron residing predominately on the more electronegative As atom. In a situation somewhat similar to 2D-Si, the lower conduction band has two local minima. However, in GaAs the direct gap at Γ is lower than the sp hybrid indirect gap near the zone edge, and thus relaxed luminescence is dipole allowed and strong in GaAs.

In 2D-Si the two gaps are almost degenerate in energy when the sheet polymer is "capped" with H atoms. Generally speaking, capping with other chemical species either more or less electronegative than H should tend to withdraw or donate electron density into the sheet on specific Si atoms. This in turn should modify the band structure. A chemical synthesis of "hydroxy" 2D-Si $(Si_2HOH)_n$ is known; in this sheet polymer half of the Si atoms are capped with hydroxyl groups and half with H atoms (Weiss, Beil, and

Meyer, 1979; Brandt et al., 1992). This material, a member of the siloxene family, shows strong, direct gap like luminescence in the green, and calculation shows that indeed the direct, charge transfer gap is lower than the sp hybrid indirect gap (Van de Walle and Northrup, 1993; Takeda and Shiraishi, 1993; Deak et al., 1992). This example shows the ability of surface chemical derivatization to change electronic properties.

2. LUMINESCENCE

Band structure does not include electron–hole interaction, and/or possible change in bond lengths and angles in the excited state. Both these phenomena will influence the nature of luminescence. 1D-Si emits strongly in the ultraviolet as expected for a direct gap material (Miller and Michl, 1989). The electron–hole interaction is on the order of 1 eV, and pairs are delocalized over some 10–20 Si atoms if the chain is dissolved in the condensed phase (Soos and Hayden, 1990; Moritomo et al., 1991; Thorne et al., 1989; Kanemitsu et al., 1992). Gas phase 1D-Si chains seem to show excited state localization onto a single Si–Si bond, with subsequent dissociation into two fragments (Allan, Delerue, and Lannoo, 1993).

2D-Si shows weak, indirect gap type luminescence with a strong Stokes shift that suggests structural localization of the excited state before emission. It is difficult to synthesize, characterize, and purify samples of 2D-Si. The hydroxy 2D-Si variant shows strong photoluminescence (PL) in the green and yellow, with nanosecond lifetime (Wilson and Weidman, 1991; Kanemitsu et al., 1993; Friedman et al., 1993; Stutzman et al., 1993; Dahn et al., 1994). Good single crystal experimental studies of band and exciton structure have not been done. Siloxene can be grown directly on wafer Si by acid attack on a surface layer of $CaSi_2$. This material may have technological value and needs to be carefully characterized. A potential problem is that siloxene is thermodynamically metastable; upon heating it evolves hydrogen gas and converts to a Si dioxide material with embedded Si particles (Dahn et al., 1994). In this condition it exhibits the slow red photoluminescence also seen in π-Si and Si nanocrystals.

3D-Si has very weak electron–hole Coulomb interaction because of the high Si dielectric constant. As a consequence the excitonic interaction in 3D-Si is about 100 times weaker than in 1D-Si. 3D-Si also has a very weak electron–phonon interaction due to the covalent, rigid nature of the lattice. As a result, direct multiphonon, nonradiative recombination of band edge electrons and holes, releasing 1.1 eV of energy in heat, is extremely slow and has never been observed in perfect Si crystals. Nonradiative recombination is always catalyzed by a local defect, or occurs by a three body Auger electronic process. In a perfect crystal at 300 K, electrons and holes can live as long as 40 ms (Yablonovitch et al., 1986). On this time scale, individual carriers are mobile over macroscopic distances. At low temperature, careful studies of luminescence (Bradfield, Brown, and Hall, 1988; Schall, Thonke, Sauer, 1986) from weakly bound Si excitons shows indirect gap type emission, with the TO phonon vibronically induced lines much stronger than the zero phonon line (hereafter, ZPL). In 3D-Si, and in Si nanocrystals as we shall see, the indirect nature of the bandgap transition is robust and not significantly affected by physical perturbation, such a finite size in the exciton. This is true because the nearest direct gap with high oscillator strength is 2.3 eV higher in energy, and across the Brillouin zone from the indirect gap. Pertubation induced mixing with the direct gap is therefore weak.

III. Passivated Silicon Nanocrystals

1. Theory of Optical Properties

Finite nanocrystal size breaks translational symmetry. As a consequence, the luminescence acquires some purely electric dipole character, and the ZPL becomes allowed. However, quantitative calculation shows that this effect is small (Takagahara and Takeda, 1992; Proot, Delerue, and Allan, 1992; Delley and Steigmeier, 1993). Figure 3 shows electron and hole distributions in k space due to spatial localization in 2.5 nm nanocrystals. There is a small overlap between the tails of the distributions — this creates the ZPL oscillator strength. In Fig. 4, the radiative lifetime of a confined pair is shown as a function of cubic nanocrystal size, from the work of Hybertsen (1994). The predicted purely electronic radiative lifetimes are around 10^{-3} s for 2.5 nm size, and 5×10^{-5} s for 1.5 nm size. In the bulk Si exciton, the coupling with the radiation field is due principally to vibronic coupling via the TO phonon. The strength of this coupling should increase with deceasing size, roughly in proportion to inverse volume (Schmitt-Rink, Miller, and Chelma, 1987). Figure 4 also shows this predicted TO vibronic radiative lifetime, which is in fact shorter than the purely electronic lifetime. Thus, theory indicates that Si nanocrystals should remain indirect gap like, with TO phonon vibronic structure near the bandgap, and radiative lifetimes 10^4 or more longer than that of an allowed transition, down to sizes of roughly 1.5 nm.

In Fig. 4 the nanocrystal bandgap versus size is estimated from an effective mass model. Several theoretical studies have shown that this

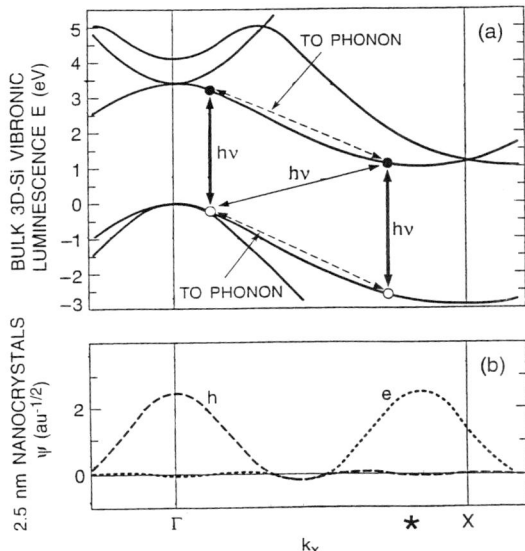

FIG. 3. Upper panel: 3D-Si band structure showing TO vibronic luminescence mechanism. Broad arrows are allowed transitions, and the narrow arrow is the weak vibronically induced transition. Lower panel: Wavevector distributions for confined electron and hole in a 2.5 nm nanocrystal. (From Hybertsen, 1994.)

approximation overestimates the increase as size decreases, especially if the surface is also taken as an infinite potential barrier. Figure 5 shows the exciton energy (including the size dependent Coulomb interaction) versus size from a representative calculation employing H atom termination, (Delerue, Allan, and Lannoo, 1993). Accurate calculations with oxide termination have not been done.

2. Nanocrystal Synthesis, Characterization, and Luminescence

Liquid phase chemical synthesis of crystalline Si particles is difficult, as the Si–Si bond energies are large, and thus high temperatures are necessary to anneal into crystalline structures. However, a high temperature, two stage aerosol method makes Si shell particles with a crystalline Si core, and a passivating surface oxide layer about 0.8 nm thick (Littau et al., 1993). In the first stage, diamond lattice Si aerosol nanocrystals are grown from dilute disilane in 1.5 atm of He gas. In roughly 30 ms at 800 or 1000°C, the

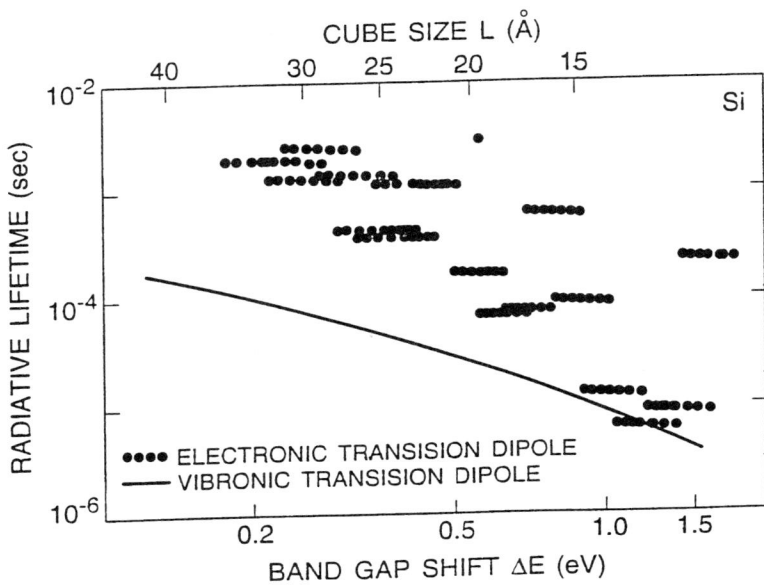

FIG. 4. Purely electronic radiative lifetimes (dots) and vibronic radiative lifetimes (line) versus size on upper axis. For a given size, several lifetimes are possible depending upon shape, as shown in Hybertsen, 1994.

particles anneal and become faceted. In the second stage, the particles are briefly (roughly 30 ms again) oxidized at 800°C with O_2. The nanocrystals are collected as a colloid in ethylene glycol.

The aerosol process creates a rather wide size distribution. High pressure liquid chromatography and size selective precipitation, chemical methods that are commonly used to purify proteins and organic molecules, are used to achieve a partial size separation. For example, Fig. 6 shows broad luminescence spectra and size distributions of two fractions created by size selective precipitation from a nanocrystal colloid that initially showed photoluminescence in the 600 to 900 nm range (Wilson, Szajowski, and Brus, 1993). The smaller fraction, as shown by the high pressure liquid chromatogram, consists of only smaller, single nanocrystals and emits at 650 nm. The larger fraction consists of larger single nanocrystals, and of aggregates; and emits near 800 nm. Both groups have luminescence shifted blue of the bulk 1.1 eV Si bandgap. The absolute luminescence quantum yield of the smaller nanocrystals in dilute solution is 5.8% at 293 K. The quantum yield increases with decreasing temperature, and reaches about 50% at and below 50 K.

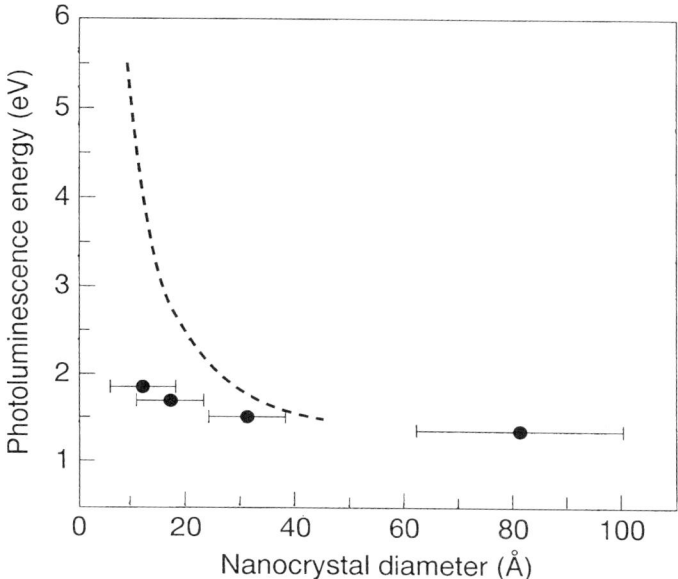

FIG. 5. Comparison of Si nanocrystals luminescence energies with calculated nanocrystal exciton energies (dashed line). (From Delerue, Allan, and Lannoo, 1993.)

Physical characterization of small nanocrystals is difficult. In larger 7 nm particles, with 5 nm Si cores, it is possible to directly observe shell structure in transmission electron microscopy (TEM), and to confirm Si core crystallinity by X-ray powder patterns (Littau et al., 1993). The X-ray powder patterns show diamond lattice Si with a lattice constant unchanged from the bulk within experimental error. However, neither method provides useful data for the smaller particles emitting near 650 nm. In very small particles, with Si cores below 2 nm, the contrast in TEM is weak, and line broadening in X-ray powder is severe. This inability to cleanly recognize small particles in TEM makes it difficult to accurately determine their presence in the presence of large Si particles, as is the case in direct π-Si characterization, and also in the aerosol colloids without chemical processing to remove the larger particles.

Si K shell X-ray near edge absorption spectra (NEXAFS) provide valuable data (Schuppler et al., 1994). Figure 7 shows this near edge structure as a function of size for partially size selected Si core-shell particles. There is a 4 eV shift between the absorption edge of Si in Si dioxide, and in elemental crystalline Si. The data show the particles are composed of both

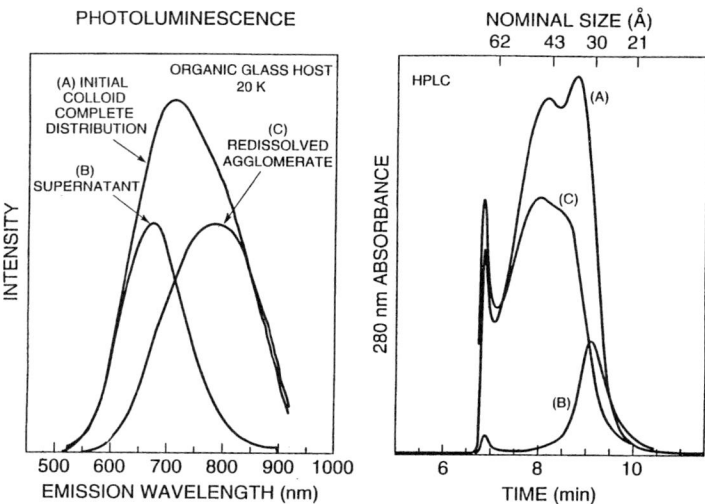

FIG. 6. Left-hand side: Spectrally corrected Si nanocrystal luminescence spectra in organic glass (350 nm excitation wavelength, 20 K). Right-hand side: Corresponding HPLC chromatograms with approximate logarithmic size calibration. Relative intensities are arbitrary. (From Wilson, Szajowski, and Brus, 1993.)

oxide and elemental Si, with an interface region about one monolayer thick. Figure 8 shows the oxidized nanocrystal structure derived from these data, and the outer oxide diameters as measured by TEM. The smallest particles emitting near 650 nm have outer diameters near 2.5 nm, and inner Si core diameters of 1.2–1.5 nm. They are actually mostly oxide in terms of volume fraction. They show PL near 650 nm that is indistinguishable in most characteristics from PL of π-Si of $\sim 80\%$ porosity.

The HF etching reaction that creates π-Si thin films initially leaves the Si surfaces with H atom termination. H atom as a nearest neighbor is invisible in Si EXAFS spectra, and so the π-Si EXAFS data show an apparent absence of Si atom nearest neighbors, when most Si atoms are part of the surface and bonded to H (Schuppler et al., 1995). In this way one can derive an average nanocrystal size in π-Si, by fitting the EXAFS data as a sum of surface atoms, and interior atoms with four Si nearest neighbors. Figure 9 is a correlation of mass weighted average size with the peak PL at room temperature, showing that as the average size decreases to about 1.4 nm the emission peak shifts blue to about 650 nm. This average size in π-Si is quite similar to the core size in the oxide shell Si nanocrystals, where an average size of about 1.2 nm has a peak near 670 nm. This correlation is further evidence that the luminescence process and emitting species are the same in both systems.

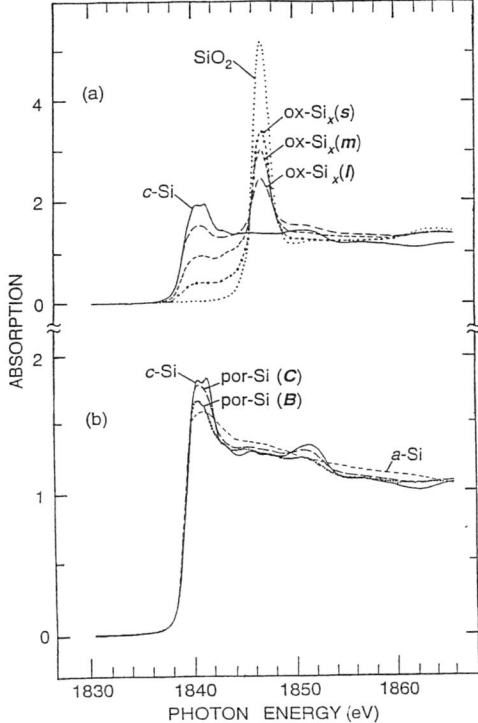

FIG. 7. (a) K-edge NEXAFS data from crystalline Si, silicon dioxide, and shell Si nanocrystals of different sizes: labeled small (s), medium (m) and large (l). (b) NEXAFS data from crystalline Si, amorphous Si, and two differently prepared π-Si samples, labeled B and C. This figure is adapted from Schuppler et al., 1994.

Dilute colloids of Si nanocrystals in ethylene glycol are optically clear, and show the pale yellow color of Si dust. Upon freezing in thin sections between sapphire plates, a relatively crack free organic glass with embedded nanocrystals is formed. The luminescence spectrum, lifetime, and excitation spectrum can be obtained in such optically thin condition. Figure 10 shows that the excitation spectrum of 600 nm luminescence at ~15 K is a featureless, monotonically increasing continuum, characteristic of an indirect material with a bandgap experimentally the same as the monitored 600 nm luminescence wavelength (Brus et al., 1995). This spectrum actually shows less structure than the equivalent, calculated spectrum for the bulk diamond lattice spectrum, in that the broad ultraviolet feature due to the 3.4 eV direct gap in bulk Si is almost washed out in Si nanocrystals.

O-passivated Si nanocrystals

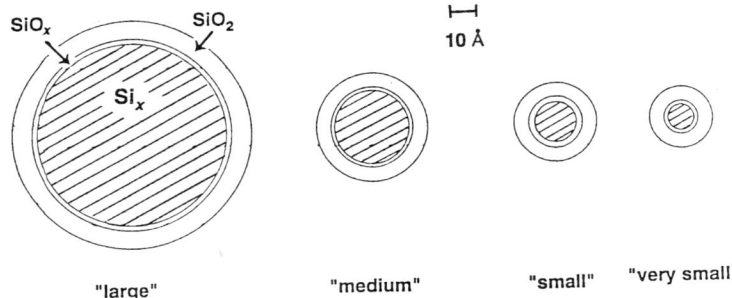

FIG. 8. Schematic illustration of different shell Si nanocrystals, drawn to relative scale. (From Schuppler et al., 1994.)

The ultraviolet excited PL is a broad featureless band, peaking near 700 nm in the particular sample shown in Fig. 11. In any one sample, smaller nanocrystals emit on the high energy side of this band, and bigger particles emit on the low energy side. Possible phonon structure in the spectra of just one nanocrystal is obscured by this size broadening. As originally shown in optical studies of π-Si luminescence (Calcott et al., 1993a, 1993b; Suemoto et al., 1993, 1994), this broadening can be probed by laser excitation of luminescence within the bandgap distribution at low temperature.

In Fig. 11, continuous wave laser excitation at 710 nm excites luminescence from just the larger nanocrystals in this sample, and shows partial TO phonon structure. The luminescence data can be fitted, as a function of excitation wavelength, by assuming a near-gaussian band gap size distribution centered at 650 nm and single nanocrystal spectra, as shown in Fig. 12. The strongest features are TO phonon transitions present in both absorption and emission. There is a weaker feature in the transverse acoustic (TA) phonon position. In absorption there is also rising continuous absorption that corresponds to absorption into the higher electronic states of an indirect gap type material. These spectra and assignments are very similar to those of π-Si (Calcott et al., 1993a, 1993b; Suemoto et al., 1993, 1994), and confirm the indirect gap, phonon induced nature of the Si nanocrystals emitting near 650 nm. The value of this size selective optical technique is that it determines the absorption and emission spectra near the bandgap, without the necessity of simultaneously measuring size or accurately calculating the exciton energy. As mentioned above, direct TEM measurement of such small sizes is quite difficult, especially in the presence of larger particles.

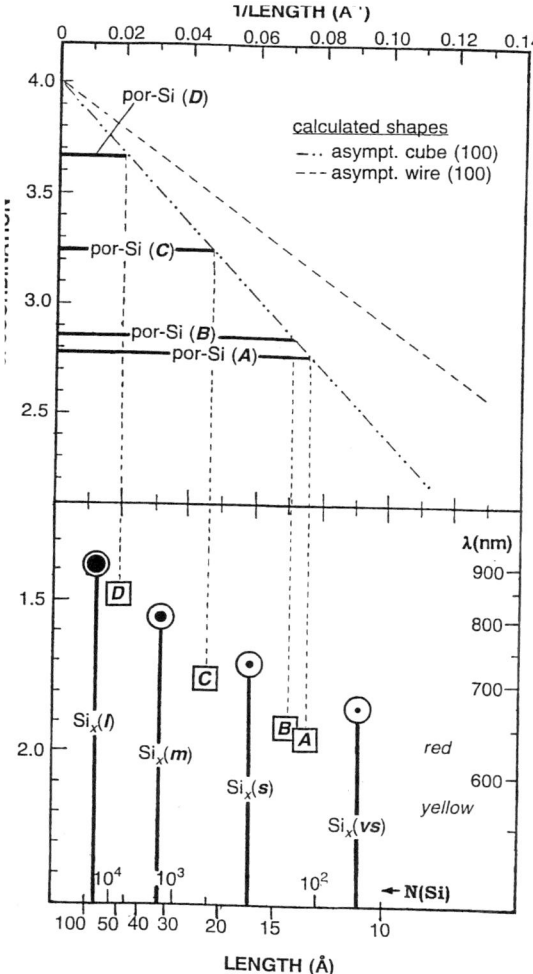

FIG. 9. Correlation between average Si coordination calculation for particles of different shapes versus their inverse characteristic length. The lower panel shows the correlation with luminescence wavelength. The average includes bulk and surface atoms, which have less than four first neighbors, explaining the trend of lower values with decreasing particle size. Experimental Si coordinations from EXAFS data of four differently prepared π-Si samples are indicated, as are upper-limit characteristic lengths. (From Schuppler et al., 1994.)

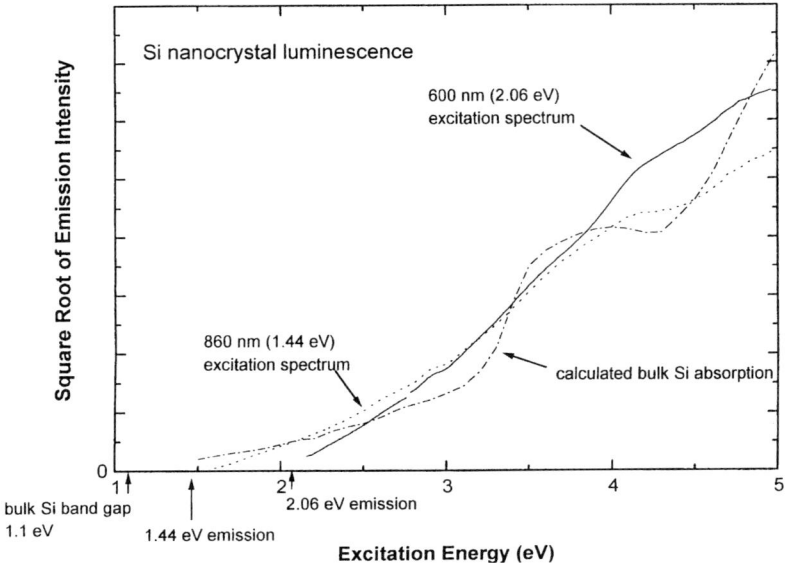

FIG. 10. Square root of 600 nm and 860 nm luminescence intensity from Si nanocrystals at ~15 K as a function of excitation radiation energy. For comparison, the figure also shows the square root of the optical absorption cross section of a small Si sphere calculated from the electric dipole term of Mie theory using bulk Si dielectric constant data. (From Brus et al., 1995.)

Figure 13 shows luminescence decay versus emission wavelength. The 630 nm lifetime τ increases from $\sim 5 \times 10^{-5}$ s at 293 K to $\sim 2.5 \times 10^{-3}$ s at 20 K. In a molecule, or in a nanocrystal with just one electron–hole pair, the decay rate (inverse lifetime) is $\tau^{-1} = \Gamma_r + \Gamma_{nr}$. The first term is the radiative rate, and the second term is the competing nonradiative rate. The observed intensity of emission is proportional to the luminescence quantum yield $QY = \Gamma_r/(\Gamma_r + \Gamma_{nr})$. From measurement of both QY and τ, both Γ_r and Γ_{nr} are determined as a function of temperature. At 20 K, Γ_{nr} is negligible with respect to Γ_r, which itself is quite slow being $\sim 10^3$ s^{-1} (Wilson, Szajowski, and Brus, 1993). Γ_r increases by an order of magnitude as temperature increases to about 150 K. Γ_{nr} increases by several orders of magnitude as temperature increases, so at room temperature the quantum yield has decreased to $\sim 5\%$. In both π-Si and nanocrystal Si, the nonradiative process is not understood. It shows some characteristics of tunneling through a barrier.

In Si nanocrystals and in π-Si, there is essentially no vibrational Franck–Condon (Stokes) shift, as this term is normally used, between the bandgap

8 SILICON POLYMERS AND NANOCRYSTALS 317

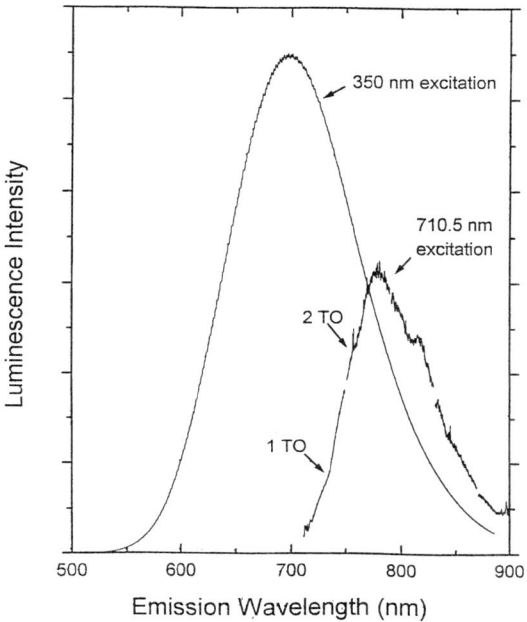

FIG. 11. Comparison of the 350 nm excited, low resolution emission spectrum of Si nanocrystals at ~15 K, and the higher resolution (0.2 nm) emission spectrum excited at 710.5 nm. Weak thresholds are observed at one and two times the TO phonon frequency in the 710.5 spectrum. (From Brus et al., 1995.)

determined by excitation spectra and the luminescence. The Franck–Condon principle indicates that strong optical transitions are vertical on the configuration coordinate diagram when the transition is electrically allowed. If the excited state oscillator is displaced with respect to the ground state oscillator along a totally symmetric coordinate, then luminescence will be Stokes shifted to lower energy than absorption. In π-Si and Si nanocrystals, the transition is electrically forbidden but vibronically allowed along the nontotally symmetric TO coordinate. While there is no real displacement of the excited state along the TO coordinate, there is a 2 TO-phonon-energy splitting between the apparent luminescence origins in absorption and luminescence, because vertical transitions are absent. In the π-Si spectroscopic data of Calcott et al. (1993a, 1993b) which shows more structure and higher resolution than the nanocrystal data, there is an additional singlet-triplet splitting of a few millielectron volts.

Figure 5 shows the peak luminescence plotted versus the calculated

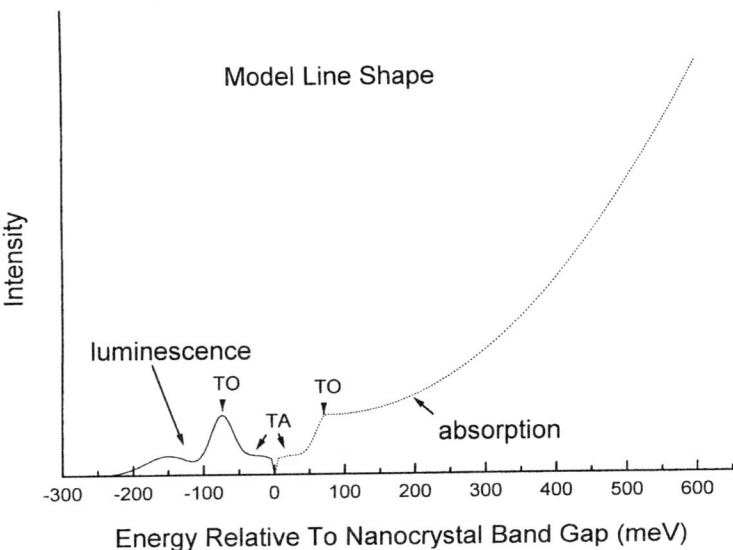

FIG. 12. Si single nanocrystal model luminescence and absorption spectra, derived from fitting spectroscopic data in Brus *et al.* (1995). The zero of energy corresponds to the exciton energy of the single nanocrystal.

exciton energy including the Coulomb interaction. While theory predicts that a 2 nm unperturbed nanocrystal has a bandgap near 2.5 eV, the luminescence of 2 nm particles appears near 1.7 eV. As discussed above, this difference does not result from a Stokes shift. While in larger nanocrystals the core Si is definitely crystalline, there is presently little structural characterization of the core Si in 2 nm and smaller nanocrystals, either in the aerosol nanocrystals or in π-Si. Perhaps such small nanocrystals have an equilibrium structure with a somewhat different bond length, or are modified in some other way. Sometime in the next decade with improved nanocrystal samples, this question of structure and bandgap versus size needs to be further explored. Nevertheless, the majority of evidence supports assignment of the emitting species to a simple volume confined electron–hole pair in a Si nanocrystal. (The pronounced polarization of π-Si luminescence proves that the transition dipole direction is determined by nanocrystal shape, and is not determined by localized traps uncorrelated with overall nanocrystal shape.) This assignment could be confirmed by observation of the expected spin- and valley-orbit electronic fine structure. We find no evidence for molecular or interface assignments. Both oxide- and H-atom terminated π-Si, as well as the oxide-terminated nanocrystals, show the same red

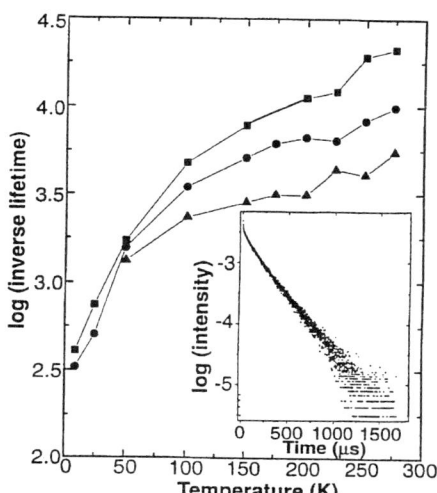

FIG. 13. Log of the inverse of the average single exponential decay time versus temperature. Emission wavelength: squares, 630 nm; circles, 730 nm; triangles, 830 nm. Insert: luminescence decay at 730 nm and 150 K. (From Wilson, Szajowski, and Brus, 1993.).

luminescence. Nanocrystals must be passivated to emit, but the chemical nature of the passivation is of secondary importance.

The smallest 1.2–1.5 nm nanocrystals, that we have been able to make, emit in the 2.0–1.7 eV range. In octasilacubane chemical compounds R_8Si_8, a cube of Si atoms is capped with organic groups R. These compounds show indirect-gap-like optical spectra, with luminescence near 2.5 eV (Matsumoto et al., 1988, Malsumoto et al., 1992). This limited evidence suggests that the smallest possible three dimensional capped Si nanocrystals emit near 2.5 eV. As size increases from this limit, the emission shifts to the red. It would be difficult to make a red-green-blue emitting display from a Si nanocrystal material such as π-Si.

3. COMPARISON BETWEEN NANOCRYSTALS AND MACROSCOPIC CRYSTALLINE Si

At liquid He temperatures the Coulombic exciton in bulk Si has similarities to the quantum confined electron–hole pair in a nanocrystal. Both are spatially confined systems, yet in both cases luminescence occurs principally by TO phonon participation rather than by emission on the ZPL. That is, in both cases the transition remains indirect, with radiative lifetimes at least

10^4 longer than that of an allowed transition. In both cases the competing multiphonon radiationless transition is negligible, and the quantum yield of luminescence is near unity once the bound pair is formed. In Si, slow multiphonon radiationless decay is a consequence of weak electron–phonon coupling, which in turn is characteristic of a covalent, strongly bound, defect free lattice.

At room temperature, however, bulk crystalline Si is fundamentally different than nanocrystal Si. In bulk Si, electron–hole pairs dissociate, and the ensemble of individual carriers interact by many body kinetics. In high quality wafers, the carriers are mobile over macroscopic distances, and recombination can be efficiency catalyzed by defects and impurities present in very low concentrations. At very low carrier densities, collisions between the free carriers are rare. In the perfect lattice, all recombination process are slow, and carriers are observed to live as long as 40×10^{-3} s at 293 K. At normal carrier densities, however, the dominate recombination process is the three body Auger process (Yablonovitch et al., 1986)

$$e + h + (e \text{ or } h) \rightarrow (e \text{ or } h) \text{ with } 1.1 \text{ eV kinetic energy} \qquad (1)$$

The third carrier carries away the bandgap recombination energy, and then rethermalizes. This process dominates the kinetics, and shortens the effective lifetime by many orders of magnitude.

The luminescence quantum yield increases in nanocrystal Si at 293 K, not because Γ_r increases, but because Γ_{nr} decreases, with respect to bulk crystalline Si (Brus, 1994; Brus et al., 1995). In a material made of well passivated Si nanocrystals, an electron–hole pair in one nanocrystal is electrically isolated from other pairs in other nanocrystals. In this situation the three body Auger process in Eq. (1) is significantly decreased. In addition, a rare impurity or lattice defect can only quench luminescence in one nanocrystal, having a volume much smaller than the macroscopic region of mobility in bulk wafers. In nanocrystal Si, an electron and hole are superimposed for long periods of time, not by their Coulomb attraction but by confinement in one nanocrystal. Because the pair remains spatially superimposed, and competing nonradiative processes are depressed, the QY is far larger in nanocrystal Si than in bulk Si. In almost any electronic material, PL increases as mobility decreases. The point that some of the increased luminescence in π-Si is due to reduced mobility has also been made in Reed et al. (1992); Koch, Petrova-Koch, and Muschik (1993).

The indirect gap semiconductor AgBr, which is used as micron sized crystals in photographic film, shows an analogous series of changes in nanocrystals. In 6–9 nm nanocrystals, the radiative lifetime increases slightly in comparison with bulk AgBr, yet the nanocrystals remain essentially indirect gap. However, experimentally there is a large increase of luminescence QY in nanocrystals because of impurity exclusion with decreasing size (Johansson, McLendon, and Marchetti, 1991, 1992, 1993; Chen et al., 1994).

4. Physical Size Regimes for Individual Nanocrystals

The work on Si and AgBr, along with the earlier work on the direct gap II–VI materials and CuCl, allows us to recognize broad size regimes in Fig. 14 (Brus, 1991; Brus et al., 1995; Efros and Efros, 1982). Spectroscopic properties can be broken into molecular, quantum dot (e.g., nanocrystal), polariton, and bulk regions. These labels represent the evolution of molecular to unit-cell structures, discrete electronic states to continuous bands, and weak dipole scattering into strong polariton electromagnetic scattering. Bare Si_{10} and Si_{45} show bonding strongly distorted from the diamond lattice, in order to stabilize the broken bonds of surface atoms. This is the molecular regime. If the surface is passivated with a species such as H that

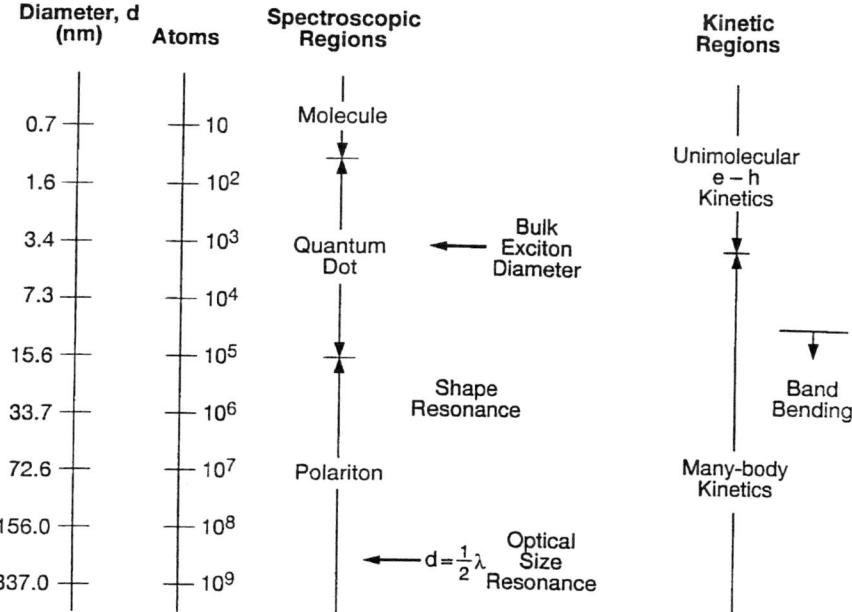

FIG. 14. Schematic representation of size regimes used for describing spectroscopic and excited state kinetic properties of semiconductor nanocrystals. (From Brus et al., 1995.)

preserves sp³ local hybridization, then the cross-over to bulk bonding in the nanocrystal regime occurs at a smaller size than otherwise. Bulk optical properties appear only when size is much larger than an optical wavelength.

In the nanocrystal or quantum dot regime, the structure is that of an excised fragment of the bulk lattice, while strong three dimensional quantum confinement is present. It is interesting to compare the indirect gap Si shell nanocrystals, and the direct gap CdSe nanocrystals. Both show a bandgap shift to higher energy. CdSe nanocrystals show discrete excited state optical absorption (Murray, Norris, and Bawendi, 1993), while in Si the optical absorption is continuous. CdSe nanocrystal spectra actually show more structure than bulk CdSe, while Si nanocrystal spectra show less structure than bulk Si. Two effects contribute to this difference: First, the spacing of quantized conduction band states is much larger in CdSe than in Si, because of the very small CdSe electron effective mass, which is $\sim 0.05\ m_e$. Second, in CdSe only a few transitions, out of the many possible, appear strongly because of electric dipole selection rules. In Si, all possible transitions appear to be present at similar intensity due to vibronic interaction. In Si, the discrete yet dense spectrum appears continuous, as individual transitions overlap with respect to their line widths, at least in the present Si nanocrystal samples.

The previous discussion of radiationless transitions in Si and AgBr indicates that size regimes exist in kinetic properties as well. At typical excitation intensities, large crystallites contain several dissociated pairs and exhibit many interacting carrier recombination kinetics. Small nanocrystals can exhibit size exclusion of bulk lattice defects and impurities, and also do not show surface band bending if dopants are excluded due to small size. They behave as intrinsic dielectric particles. At typical excitation intensities, unimolecular decay of a single, quantum confined electron–hole pair is observed.

IV. Electron Transport in Porous Nanocrystal Materials

Transport and electroluminescence (EL) in π-Si samples of high porosity are very sensitive to polar fluids in the pore structure. Dry π-Si is extremely insulating; if dilute methanol vapor is admitted, the conductivity rises by 10^4 (Ben-Chorin, Kux, and Schechter 1994; Ben-Chorin et al., 1995). The methanol adsorbs onto and wets the interior Si-H covered surfaces, but at such low pressures it does not condense and fill the pore with liquid. Additionally, the quantum efficiency of diode EL is significantly higher if a liquid electrolyte junction is used (Halimaoui, 1991; Bressers et al., 1992;

Bsiesy et al., 1993; Ligeon et al., 1993; Kooij, Despo, and Kelly, 1995). While transport is enhanced, red PL is quenched by polar liquids and vapors (Li et al., 1993; Dubin, Ozanan, and Chazalviel, 1994; Ichinolhe et al., 1995; Lauerhass et al., 1992; Lauerhaas and Sailor, 1993). These effects are reversible and are not due to a permanent surface chemical reaction, or to electrical doping of the nanocrystals, by the polar molecules. Generally speaking, these observations are similar to kinetic solvation effects that occur for molecules and proteins.

If a nanocrystal contains a hole or an electron and thus has a net charge, an electric field exits the nanocrystal and interacts with the local environment. In π-Si of high porosity this external field dominates both single nanocrystal energetics (Brus, 1983, 1984; Babic, Tsu, and Greene, 1992; Tsu and Babic, 1994; Lannoo, Delerue, and Allan, 1995; Chazalviel, Ozanam, and Dubin, 1994; Martin et al., 1994) and transport phenomena (Brus, 1996), and provides a way of "tuning" the transport behavior via the polarity of the media in the pores. The electron affinity of a Si nanocrystal decreases by ΔA from the approximate 4 eV value of an electron on the conduction band edge in bulk Si:

$$\Delta A = KE(d) + \langle 1S|P(r)|1S \rangle$$

where $KE(d)$ is the kinetic energy of localization in a nanocrystal of size d. The second term is the average of the dielectric polarization energy over a 1S wave function in the nanocrystal; it tends to be larger than $KE(d)$ in π-Si of high porosity. The second term can be approximated as (Lannoo, Delerue and Allan, 1995)

$$\langle 1S|P(r)|1S \rangle \simeq (e^2/d)(1/\varepsilon_{out} - 1/\varepsilon_{Si}) + \delta\Sigma$$

Here ε_{out} is the dielectric constant of the medium outside the nanocrystal, and

$$\delta\Sigma = \frac{0.94e^2}{\varepsilon_{Si}d}\left(\frac{\varepsilon_{out} - \varepsilon_{Si}}{\varepsilon_{out} - \varepsilon_{Si}}\right)$$

Transport in π-Si can be modeled as the hopping of electrons in a system of touching nanocrystals of variable diameter (Brus, 1996). If an electron jumps from a 2 nm Si nanocrystal to a touching 4 nm nanocrystal in vacuum, the exothermicity is approximately $\Delta G = \Delta A(4\,\text{nm}) - \Delta A(2\,\text{nm}) = -0.5\,\text{eV}$. The electron energy levels in the 4 nm nanocrystal are discrete, and in general no state in the 4nm particle is resonant with the electron in the 2 nm particle. If the electron hops directly to the lowest, bandgap state in the 4 nm particle, then ΔG must be dissipated in some coupled "vibrational" degree of freedom.

Reorganization energies λ are a measure of the electron's coupling to various "slow" degrees of freedom (Marcus, 1956, 1960; Levich and Dogonadze, 1959; Jortner, 1976; German and Kuznetsov, 1981; Kharkats, 1976; Miller et al., 1995; Newton and Sutin, 1984; Holstein, 1959; Henry and Lang, 1977; Ridley, 1978). For example, the electron is weakly coupled to Si acoustic modes by the deformation potential. λ is the vertical Franck–Condon shift between the acoustic mode harmonic potential with the electron on and off the nanocrystal. For 2 nm Si nanocrystals, this λ is only ~12 meV. A polar liquid, for example water, outside the nanocrystal is orientationally polarized by the electron's field, and the λ associated with this polarization is much larger ~400 meV (Brus, 1996). The "electron–phonon" coupling to the nearby water is far stronger than the coupling to the internal Si vibrations; this is the source of the solvent effect on transport kinetics. At room temperature the electron hoping rate from one nanocrystal to another is $k(s^{-1}) \propto \exp(-G_{act}/kT)$, where the activation energy $G_{act} = (\lambda + \Delta G)^2/4\lambda$. For fast, activationless transfer, the negative exothermicity ΔG must equal the total λ, which is positive. Thus for fast electron transfer from a 2 nm to a 4 nm nanocrystal, λ must be ~0.5 eV. This only happens in a polar fluid; the orientational motions of the polar molecules take up the exothermicity.

Figure 15 is a plot of $\log k$ versus the size of the electron accepting nanocrystal, for a fixed 2 nm donor size. For a 2 nm acceptor in vacuum, the transfer is resonant and the rate is fast. It is also fast for specific larger sizes where resonant transfer to excited 2S and 3S states is possible. For other sizes transfer is very slow due to a huge activation energy, as λ is small. In water where λ is large, the resonant nature of transfer is lost. Transfer is fast for all cases *except* resonant 2 nm–2nm transfer. In this case, the polarization of the water around the initial nanocrystal tends to "self-trap", or stabilize, the carrier. This situation is analogous to strong lattice deformation around a carrier in an insulator.

In a nanocrystal material, and in π-Si specifically, a polar liquid tends to assist nonresonant transport when the porosity is high and the size distribution is wide. It also plays a critical role in electron–hole injection, where an electron in one nanocrystal hops into a neighboring nanocrystal containing a hole. In this situation, even if the nanocrystals have the same size, the exothermicity is high because two charged nanocrystals convert into two neutral nanocrystals. Thus electron–hole injection is fast only in the presence of a polar liquid. These results help to explain why conductivity increases, and EL improves, in the presence of methanol or aqueous

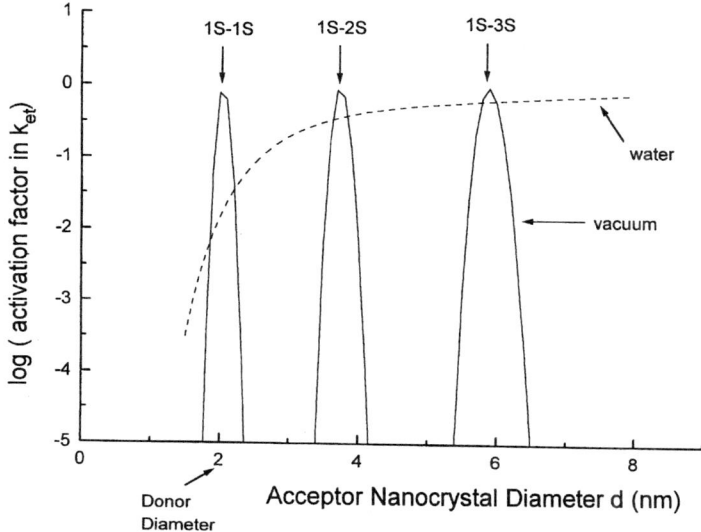

FIG. 15. Log of the unimolecular electron hoping rate in water and in vacuum, for a 2 nm Si donor nanocrystal, as a function of acceptor nanocrystal size. (From Brus, 1996.)

electrolyte. Polar solvent also quenches red PL. This experimental observation appears to be related to the fact that some preparations of π-Si make surfaces that are not completely terminated with H. If a nanocrystal has a mid-gap dangling bond state due to a missing H atom, then a photoexcited electron could trap on this state. Such nonradiative trapping competes with luminescence. This trapping step has an exothermicity of 1 eV in a 1.2.– 1.5 nm particle with a bandgap near 2.0 eV. The rate of this process is faster by many orders of magnitude in the presence of a polar liquid.

Generally speaking, dry π-Si behaves electrically like a resonant tunneling device. However, wet π-Si show fast, nonresonant electron transfer similar to that commonly seen in biological and molecular processes in polar environments. This solvation effect on kinetics demonstrates yet another molecule-like property of nanocrystals.

Acknowledgments

This article reviews research conducted while the author was a member of Bell Laboratories, now part of Lucent Technologies. I express deep appreciation to former post-doctoral students A. P. Alivisatos, M. G.

Bawendi, and K. A. Littau. Many colleagues have collaborated in the work described here: M. L. Steigerwald, T. D. Harris, A. Muller, A. R. Kortan, A. Harris, D. C. Douglass, A. R. Hull, R. L. Opila, P. J. Carroll, P. J. Szajowski, P. H. Citrin, W. L. Wilson, and L. Rothberg. I have benefited from the advice of D. Monroe, M. L. Mandich, M. L. Hybertsen, F. Stillinger, and J. Trautman. Physical and materials chemistry thrived under the leadership of J. C. Tully, M. J. Cardillo, and R. H. Laudise.

REFERENCES

Allan, G., Delerue, C., and Lannoo, M.(1993). *Phys. Rev. B* **48**, 7951.
Babic, D. Tsu, R., and Greene, R. F. (1992). *Phys. Rev. B* **45**, 14150.
Banyai, L., and Koch, S. W. (1993). *Semiconductor Quantum Dots*. World Scientific, Singapore.
Ben-Chorin, M., Kux, A., and Schechter, I. (1994). *Appl. Phys. Lett.* **64**, 481.
Ben-Chorin, M., Moller, F., Koch, F., Schirmacher, W., and Eberhard, M. (1995). *Phys. Rev. B* 51, 2199.
Bradfield, P. L., Brown, T. G., and Hall, D. G. (1988). *Phys. Rev. B* **38**, 3533.
Brandt, M. S., Breitschwerdt, A., Fuchs, H. D., Hopner, A., Rosenbauer, M. Stutzmann, M., and Weber M. (1992). *Appl. Phys. A*, **54**, 567.
Bressers, P., Knapen, J., Meulenkamp, E., and Kelly, J. (1992). *Appl. Phys. Lett.* **61**, 108.
Brus, L. E. (1983, 1984). *J. Chem. Phys.* **79**, 5566; *J. Chem. Phys.* **80**, 4403.
Brus, L. E. (1991). *Appl. Phys. A* **53**, 465.
Brus, L. E. (1994). *J. Phys. Chem.* **98**, 3575.
Brus, L. E., Szajowski, P., Wilson, W., Harris, T., Schuppler, S., and Citrin, P. (1995). *J. Am. Chem. Soc.* **117**, 2915.
Brus, L. E. (1996). *Phys. Rev. B* **53**, 4649.
Bsiesy, A., and co-workers (1993). *Phys. Rev. Lett.* **71**, 637.
Calcott, P. D. J., Nash, K. J., Canham, L. T., Kane, M. J., and Brumhead, D. (1993a). *J. Phys. Condens. Matter* **5**, L9; (1993b) *J. Lumin* **57**, 257.
Canham, L. T. (1990). *Appl Phys. Lett.* **57**, 1046.
Chazalviel, J., Ozanam, F., and Dubin, V. (1994). *J. Phys. I France* **4**, 1325.
Chen, W., McLendon, G., Marchetti, A., Rehm, J. M., Freedhoff, M., and Myers, C. (1994). *J. Am. Chem. Soc.* **116**, 1585.
Cohen, M. L., and Chelikowsky, J. R. (1988). *Electronic and Optical Properties of Semiconductors*. Springer Verlag, Berlin.
Cuthbert, J. D. (1970). *Phys. Rev. B*1, 1552.
Dahn, J. R., and co-workers (1994). *J. Appl. Phys.* **75**, 1946.
Deak, P., Rosenbauer, M., Stutzmann, M., Weber, J., and Brandt, M. S. (1992). *Phys. Rev. Lett.* **69**, 2531.
Delerue, C., Allan, G., and Lannoo, M. (1993). *Phys. Rev. B* **48**, 11024.
Delley, B., and Steigmeier, E. F. (1993). *Phys. Rev. B* **47**, 1397.
Dubin, V., Ozanam, F., and Chazalviel, J. (1994). *Phys. Rev. B* **50**, 14867.
Efros, Al. L., and Efros, A. L. (1982). *Sov. Phys. Semicond.* **16**, 1209.
Friedman, S. L., Marcus, M. A., Adler, D. L., Xie, Y.-H., Harris, T. D., and Citrin, P. H. (1993). *Appl. Phys. Lett.* **62**, 1934.

German, E., and Kuznetsov, A. (1981). A review of dielectric continuum reorganisation energy. *Spectrochim. Acta* **26**, 1595.
Halimaoui, A., and co-workers (1991). *Appl. Phys. Lett.* **59**, 304.
Haynes, J. R., Lax, M., and Flood, W. F. (1961). *Proc. Int. Conf. Semicond. Phys. Prague 423.*
Henry, C., and Lang, D. (1977). *Phys. Rev. B* **15**, 989.
Hybertsen, M. S. (1994). *Phys. Rev. Lett.* **72**, 1514.
Ichinolhe, T., Nozaki, S., Ono, H., and Morisaki, H. (1995). *Appl. Phys. Lett.* **66**, 1644.
Johansson, K. P., McLendon, G. P., and Marchetti, A. P. (1991). *Chem Phys. Lett.* **179**, 32.
Johansson, K. P., Marchetti, A. P., and McLendon, G. P. (1992). *J. Phys. Chem.* **96**, 2873.
Jortner, J. (1976). *J. Chem. Phys.* **64**, 4860.
Kanemitsu, Y., Suzuki, K., Nakayoshi, Y., and Masumoto, Y. (1992). *Phys. Rev. B* **46**, 3916.
Kanemitsu, Y., Suzuki, K., Masumoto, Y., Komatsu, T., Sato, K., Kyushin, S., and Matusmoto, H. (1993). *Sol. St. Comm.* **86**, 545.
Kanzaki, H., and Tadakuma, Y. (1991). *Sol. St. Comm.* **80**, 33.
Kharkats, Yu. (1976). *Elektrokhimiya* **12**, 1866.
Koch, F. Petrova-Koch, V., and Muschik, T. (1993). *J. Lumines.* **57**, 271.
Kooij, E., Despo, R., and Kelly, J. (1995). *Appl. Phys. Lett.* **66**, 2552.
Koyama, H., Araki, M., Yamamoto, Y., and Koshida, N. (1991). *Jpn. J. Appl. Phys.* **30**, 3606.
Lannoo, M., Delerue, C., and Allan, G. (1995). *Phys. Rev. Lett.* **74**, 3415.
Lauerhass, J., Credo, G., Heinrich, J., and Sailor, M. (1992). *J. Am. Chem Soc.* **114**, 1911.
Lauerhaas, J., and Sailor, M. (1993). *Science* **261**, 1567.
Lehmann, V., and Gösele, U. (1991). *Appl. Phys. Lett.* **58**, 856.
Levich, V., and Dogonadze, R. (1959). *Dokl. Akad. Nauk. SSR* **124**, 123.
Li, K., Tsai, C., Sarathy, J., and Campbell, J. (1993). *Appl. Phys. Lett.* **62**, 3192.
Ligeon, M., and co-workers (1993). *J. Appl. Phys.* **74**, 1265.
Littau, K. A., Szajowski, P. F., Muller, A. J., Kortan, R. F., and Brus, L. E. (1993). *J. Phys. Chem.* **97**, 1224.
Marchetti, A. P., Johansson, K. P., and McLendon, G. P. (1993). *Phys. Rev. B* **47**, 4268.
Marcus, R. (1956). *J. Chem. Phys.* **24**, 966.
Marcus, R. (1960). *Faraday Disc. Chem. Soc.* **29**, 21.
Martin, E., Delerue, C., Allan, G., and Lannoo, M. (1994). *Phys. Rev. B* **50**, 18258.
Matsumoto, H., Higuchi, K., Hoshino, Y., Koike, H., Naoi, Y., and Nagai, Y. (1988). *J. Chem. Soc., Chem. Commun.* 1083.
Matsumoto, H., Higuchi, K., Kyushin, S., and Goto, M. (1992). *Angew. Chem. Int. Ed. Engl.* **31**, 1354.
Miller, R. J. D., and Michl, J. (1989). A review of polysilanes. *Chem. Rev.* **89**, 1359.
Miller, R. J. D., and co-workers (1995). *Surface Electron Transfer Processes.* VCH Publishers, New York, chapters 1 and 4.
Moritomo, Y., Tokura, Y., Tachibana, H., Kawabata, Y., and Miller, R. D. (1991). *Phys. Rev. B* **43**, 14746.
Murray, C. B., Norris, D. J., and Bawendi, M. G. (1993). *J. Am. Chem. Soc.* **115**, 8706.
Newton, M., and Sutin, N. (1984). A review of molecular dynamics applications. *Annu. Rev. Phys. Chem.* **35**, 437.
Petrova-Koch, V., and co-workers (1992). *J. Appl. Phys.* **61**, 943.
Proot, J. P., Delerue, C., and Allan, G. (1992). *Appl. Phys. Lett.* **61**, 1948.
Reed, A. J., Needls, O. R. J., Nash, K. J., Canham, L. T., Calcott, P. D. J., and Qteish, A. (1992). *Phys. Rev. Lett.* **69**, 1232.
Ridley, B. (1978). *J. Phys. C* **11**, 2323.
Schall, U., Thonke, K., and Sauer, R. (1986). *Phys. Stat. Sol.(b)* **137**, 305.
Schmitt-Rink, S., Miller, D. A. B., and Chemla, D. S. (1987). *Phys. Rev. B* **35**, 8113.

Schuppler, S., Friedman, S. L., Marcus, M. A., Adler, D. L., Xie, Y.-H., Ross, F. M., Harris, T. D., Brown, W. L., Chabal, Y. J., Brus, L. E., and Citrin, P. H. (1994). *Phys. Rev. Lett.* **72**, 2648; (1995). *Phys. Rev. B* **52**, 4910.

Soos, Z. G., and Hayden, G. W. (1990). *Chem. Phys.* **143**, 199.

Stutzman, M., Brandt, M. S., Rosenbauer, M., Weber, J., and Fuchs, H. D. (1993). *Phys. Rev. B* **47**, 4806.

Suemoto, T., Tanaka, K., Nakajima, A., and Itakura, T. (1993). *Phys. Rev. Lett.* **70**, 3659.

Suemoto, T., Tanaka, K., and Nakajima, A. (1994). *J. Phys. Soc. Jpn.* (Suppl. B) **63**, 190.

Takagahara, T., and Takeda, K. (1992). *Phys. Rev. B* **46**, 15578.

Takeda, K., and Shiraishi, K. (1989). *Phys. Rev. B* **39**, 11028.

Takeda, K., Shiraishi, K., and Matsumoto, N. (1990). *J. Am. Chem. Soc.* **112**, 5043.

Takeda, K., and Shiraishi, K. (1993). *Sol. St. Comm.* **85**, 301.

Thorne, J. R., Ohsako, Y., Zeigler, J. M., and Hochstrasser, R. M. (1989). *Chem. Phys. Lett.* **162**, 455.

Tsu, R., and Babic, D. (1994). *Appl. Phys. Lett.* **64**, 1806.

Van de Walle, C. G., and Northrup, J. E. (1993). *Phys. Rev. Lett.* **70**, 1116.

Wang, Y., and Herron, N. (1991). *J. Phys. Chem.* **95**, 525.

Weiss, A., Beil, G., and Meyer, H. (1979). *Z. Naturforsch.* **34b**, 25.

Weller, H. (1993). *Angew. Chem. Int. Ed. Engl.* **105**, 41.

Wilson, W. L., and Weidman, T. W. (1991). *J. Phys. Chem.* **95**, 4568.

Wilson, W. L., Szajowski, P. F., and Brus, L. E. (1993). *Science* **262**, 1242.

Yablonovitch, E., Allara, D. L., Chang, C. C., Gmitter, T., and Bright, T. B. (1986). *Phys. Rev. Lett.* **57**, 249.

Index

A

Ab initio calculations, 161, 179, 275, 292
Alloying for band structure engineering, 9–11
Alloy layers, silicon-germanium, 9–11, 54–58
Amorphous silicon (a-Si), 16, 20
Annealing experiments, 61–62
ARROW (Anti-Resonant Reflecting Optical Waveguide), 105
Atomic force microscopy (AFM), 236
Atomic layer superlattices, silicon germanium, 6–9, 59–63
Auger process, 13, 80
Auger recombination, nonradiative, 286–292

B

Band gaps. See Energy band gaps
Band structure engineering, 9–11
Band-to-band recombination, 16, 21
Barrier heights, 188
Beryllium-doped SiGe alloys, 94
  epitaxial growth and, 99–103
  isoelectronic bound exciton emission from, 94–103
  photoluminescence from SiGeSi QWs, 97–99
  photoluminescence from thick, 95–96
Blue PL band, 16, 212–213, 222–224
Bohr radius, free exciton, 14
Bond-and-etchback process, 105
Brillouin zone folding in atomic layer superlattices, 6–9, 52–54
Buffer layer, graded, 8, 26, 41

C

Capacitance-voltage (CV) profiling, 136, 137
Carrier mobility, 234
Chemical etching, 14, 26
Chemical methods used to fabricate Ge nanocrystals, 168–170
Chemical vapor deposition (CVD), 55, 118–120
Conduction band structure in semiconductors, 4
Co-sputtering, used to fabricate Si and Ge nanocrystals, 166–168
Coulomb charging effects, 290, 295–298
Crystalline silicon (c-Si)
  electroluminescence and, 85, 93–94
  isoelectronic bound exciton emission from, 83–94
  isoelectronic impurities in, 83
  photoluminescence from chalcogen-related centers, 91–93
  photoluminescence from Si:In, Si:Al, and Si:Be, 85–91
  quantum efficiency, 3
  sample preparation and processing, 83–85

D

Dangling bonds, 16, 24, 225, 259, 263, 272
  nonradiative recombination of, 279–286
Deep level transient spectroscopy (DLTS), 136, 137
Defect engineering, 206–207
Density-functional theory, 143

329

Diamond structure semiconductors, 4
Donor binding energy, 294–295

E

Effective dielectric constant, 295–298
Effective mass approximation (EMA), 17–20, 254, 255–256
Effective masses, electron and hole, 20–21
Electrical properties
  erbium doped Si, 136–139
  porous Si, 233–236
Electrochemical etching of porous Si and Ge, 163–165, 207
Electroluminescence (EL)
  c-Si and, 85, 93–94
  GeSiGE coupled QWs and, 64
  impurity centers introduced to increase efficiency, 11–13
  multi-quantum wells and, 55, 57
  nanoparticle, 19
  porous silicon and, 14–15, 238–246
  quantum dots and, 23–24
  quantum wires and, 22–23
  Si-Ge and, 8, 10, 55, 57–58
  Si:In and, 11, 85
Electron cyclotron resonance plasma enhanced CVD (ECR-PECVD), 118–119
Electronic properties
  effective mass approximation, 17–20, 254, 255–256
  empirical pseudopotential, 254, 255–258
  empirical tight-binding, 254, 255–258, 272, 274–277
  local density approximation, 254, 255–258, 272, 275–277
Electronic structure, erbium, 142–149
Electron paramagnetic resonance (EPR) studies, 143
Electron transport in porous Si, 322–325
Empirical pseudopotential (EPS), 254, 255–258
Empirical tight-binding (ETB), 254, 255–258, 272, 274–277
Energy band gaps, 4, 6
  quasi-direct, 7, 39, 52–53

Energy band gaps and light emission in atomic layer structures
  background research into, 37–40
  bandgaps determined, 44–48
  band minima, 44–45
  band offsets, 48–52
  Brillouin zone folding, 52–54
  germanium-silicon-germanium structures and interfaces, 64–67
  hydrostatic stress and conduction band shifting, 46
  hydrostatic stress and valence band shifting, 46–47
  lateral confinement in silicon-germanium QWs, 67–70
  minibands, formation of, 50, 52
  photoluminescence and electroluminescence, 8, 10, 54–70
  silicon-germanium alloy layers and quantum wells, 9–11, 54–58
  silicon-germanium short period superlattices, 6–9, 59–63
Energy band structure
  in polymers, 304–307
  in semiconductors, 4
Epitaxial growth, beryllium doping and, 99–103
Epitaxial lift-off technique, 26
Epitaxial SiGe alloy layer waveguide, 106
Epitaxial Si waveguide, 105
Epitaxy
  ion beam, 120–121
  molecular beam, 7, 19, 38, 55, 95, 118
  solid phase, 117
Erbium (Er) in Si, 13, 27, 216–217
  advantages of, 112
  complexes, 148–149
  diffusivity and solubility, 121–142
  electrical properties, 136–139
  electronic structure, 142–149
  excitation and de-excitation processes, 139–142
  isolated impurities, 145–148
  ligands, 133–135
  light emitting diodes designed from, 150–153
  physics of light emission, 127–132
Erbium doping of Si
  chemical vapor deposition, 118–120

ion beam epitaxy, 120–121
ion implantation, 113–115
molecular beam epitaxy, 118
solid phase epitaxy, 117
EXAFS, 312
Excess-stress concept, 41
Exchange splitting, 262–269
Exciton binding, 79–82
Excitons, 6
　self-trapped, 271–279

F

Fermi golden rule, 258, 288
Fourier-transform infrared (FTIR), 162, 167
Franck-Condon shift, 271, 281, 288, 316–317

G

Gallium arsenide, 2, 26
　physical properties of, 4–6
GaP, 11, 79
　isoelectronic impurities in, 82–83
Germanium rich islands, 68–70
Germanium-silicon-germanium structures and interfaces, 64–67
Glass optical waveguide, 104–105

H

Hamiltonian matrix, 263–264
Hardness, micro-, 236–237
Hartree-Fock (HF) theory, 143
Huang-Rhys factor, 281
Hybrid methods for integrating direct gap materials with Si, 2, 25–26
Hydrogenic impurities, 292–295
Hydrogen passivation, 199–200
Hydrostatic stress
　conduction band shifting and, 46
　valence band shifting and, 46–47

I

Indirect bandgap limitations, methods for handling

alloying for band structure engineering, 9–11
Brillouin zone folding in atomic layer superlattices, 6–9
hybrid methods, use of, 25–26
impurity centers introduced to increase luminescence efficiency, 11–13
nanostructures, 14–24
polymers and molecules containing silicon, 24–25
Indium gallium arsenide multiple QW laser on Si, 26
Indium phosphide (InP), 26
Infrared PL bands, 16, 213–216, 224–225
Ion beam epitaxy (IBE), 120–121
Ion implantation, 113–115
　used to fabricate Si and Ge nanocrystals, 166–168
Isoelectronic bound exciton emission, Be-doped SiGe alloys, 94
　epitaxial growth and, 99–103
　photoluminescence from SiGeSi QWs, 97–99
　photoluminescence from thick, 95–96
Isoelectronic bound exciton emission, c–Si
　electroluminescence and, 85, 93–94
　photoluminescence from chalcogen-related centers, 91–93
　photoluminescence from Si:In, Si:Al, and Si:Be, 85–91
　sample preparation and processing, 83–85
Isoelectronic bound excitons (IBEs), radiative decay of, 77
Isoelectronic centers, 11–12
Isoelectronic complexes, annealing for formation of, 85
　heat treatment and rapid quenching for, 84–85
Isoelectronic impurities
　atoms and complexes, 11–12, 78–79
　in crystalline silicon, 11–12, 83
　device considerations, 103–106
　energy levels with bound excitons, 80
　exciton binding, 79–82
　in GaP, 82–83

J

Junction photocurrent spectroscopy (JPCS), 141–142

K

k-conservation selection rule, 6, 77, 158

L

Landé interval rule, 128, 129
Laser evaporation methods, pulsed, 158
Lasers, Si-based, 3, 26, 27, 29
Lateral confinement in silicon-germanium QWs, 67–70
Lattice mismatch, 26
Lifetime, radiative, 258–261
Ligands, 133–135
Light emission, physics of
  in REs, 127–132
  in Si, 6
Light emission in silicon, 206–207
  See also Energy band gaps and light emission in atomic layer structures
  alloying for band structure engineering, 9–11
  Brillouin zone folding in atomic layer superlattices, 6–9
  hybrid methods, use of, 25–26
  impurity centers introduced to increase luminescence efficiency, 11–13
  methods for handling indirect bandgap limitations, 6–26
  nanostructures, 14–24
  optoelectronic devices problems with, 2–4
  polymers and molecules containing silicon, 24–25
  prospects for optoelectronic devices using, 26
Light emitting diodes (LEDs), 3
  compatibility with microelectronics, 244–246
  designed from Er-doped Si, 27, 150–153
  GaAs, 26
  GaP, 82
  porous silicon, 17, 27, 238–246
  power efficiency, 240–241
  quantum dot, 23, 28
  response time, 241–243
  silicon-germanium, 11
  spectral coverage, 243–244
  stability, 239–240

Linear combination of atomic orbitals (LCAO), 292, 297
Local density approximation (LDA), 254, 255–258, 272, 275–277
Luminescence wavelengths, changing, 194–200

M

Materials engineering, 6
Molecular beam epitaxy (MBE), 7, 19, 38, 55, 95, 118

N

Nanoclusters, 18–19
Nanocrystals
  See also Nonradiative recombination; Radiative recombination
  chemical methods used to fabricate Ge, 168–170
  co-sputtering and ion implantation used to fabricate Si and Ge, 166–168
  decomposition of silane gas, 162
  defined, 157
  electrochemical etching of porous Si and Ge, 163–165
  exchange splitting and symmetry of, 262–269
  excitation power dependence of red and blue PL in porous Si, 194–200
  fabrication of Si and Ge, 158–170
  nonlinear optical properties of porous Si, 18, 189–194
  optical absorption, 159, 171–172
  organic synthesis and purification techniques used to produce Si, 158–162
  passivated silicon, 308–322
Nanocrystals, passivated silicon
  compared with macroscopic crystalline Si, 319–321
  optical properties, 18, 308–309
  size regimes for, 321–322
  synthesis, characterization, and photoluminescence, 309–319
Nanocrystals, photoluminescence and surface-oxidized, 18, 170
  decay dynamics, 185–189

resonantly excited luminescence
  spectrum, 174–177
size dependence of peak energy,
  171–174
size dependence of PL intensity, 178
three region model, 177–185
Nanostructures
to handle indirect bandgap limitations,
  14–24
  nanoclusters, 18–19
  porous silicon, 14–18
  quantum dots, 23–24, 28
  quantum wells, 19–22, 28
  quantum wires, 22–23
NEXAFS, 311–312
Nonbridging oxygen hole center (NBOHC),
  221–222
Nonlinear optical properties of porous Si,
  18, 189–194
Nonoptical properties of porous Si, 233–237
Nonradiative recombination, 11–12, 206
  Auger recombination, 286–292
  dangling bonds and, 279–286
No phonon (NP) lines
  GeSiGe coupled QWs and, 64, 67
  isoelectronic bound excitons and, 77
  Si:Al-N and, 87
  Si-Ge alloy layers/quantum wells and, 55,
    56
  Si-Ge short period superlattices and, 8,
    59–62
  Si:In and, 11, 85, 86

O

Optical absorption, 14, 16–18, 159, 171–172, 260–261, 273
Optical amplifiers, 3
Optical detectors, 3, 26, 28
Optical fibers, 3, 26
Optical microcavity, 17, 21, 28
Optical modulators, 2–3, 28
Optical properties, passivated silicon, 308–309
Optical properties, porous Si nonlinear, 18,
  189–194
Optical waveguides, 3, 26–28, 103–106
Optoelectronic devices, 2–4
  prospects for silicon based, 26–29

Optoelectronic integrated circuits (OEICs),
  3
Organic synthesis and purification
  techniques used to produce Si
  nanocrystals, 158–162

P

Photoelectron spectroscopy (PS), 230
Photoluminescence (PL), 8, 10
  bands, origin of intrinsic, 218–233
  bands, properties of, 210–217
  blue band, 16, 212–213, 222–224
  from chalcogen-related centers, 91–93
  excitation power dependence of red and
    blue PL in porous Si, 194–200
  extrinsic bands, 216–217
  germanium rich QWs and, 69–70
  GeSiGe coupled quantum wells and,
    64–67
  infrared bands, 16, 213–216, 224–225
  nanoparticle, 18–19
  polymers and, 24–25, 307–308
  porous silicon and, 14, 15–17, 210–233
  quantum confinement and surface states,
    226–233
  quantum dots and, 23
  quantum wells and, 20–22
  quantum wires and, 22
  quenching, 13
  red band, 16, 210–211, 218–222
  Si:Al-N and, 87
  Si:Be and, 87, 89–90
  Si-Ge alloy layers and quantum wells
    and, 10, 54–58
  Si-Ge short period superlattices and, 8,
    59–63
  from SiGeSi QWs, 97–99
  Si:In and, 11, 85–87
  siloxene and, 24
  Si-SiO_2 superlattices and, 20–21
  from thick Be-doped SiGe alloys, 95–96
  voltage quenching, 290
Photoluminescence, surface-oxidized
  nanocrystals and, 170
  decay dynamics, 185–189
  resonantly excited luminescence
    spectrum, 174–177
  size dependence of peak energy, 171–174
  size dependence of PL intensity, 178
  three region model, 177–185

Photonics, 2
p-i-n diode, 3, 11, 19
Polymers, silicon
Polymers, silicon
  band structure, 304–307
  photoluminescence, 24–25, 307–308
Polysilane compounds, 24
Porous germanium, electrochemical etching, 163–165
Porous silicon (π-Si), 4
  defects in, 221–222
  discovery of, 14, 163, 207
  electrical properties, 233–236
  electrochemical etching, 14, 163–165, 207
  electroluminescence in, 14–15, 238–246
  electron transport in, 322–325
  excitation power dependence of red and blue PL in, 194–200
  fabrication of, 207–209
  to handle indirect bandgap limitations, 14–18
  light emitting diodes, 17, 27, 238–246
  nonlinear optical properties of, 18, 189–194
  nonoptical properties of, 233–237
  photoluminescence and, 14, 15–17, 210–233
  photoluminescence bands, origin of intrinsic, 218–233
  photoluminescence bands, properties of, 210–217
  structural properties, 236–237
  uses of, 207
P-type doping, 53–54, 58, 64

Q

Quantum confinement, 14
  effective mass approximation, 17–20, 254, 255–256
  effects in porous germanium, 163–164
  effects in porous silicon, 16, 163–164
  empirical pseudopotential, 254, 255–258
  empirical tight-binding, 254, 255–258, 272, 274–277
  local density approximation, 254, 255–258, 272, 275–277
  optical transitions and radiative lifetime, 258–261
  red PL band and, 16, 218–222
  surface states and pure, 226–233

Quantum dots, 16, 19, 23–24, 28
Quantum efficiency of Si, 3
Quantum wells (QWs), 6, 19–22, 28
  band offsets in, 80–82
  GeSiGe coupled, 64–67
  lateral confinement in Si-Ge QWs, 67–70
  multi, 55, 57
  photoluminescence and, 20–22
  photoluminescence from SiGeSi QWs, 97–99
  Si-Ge and, 54–58
Quantum wires, 14, 19, 22–23

R

Radiative recombination, 6, 206
  exchange splitting and crystallite symmetry, 262–269
  optical transitions and radiative lifetime, 258–261
  self-trapped excitons, 271–279
  Stokes shift for delocalized states, 270–271
  two-level model, 264–265
Raman spectrum, 219–220
Rapid-thermal-oxidization (RTO) processes, 196–198
Rare earth (RE), 12
  energy levels, 128–129
  extrinsic centers, 11, 12–13
  ligands, 133–135
  physics of light emission, 127–132
Recrystallization, 224
Red PL band, 16, 210–211, 218–222
Reflective high energy electron diffraction (RHEED), 121
Resonantly excited luminescence spectrum, 174–177
Russell-Saunders (RS) approximation, 127
Scanning electron microscopy (SEM), 121, 236
Scanning tunneling microscopy (STM), 199
Schredinger equation, 143, 144
Secondary ion mass spectroscopy (SIMS), use of, 115, 121
Self-trapped excitons, 271–279
Shadow masks, 67–68
Silane gas, decomposition of, 162
Silicon (Si)
  backbone structure, 24–25, 159, 160, 161

erbium doping of, 113–121
indirect energy gaps in, 6
methods for handling indirect bandgap limitations, 6–26
optoelectronic devices and light emission problems, 2–4
physical properties of, 4–6
prospects for optoelectronic devices using, 26–29
quantum efficiency, 3
Silicon:aluminum-nitrogen (Si:Al-N), 87
Silicon:beryllium (Si:Be), 87, 89–90
Silicon-carbon (Si-C), 10
Silicon-germanium (Si-Ge)
  See also Energy band gaps and light emission in atomic layer structures
  alloy layers and quantum wells, 9–11, 54–58
  energy gaps of unstrained, 45
  heterostructures, 22
  interface sharpness, 42
  lattice constants, 40
  short period superlattices, 6–9, 59–63
  strain relaxation, 40–42
  structural properties, 40–44
  virtual substrates, 41, 42
  wires, 22
Silicon-germanium-silicon (SiGeSi) QWs, photoluminescence from, 97–99
Silicon:indium (Si:In), photoluminescence from, 11, 85–87
Silicon-on-insulator (SOI) optical waveguide, 105–106, 150, 207
Silicon oxide, blue PL band and, 222–224
Silicon:selenium (Si:Se), 92–93
Silicon-silicon dioxide (Si-SiO$_2$) superlattices, 19–22
Silicon:sulfur (Si:S), 91
Siloxene, 24, 220–221
SIMOX (Separation by IMplantation of OXygen), 105, 181
Size dependence of PL peak energy, 171–174
Slater determinants, 263, 264, 288
Sol-gel process, 168
Solid phase epitaxy (SPE), 117
Spin-unrestricted formalism, 147
Stark levels, 129
Stokes shift, 230, 232, 233
  for delocalized states, 270–271

Stranski-Krastanow growth mode, 23, 41, 68
Structural properties, porous Si, 236–237
Superlattices,
  silicon-germanium short period, 6–9, 59–63
  silicon-silicon dioxide, 19–22
Surface state models, 16, 19, 226–233

T

Third harmonic generation (THG) Maker fringe method, 189–190, 191
Transmission electron microscopy (TEM), use of, 115, 117, 119, 121, 162, 167, 219, 220, 236, 311, 314

U

Ultra-high vacuum CVD (UHV-CVD), 118–120

V

Valence band offset, 48–50
Valence band shifting, hydrostatic stress and, 46–47
Valence band structure in semiconductors, 4
van der Waals force, 26
Vickers test, 236–237

W

Waveguide modulators, 3

X

X-ray diffraction studies, 162
X-ray photoemission spectroscopy (XPS), 162, 167
X-ray scattering, 236

Z

Zero phonon line (ZPL), 308, 319
Zinc-blende structure semiconductors, 4

Contents of Volumes in This Series

Volume 1 **Physics of III–V Compounds**

C. Hilsum, Some Key Features of III–V Compounds
Franco Bassani, Methods of Band Calculations Applicable to III–V Compounds
E. O. Kane, The k-p Method
V. L. Bonch-Bruevich, Effect of Heavy Doping on the Semiconductor Band Structure
Donald Long, Energy Band Structures of Mixed Crystals of III–V Compounds
Laura M. Roth and Petros N. Argyres, Magnetic Quantum Effects
S. M. Puri and T. H. Geballe, Thermomagnetic Effects in the Quantum Region
W. M. Becker, Band Characteristics near Principal Minima from Magnetoresistance
E. H. Putley, Freeze-Out Effects, Hot Electron Effects, and Submillimeter Photoconductivity in InSb
H. Weiss, Magnetoresistance
Betsy Ancker-Johnson, Plasma in Semiconductors and Semimetals

Volume 2 **Physics of III–V Compounds**

M. G. Holland, Thermal Conductivity
S. I. Novkova, Thermal Expansion
U. Piesbergen, Heat Capacity and Debye Temperatures
G. Giesecke, Lattice Constants
J. R. Drabble, Elastic Properties
A. U. Mac Rae and G. W. Gobeli, Low Energy Electron Diffraction Studies
Robert Lee Mieher, Nuclear Magnetic Resonance
Bernard Goldstein, Electron Paramagnetic Resonance
T. S. Moss, Photoconduction in III–V Compounds
E. Antoncik ad J. Tauc, Quantum Efficiency of the Internal Photoelectric Effect in InSb
G. W. Gobeli and I. G. Allen, Photoelectric Threshold and Work Function
P. S. Pershan, Nonlinear Optics in III–V Compounds
M. Gershenzon, Radiative Recombination in the III–V Compounds
Frank Stern, Stimulated Emission in Semiconductors

Volume 3 Optical of Properties III–V Compounds

Marvin Hass, Lattice Reflection
William G. Spitzer, Multiphonon Lattice Absorption
D. L. Stierwalt and R. F. Potter, Emittance Studies
H. R. Philipp and H. Ehrenveich, Ultraviolet Optical Properties
Manuel Cardona, Optical Absorption above the Fundamental Edge
Earnest J. Johnson, Absorption near the Fundamental Edge
John O. Dimmock, Introduction to the Theory of Exciton States in Semiconductors
B. Lax and J. G. Mavroides, Interband Magnetooptical Effects
H. Y. Fan, Effects of Free Carries on Optical Properties
Edward D. Palik and George B. Wright, Free-Carrier Magnetooptical Effects
Richard H. Bube, Photoelectronic Analysis
B. O. Seraphin and H. E. Bennett, Optical Constants

Volume 4 Physics of III–V Compounds

N. A. Goryunova, A. S. Borschevskii, and D. N. Tretiakov, Hardness
N. N. Sirota, Heats of Formation and Temperatures and Heats of Fusion of Compounds $A^{III}B^{V}$
Don L. Kendall, Diffusion
A. G. Chynoweth, Charge Multiplication Phenomena
Robert W. Keyes, The Effects of Hydrostatic Pressure on the Properties of III–V Semiconductors
L. W. Aukerman, Radiation Effects
N. A. Goryunova, F. P. Kesamanly, and D. N. Nasledov, Phenomena in Solid Solutions
R. T. Bate, Electrical Properties of Nonuniform Crystals

Volume 5 Infrared Detectors

Henry Levinstein, Characterization of Infrared Detectors
Paul W. Kruse, Indium Antimonide Photoconductive and Photoelectromagnetic Detectors
M. B. Prince, Narrowband Self-Filtering Detectors
Ivars Melngalis and T. C. Harman, Single-Crystal Lead-Tin Chalcogenides
Donald Long and Joseph L. Schmidt, Mercury-Cadmium Telluride and Closely Related Alloys
E. H. Putley, The Pyroelectric Detector
Norman B. Stevens, Radiation Thermopiles
R. J. Keyes and T. M. Quist, Low Level Coherent and Incoherent Detection in the Infrared
M. C. Teich, Coherent Detection in the Infrared
F. R. Arams, E. W. Sard, B. J. Peyton, and F. P. Pace, Infrared Heterodyne Detection with Gigahertz IF Response
H. S. Sommers, Jr., Macrowave-Based Photoconductive Detector
Robert Sehr and Rainer Zuleeg, Imaging and Display

Volume 6 Injection Phenomena

Murray A. Lampert and Ronald B. Schilling, Current Injection in Solids: The Regional Approximation Method
Richard Williams, Injection by Internal Photoemission
Allen M. Barnett, Current Filament Formation

R. Baron and J. W. Mayer, Double Injection in Semiconductors
W. Ruppel, The Photoconductor-Metal Contact

Volume 7 Application and Devices
Part A

John A. Copeland and Stephen Knight, Applications Utilizing Bulk Negative Resistance
F. A. Padovani, The Voltage-Current Characteristics of Metal-Semiconductor Contacts
P. L. Hower, W. W. Hooper, B. R. Cairns, R. D. Fairman, and D. A. Tremere, The GaAs Field-Effect Transistor
Marvin H. White, MOS Transistors
G. R. Antell, Gallium Arsenide Transistors
T. L. Tansley, Heterojunction Properties

Part B

T. Misawa, IMPATT Diodes
H. C. Okean, Tunnel Diodes
Robert B. Campbell and Hung-Chi Chang, Silicon Junction Carbide Devices
R. E. Enstrom, H. Kressel, and L. Krassner, High-Temperature Power Rectifiers of $GaAs_{1-x}P_x$

Volume 8 Transport and Optical Phenomena

Richard J. Stirn, Band Structure and Galvanomagnetic Effects in III–V Compounds with Indirect Band Gaps
Roland W. Ure, Jr., Thermoelectric Effects in III–V Compounds
Herbert Piller, Faraday Rotation
H. Barry Bebb and E. W. Williams, Photoluminescence I: Theory
E. W. Williams and H. Barry Bebb, Photoluminescence II: Gallium Arsenide

Volume 9 Modulation Techniques

B. O. Seraphin, Electroreflectance
R. L. Aggarwal, Modulated Interband Magnetooptics
Daniel F. Blossey and Paul Handler, Electroabsorption
Bruno Batz, Thermal and Wavelength Modulation Spectroscopy
Ivar Balslev, Piezopptical Effects
D. E. Aspnes and N. Bottka, Electric-Field Effects on the Dielectric Function of Semiconductors and Insulators

Volume 10 Transport Phenomena

R. L. Rhode, Low-Field Electron Transport
J. D. Wiley, Mobility of Holes in III–V Compounds
C. M. Wolfe and G. E. Stillman, Apparent Mobility Enhancement in Inhomogeneous Crystals
Robert L. Petersen, The Magnetophonon Effect

Volume 11 Solar Cells

Harold J. Hovel, Introduction; Carrier Collection, Spectral Response, and Photocurrent; Solar Cell Electrical Characteristics; Efficiency; Thickness; Other Solar Cell Devices; Radiation Effects; Temperature and Intensity; Solar Cell Technology

Volume 12 Infrared Detectors (II)

W. L. Eiseman, J. D. Merriam, and R. F. Potter, Operational Characteristics of Infrared Photodetectors
Peter R. Bratt, Impurity Germanium and Silicon Infrared Detectors
E. H. Putley, InSb Submillimeter Photoconductive Detectors
G. E. Stillman, C. M. Wolfe, and J. O. Dimmock, Far-Infrared Photoconductivity in High Purity GaAs
G. E. Stillman and C. M. Wolfe, Avalanche Photodiodes
P. L. Richards, The Josephson Junction as a Detector of Microwave and Far-Infrared Radiation
E. H. Putley, The Pyroelectric Detector–An Update

Volume 13 Cadmium Telluride

Kenneth Zanio, Materials Preparations; Physics; Defects; Applications

Volume 14 Lasers, Junctions, Transport

N. Holonyak, Jr. and M. H. Lee, Photopumped III–V Semiconductor Lasers
Henry Kressel and Jerome K. Butler, Heterojunction Laser Diodes
A Van der Ziel, Space-Charge-Limited Solid-State Diodes
Peter J. Price, Monte Carlo Calculation of Electron Transport in Solids

Volume 15 Contacts, Junctions, Emitters

B. L. Sharma, Ohmic Contacts to III–V Compounds Semiconductors
Allen Nussbaum, The Theory of Semiconducting Junctions
John S. Escher, NEA Semiconductor Photoemitters

Volume 16 Defects, (HgCd)Se, (HgCd)Te

Henry Kressel, The Effect of Crystal Defects on Optoelectronic Devices
C. R. Whitsett, J. G. Broerman, and C. J. Summers, Crystal Growth and Properties of $Hg_{1-x}Cd_xSe$ alloys
M. H. Weiler, Magnetooptical Properties of $Hg_{1-x}Cd_xTe$ Alloys
Paul W. Kruse and John G. Ready, Nonlinear Optical Effects in $Hg_{1-x}Cd_xTe$

Volume 17 CW Processing of Silicon and Other Semiconductors

James F. Gibbons, Beam Processing of Silicon
Arto Lietoila, Richard B. Gold, James F. Gibbons, and Lee A. Christel, Temperature Distribu-

tions and Solid Phase Reaction Rates Produced by Scanning CW Beams
Arto Leitoila and James F. Gibbons, Applications of CW Beam Processing to Ion Implanted Crystalline Silicon
N. M. Johnson, Electronic Defects in CW Transient Thermal Processed Silicon
K. F. Lee, T. J. Stultz, and James F. Gibbons, Beam Recrystallized Polycrystalline Silicon: Properties, Applications, and Techniques
T. Shibata, A. Wakita, T. W. Sigmon, and James F. Gibbons, Metal-Silicon Reactions and Silicide
Yves I. Nissim and James F. Gibbons, CW Beam Processing of Gallium Arsenide

Volume 18 Mercury Cadmium Telluride

Paul W. Kruse, The Emergence of $(Hg_{1-x}Cd_x)Te$ as a Modern Infrared Sensitive Material
H. E. Hirsch, S. C. Liang, and A. G. White, Preparation of High-Purity Cadmium, Mercury, and Tellurium
W. F. H. Micklethwaite, The Crystal Growth of Cadmium Mercury Telluride
Paul E. Petersen, Auger Recombination in Mercury Cadmium Telluride
R. M. Broudy and V. J. Mazurczyck, (HgCd)Te Photoconductive Detectors
M. B. Reine, A. K. Soad, and T. J. Tredwell, Photovoltaic Infrared Detectors
M. A. Kinch, Metal-Insulator-Semiconductor Infrared Detectors

Volume 19 Deep Levels, GaAs, Alloys, Photochemistry

G. F. Neumark and K. Kosai, Deep Levels in Wide Band-Gap III–V Semiconductors
David C. Look, The Electrical and Photoelectronic Properties of Semi-Insulating GaAs
R. F. Brebrick, Ching-Hua Su, and Pok-Kai Liao, Associated Solution Model for Ga-In-Sb and Hg-Cd-Te
Yu. Ya. Gurevich and Yu. V. Pleskon, Photoelectrochemistry of Semiconductors

Volume 20 Semi-Insulating GaAs

R. N. Thomas, H. M. Hobgood, G. W. Eldridge, D. L. Barrett, T. T. Braggins, L. B. Ta, and S. K. Wang, High-Purity LEC Growth and Direct Implantation of GaAs for Monolithic Microwave Circuits
C. A. Stolte, Ion Implantation and Materials for GaAs Integrated Circuits
C. G. Kirkpatrick, R. T. Chen, D. E. Holmes, P. M. Asbeck, K. R. Elliott, R. D. Fairman, and J. R. Oliver, LEC GaAs for Integrated Circuit Applications
J. S. Blakemore and S. Rahimi, Models for Mid-Gap Centers in Gallium Arsenide

Volume 21 Hydrogenated Amorphous Silicon
Part A

Jacques I. Pankove, Introduction
Masataka Hirose, Glow Discharge; Chemical Vapor Deposition
Yoshiyuki Uchida, di Glow Discharge
T. D. Moustakas, Sputtering
Isao Yamada, Ionized-Cluster Beam Deposition
Bruce A. Scott, Homogeneous Chemical Vapor Deposition

Frank J. Kampas, Chemical Reactions in Plasma Deposition
Paul A. Longeway, Plasma Kinetics
Herbert A. Weakliem, Diagnostics of Silane Glow Discharges Using Probes and Mass Spectroscopy
Lester Gluttman, Relation between the Atomic and the Electronic Structures
A. Chenevas-Paule, Experiment Determination of Structure
S. Minomura, Pressure Effects on the Local Atomic Structure
David Adler, Defects and Density of Localized States

Part B

Jacques I. Pankove, Introduction
G. D. Cody, The Optical Absorption Edge of a-Si:H
Nabil M. Amer and Warren B. Jackson, Optical Properties of Defect States in a-Si:H
P. J. Zanzucchi, The Vibrational Spectra of a-Si:H
Yoshihiro Hamakawa, Electroreflectance and Electroabsorption
Jeffrey S. Lannin, Raman Scattering of Amorphous Si, Ge, and Their Alloys
R. A. Street, Luminescence in a-Si:H
Richard S. Crandall, Photoconductivity
J. Tauc, Time-Resolved Spectroscopy of Electronic Relaxation Processes
P. E. Vanier, IR-Induced Quenching and Enhancement of Photoconductivity and Photoluminescence
H. Schade, Irradiation-Induced Metastable Effects
L. Ley, Photoelectron Emission Studies

Part C

Jacques I. Pankove, Introduction
J. David Cohen, Density of States from Junction Measurements in Hydrogenated Amorphous Silicon
P. C. Taylor, Magnetic Resonance Measurements in a-Si:H
K. Morigaki, Optically Detected Magnetic Resonance
J. Dresner, Carrier Mobility in a-Si:H
T. Tiedje, Information about band-Tail States from Time-of-Flight Experiments
Arnold R. Moore, Diffusion Length in Undoped a-Si:H
W. Beyer and J. Overhof, Doping Effects in a-Si:H
H. Fritzche, Electronic Properties of Surfaces in a-Si:H
C. R. Wronski, The Staebler-Wronski Effect
R. J. Nemanich, Schottky Barriers on a-Si:H
B. Abeles and T. Tiedje, Amorphous Semiconductor Superlattices

Part D

Jacques I. Pankove, Introduction
D. E. Carlson, Solar Cells
G. A. Swartz, Closed-Form Solution of I–V Characteristic for a-Si:H Solar Cells
Isamu Shimizu, Electrophotography
Sachio Ishioka, Image Pickup Tubes

P. G. LeComber and W. E. Spear, The Development of the a-Si:H Field-Effect Transistor and Its Possible Applications
D. G. Ast, a-Si:H FET-Addressed LCD Panel
S. Kaneko, Solid-State Image Sensor
Masakiyo Matsumura, Charge-Coupled Devices
M. A. Bosch, Optical Recording
A. D'Amico and G. Fortunato, Ambient Sensors
Hiroshi Kukimoto, Amorphous Light-Emitting Devices
Robert J. Phelan, Jr., Fast Detectors and Modulators
Jacques I. Pankove, Hybrid Structures
P. G. LeComber, A. E. Owen, W. E. Spear, J. Hajto, and W. K. Choi, Electronic Switching in Amorphous Silicon Junction Devices

Volume 22 Lightwave Communications Technology
Part A

Kazuo Nakajima, The Liquid-Phase Epitaxial Growth of IngaAsp
W. T. Tsang, Molecular Beam Epitaxy for III–V Compound Semiconductors
G. B. Stringfellow, Organometallic Vapor-Phase Epitaxial Growth of III–V Semiconductors
G. Beuchet, Halide and Chloride Transport Vapor-Phase Deposition of InGaAsP and GaAs
Manijeh Razeghi, Low-Pressure Metallo-Organic Chemical Vapor Deposition of $Ga_x in_{1-x} As P_{1-y}$ Alloys
P. M. Petroff, Defects in III–V Compound Semiconductors

Part B

J. P. van der Ziel, Mode Locking of Semiconductor Lasers
Kam Y. Lau and Ammon Yariv, High-Frequency Current Modulation of Semiconductor Injection Lasers
Charles H. Henry, Special Properties of Semiconductor Lasers
Yasuharu Suematsu, Katsumi Kishino, Shigehisa Arai, and Fumio Koyama. Dynamic Single-Mode Semiconductor Lasers with a Distributed Reflector
W. T. Tsang, The Cleaved-Coupled-Cavity (C^3) Laser

Part C

R. J. Nelson and N. K. Dutta, Review of InGaAsP InP Laser Structures and Comparison of Their Performance
N. Chinone and M. Nakamura, Mode-Stabilized Semiconductor Lasers for 0.7–0.8- and 1.1–1.6-μm Regions
Yoshiji Horikoshi, Semiconductor Lasers with Wavelengths Exceeding 2 μm
B. A. Dean and M. Dixon, The Functional Reliability of Semiconductor Lasers as Optical Transmitters
R. H. Saul, T. P. Lee, and C. A. Burus, Light-Emitting Device Design
C. L. Zipfel, Light-Emitting Diode-Reliability
Tien Pei Lee and Tingye Li, LED-Based Multimode Lightwave Systems
Kinichiro Ogawa, Semiconductor Noise-Mode Partition Noise

Part D

Federico Capasso, The Physics of Avalanche Photodiodes
T. P. Pearsall and M. A. Pollack, Compound Semiconductor Photodiodes
Takao Kaneda, Silicon and Germanium Avalanche Photodiodes
S. R. Forrest, Sensitivity of Avalanche Photodetector Receivers for High-Bit-Rate Long-Wavelength Optical Communication Systems
J. C. Campbell, Phototransistors for Lightwave Communications

Part E

Shyh Wang, Principles and Characteristics of Integrable Active and Passive Optical Devices
Shlomo Margalit and Amnon Yariv, Integrated Electronic and Photonic Devices
Takaoki Mukai, Yoshihisa Yamamoto, and Tatsuya Kimura, Optical Amplification by Semiconductor Lasers

Volume 23   Pulsed Laser Processing of Semiconductors

R. F. Wood, C. W. White, and R. T. Young, Laser Processing of Semiconductors: An Overview
C. W. White, Segregation, Solute Trapping, and Supersaturated Alloys
G. E. Jellison, Jr., Optical and Electrical Properties of Pulsed Laser-Annealed Silicon
R. F. Wood and G. E. Jellison, Jr., Melting Model of Pulsed Laser Processing
R. F. Wood and F. W. Young, Jr., Nonequilibrium Solidification Following Pulsed Laser Melting
D. H. Lowndes and G. E. Jellison, Jr., Time-Resolved Measurement During Pulsed Laser Irradiation of Silicon
D. M. Zebner, Surface Studies of Pulsed Laser Irradiated Semiconductors
D. H. Lowndes, Pulsed Beam Processing of Gallium Arsenide
R. B. James, Pulsed CO_2 Laser Annealing of Semiconductors
R. T. Young and R. F. Wood, Applications of Pulsed Laser Processing

Volume 24   Applications of Multiquantum Wells, Selective Doping, and Superlattices

C. Weisbuch, Fundamental Properties of III–V Semiconductor Two-Dimensional Quantized Structures: The Basis for Optical and Electronic Device Applications
H. Morkoc and H. Unlu, Factors Affecting the Performance of (Al,Ga)As/GaAs and (Al,Ga)As/InGaAs Modulation-Doped Field-Effect Transistors: Microwave and Digital Applications
N. T. Linh, Two-Dimensional Electron Gas FETs: Microwave Applications
M. Abe et al., Ultra-High-Speed HEMT Integrated Circuits
D. S. Chemla, D. A. B. Miller, and P. W. Smith, Nonlinear Optical Properties of Multiple Quantum Well Structures for Optical Signal Processing
F. Capasso, Graded-Gap and Superlattice Devices by Band-Gap Engineering
W. T. Tsang, Quantum Confinement Heterostructure Semiconductor Lasers
G. C. Osbourn et al., Principles and Applications of Semiconductor Strained-Layer Superlattices

Volume 25 Diluted Magnetic Semiconductors

W. Giriat and J. K. Furdyna, Crystal Structure, Composition, and Materials Preparation of Diluted Magnetic Semiconductors
W. M. Becker, Band Structure and Optical Properties of Wide-Gap $A_{1-x}^{II}Mn_xB^{IV}$ Alloys at Zero Magnetic Field
Saul Oseroff and Pieter H. Keesom, Magnetic Properties: Macroscopic Studies
Giebultowicz and T. M. Holden, Neutron Scattering Studies of the Magnetic Structure and Dynamics of Diluted Magnetic Semiconductors
J. Kossut, Band Structure and Quantum Transport Phenomena in Narrow-Gap Diluted Magnetic Semiconductors
C. Riquaux, Magnetooptical Properties of Large-Gap Diluted Magnetic Semiconductors
J. A. Gaj, Magnetooptical Properties of Large-Gap Diluted Magnetic Semiconductors
J. Mycielski, Shallow Acceptors in Diluted Magnetic Semiconductors: Splitting, Boil-off, Giant Negative Magnetoresistance
A. K. Ramadas and R. Rodriquez, Raman Scattering in Diluted Magnetic Semiconductors
P. A. Wolff, Theory of Bound Magnetic Polarons in Semimagnetic Semiconductors

Volume 26 III–V Compound Semiconductors and Semiconductor Properties of Superionic Materials

Zou Yuanxi, III–V Compounds
H. V. Winston, A. T. Hunter, H. Kimura, and R. E. Lee, InAs-Alloyed GaAs Substrates for Direct Implantation
P. K. Bhattachary and S. Dhar, Deep Levels in III–V Compound Semiconductors Grown by MBE
Yu. Yu. Gurevich and A. K. Ivanov-Shits, Semiconductor Properties of Supersonic Materials

Volume 27 High Conducting Quasi-One-Dimensional Organic Crystals

E. M. Conwell, Introduction to Highly Conducting Quasi-One-Dimensional Organic Crystals
I. A. Howard, A Reference Guide to the Conducting Quasi-One-Dimensional Organic Molecular Crystals
J. P. Pouquet, Structural Instabilities
E. M. Conwell, Transport Properties
C. S. Jacobsen, Optical Properties
J. C. Scott, Magnetic Properties
L. Zuppiroli, Irradiation Effects: Perfect Crystals and Real Crystals

Volume 28 Measurement of High-Speed Signals in Solid State Devices

J. Frey and D. Ioannou, Materials and Devices for High-Speed and Optoelectronic Applications
H. Schumacher and E. Strid, Electronic Wafer Probing Techniques
D. H. Auston, Picosecond Photoconductivity: High-Speed Measurements of Devices and Materials
J. A. Valdmanis, Electro-Optic Measurement Techniques for Picosecond Materials, Devices, and Integrated Circuits.
J. M. Wiesenfeld and R. K. Jain, Direct Optical Probing of Integrated Circuits and High-Speed Devices
G. Plows, Electron-Beam Probing
A. M. Weiner and R. B. Marcus, Photoemissive Probing

Volume 29 Very High Speed Integrated Circuits: Gallium Arsenide LSI

M. Kuzuhara and T. Nazaki, Active Layer Formation by Ion Implantation
H. Hasimoto, Focused Ion Beam Implantation Technology
T. Nozaki and A. Higashisaka, Device Fabrication Process Technology
M. Ino and T. Takada, GaAs LSI Circuit Design
M. Hirayama, M. Ohmori, and K. Yamasaki, GaAs LSI Fabrication and Performance

Volume 30 Very High Speed Integrated Circuits: Heterostructure

H. Watanabe, T. Mizutani, and A. Usui, Fundamentals of Epitaxial Growth and Atomic Layer Epitaxy
S. Hiyamizu, Characteristics of Two-Dimensional Electron Gas in III–V Compound Heterostructures Grown by MBE
T. Nakanisi, Metalorganic Vapor Phase Epitaxy for High-Quality Active Layers
T. Nimura, High Electron Mobility Transistor and LSI Applications
T. Sugeta and T. Ishibashi, Hetero-Bipolar Transistor and LSI Application
H. Matsueda, T. Tanaka, and M. Nakamura, Optoelectronic Integrated Circuits

Volume 31 Indium Phosphide: Crystal Growth and Characterization

J. P. Farges, Growth of Discoloration-free InP
M. J. McCollum and G. E. Stillman, High Purity InP Grown by Hydride Vapor Phase Epitaxy
T. Inada and T. Fukuda, Direct Synthesis and Growth of Indium Phosphide by the Liquid Phosphorous Encapsulated Czochralski Method
O. Oda, K. Katagiri, K. Shinohara, S. Katsura, Y. Takahashi, K. Kainosho, K. Kohiro, and R. Hirano, InP Crystal Growth, Substrate Preparation and Evaluation
K. Tada, M. Tatsumi, M. Morioka, T. Araki, and T. Kawase, InP Substrates: Production and Quality Control
M. Razeghi, LP-MOCVD Growth, Characterization, and Application of InP Material
T. A. Kennedy and P. J. Lin-Chung, Stoichiometric Defects in InP

Volme 32 Strained-Layer Superlattices: Physics

T. P. Pearsall, Strained-Layer Superlattices
Fred H. Pollack, Effects of Homogeneous Strain on the Electronic and Vibrational Levels in Semiconductors
J. Y. Marzin, J. M. Gerárd, P. Voisin, and J. A. Brum, Optical Studies of Strained III–V Heterolayers
R. People and S. A. Jackson, Structurally Induced States from Strain and Confinement
M. Jaros, Microscopic Phenomena in Ordered Suprlattices

Volume 33 Strained-Layer Superlattices: Materials Science and Technology

R. Hull and J. C. Bean, Principles and Concepts of Strained-Layer Epitaxy
William J. Schaff, Paul J. Tasker, Marc C. Foisy, and Lester F. Eastman, Device Applications of Strained-Layer Epitaxy

S. T. Picraux, B. L. Doyle, and J. Y. Tsao, Structure and Characterization of Strained-Layer Superlattices
E. Kasper and F. Schäffer, Group IV Compounds
Dale L. Martin, Molecular Beam Epitaxy of IV–VI Compounds Heterojunction
Robert L. Gunshor, Leslie A. Kolodziejski, Arto V. Nurmikko, and Nobuo Otsuka, Molecular Beam Epitaxy of II–VI Semiconductor Microstructures

Volume 34   Hydrogen in Semiconductors

J. I. Pankove and N. M. Johnson, Introduction to Hydrogen in Semiconductors
C. H. Seager, Hydrogenation Methods
J. I. Pankove, Hydrogenation of Defects in Crystalline Silicon
J. W. Corbett, P. Deák, U. V. Desnica, and S. J. Pearton, Hydrogen Passivation of Damage Centers in Semiconductors
S. J. Pearton, Neutralization of Deep Levels in Silicon
J. I. Pankove, Neutralization of Shallow Acceptors in Silicon
N. M. Johnson, Neutralization of Donor Dopants and Formation of Hydrogen-Induced Defects in n-Type Silicon
M. Stavola and S. J. Pearton, Vibrational Spectroscopy of Hydrogen-Related Defects in Silicon
A. D. Marwick, Hydrogen in Semiconductors: Ion Beam Techniques
C. Herring and N. M. Johnson, Hydrogen Migration and Solubility in Silicon
E. E. Haller, Hydrogen-Related Phenomena in Crystalline Germanium
J. Kakalios, Hydrogen Diffusion in Amorphous Silicon
J. Chevalier, B. Clerjaud, and B. Pajot, Neutralization of Defects and Dopants in III–V Semiconductors
G. G. DeLeo and W. B. Fowler, Computational Studies of Hydrogen-Containing Complexes in Semiconductors
R. F. Kiefl and T. L. Estle, Muonium in Semiconductors
C. G. Van de Walle, Theory of Isolated Interstitial Hydrogen and Muonium in Crystalline Semiconductors

Volume 35   Nanostructured Systems

Mark Reed, Introduction
H. van Houten, C. W. J. Beenakker, and B. J. van Wees, Quantum Point Contacts
G. Timp, When Does a Wire Become an Electron Waveguide?
M. Büttiker, The Quantum Hall Effects in Open Conductors
W. Hansen, J. P. Kotthaus, and U. Merkt, Electrons in Laterally Periodic Nanostructures

Volume 36   The Spectroscopy of Semiconductors

D. Heiman, Spectroscopy of Semiconductors at Low Temperatures and High Magnetic Fields
Arto V. Nurmikko, Transient Spectroscopy by Ultrashort Laser Pulse Techniques
A. K. Ramdas and S. Rodriguez, Piezospectroscopy of Semiconductors
Orest J. Glembocki and Benjamin V. Shanabrook, Photoreflectance Spectroscopy of Microstructures
David G. Seiler, Christopher L. Littler, and Margaret H. Wiler, One- and Two-Photon Magneto-Optical Spectroscopy of InSb and $Hg_{1-x}Cd_xTe$

Volume 37 The Mechanical Properties of Semiconductors

A.-B. Chen, Arden Sher and W. T. Yost, Elastic Constants and Related Properties of Semiconductor Compounds and Their Alloys
David R. Clarke, Fracture of Silicon and Other Semiconductors
Hans Siethoff, The Plasticity of Elemental and Compound Semiconductors
Sivaraman Guruswamy, Katherine T. Faber and John P. Hirth, Mechanical Behavior of Compound Semiconductors
Subhanh Mahajan, Deformation Behavior of Compound Semiconductors
John P. Hirth, Injection of Dislocations into Strained Multilayer Structures
Don Kendall, Charles B. Fleddermann, and Kevin J. Malloy, Critical Technologies for the Micromachining of Silicon
Ikuo Matsuba and Kinji Mokuya, Processing and Semiconductor Thermoelastic Behavior

Volume 38 Imperfections in III/V Materials

Udo Scherz and Matthias Scheffler, Density-Functional Theory of sp-Bonded Defects in III/V Semiconductors
Maria Kaminska and Eicke R. Weber, El2 Defect in GaAs
David C. Look, Defects Relevant for Compensation in Semi-Insulating GaAs
R. C. Newman, Local Vibrational Mode Spectroscopy of Defects in III/V Compounds
Andrzej M. Hennel, Transition Metals in III/V Compounds
Kevin J. Malloy and Ken Khachaturyan, DX and Related Defects in Semiconductors
V. Swaminathan and Andrew S. Jordan, Dislocations in III/V Compounds
Krzysztof W. Nauka, Deep Level Defects in the Epitaxial III/V Materials

Volume 39 Minority Carriers in III–V Semiconductors: Physics and Applications

Niloy K. Dutta, Radiative Transitions in GaAs and Other III–V Compounds
Richard K. Ahrenkiel, Minority-Carrier Lifetime in III–V Semiconductors
Tomofumi Furuta, High Field Minority Electron Transport in p-GaAs
Mark S. Lundstrom, Minority-Carrier Transport in III–V Semiconductors
Richard A. Abram, Effects of Heavy Doping and High Excitation on the Band Structure of GaAs
David Yevick and Witold Bardyszewski, An Introduction to Non-Equilibrium Many-Body Analyses of Optical Processes in III–V Semiconductors

Volume 40 Epitaxial Microstructures

E. F. Schubert, Delta-Doping of Semiconductors: Electronic, Optical, and Structural Properties of Materials and Devices
A. Gossard, M. Sundaram, and P. Hopkins, Wide Graded Potential Wells
P. Petroff, Direct Growth of Nanometer-Size Quantum Wire Superlattices
E. Kapon, Lateral Patterning of Quantum Well Heterostructures by Growth of Nonplanar Substrates
H. Temkin, D. Gershoni, and M. Panish, Optical Properties of Ga$_{1-x}$In$_x$As/InP Quantum Wells

CONTENTS OF VOLUMES IN THIS SERIES

Volume 41 High Speed Heterostructure Devices

F. Capasso, F. Beltram, S. Sen, A. Pahlevi, and A. Y. Cho, Quantum Electron Devices: Physics and Applications
P. Solomon, D. J. Frank, S. L. Wright, and F. Canora, GaAs-Gate Semiconductor–Insulator–Semiconductor FET
M. H. Hashemi and U. K. Mishra, Unipolar InP-Based Transistors
R. Kiehl, Complementary Heterostructure FET Integrated Circuits
T. Ishibashi, GaAs-Based and InP-Based Heterostructure Bipolar Transistors
H. C. Liu and T. C. L. G. Sollner, High-Frequency-Tunneling Devices
H. Ohnishi, T. More, M. Takatsu, K. Imamura, and N. Yokoyama, Resonant-Tunneling Hot-Electron Transistors and Circuits

Volume 42 Oxygen in Silicon

F. Shimura, Introduction to Oxygen in Silicon
W. Lin, The Incorporation of Oxygen into Silicon Crystals
T. J. Schaffner and D. K. Schroder, Characterization Techniques for Oxygen in Silicon
W. M. Bullis, Oxygen Concentration Measurement
S. M. Hu, Intrinsic Point Defects in Silicon
B. Pajot, Some Atomic Configurations of Oxygen
J. Michel and L. C. Kimerling, Electical Properties of Oxygen in Silicon
R. C. Newman and R. Jones, Diffusion of Oxygen in Silicon
T. Y. Tan and W. J. Taylor, Mechanisms of Oxygen Precipitation: Some Quantitative Aspects
M. Schrems, Simulation of Oxygen Precipitation
K. Simino and I. Yonenaga, Oxygen Effect on Mechanical Properties
W. Bergholz, Grown-in and Process-Induced Effects
F. Shimura, Intrinsic/Internal Gettering
H. Tsuya, Oxygen Effect on Electronic Device Performance

Volume 43 Semiconductors for Room Temperature Nuclear Detector Applications

R. B. James and T. E. Schlesinger, Introduction and Overview
L. S. Darken and C. E. Cox, High-Purity Germanium Detectors
A. Burger, D. Nason, L. Van den Berg, and M. Schieber, Growth of Mercuric Iodide
X. J. Bao, T. E. Schlesinger, and R. B. James, Electrical Properties of Mercuric Iodide
X. J. Bao, R. B. James, and T. E. Schlesinger, Optical Properties of Red Mercuric Iodide
M. Hage-Ali and P. Siffert, Growth Methods of CdTe Nuclear Detector Materials
M. Hage-Ali and P Siffert, Characterization of CdTe Nuclear Detector Materials
M. Hage-Ali and P. Siffert, CdTe Nuclear Detectors and Applications
R. B. James, T. E. Schlesinger, J. Lund, and M. Schieber, $Cd_{1-x}Zn_xTe$ Spectrometers for Gamma and X-Ray Applications
D. S. McGregor, J. E. Kammeraad, Gallium Arsenide Radiation Detectors and Spectrometers
J. C. Lund, F. Olschner, and A. Burger, Lead Iodide
M. R. Squillante, and K. S. Shah, Other Materials: Status and Prospects
V. M. Gerrish, Characterization and Quantification of Detector Performance
J. S. Iwanczyk and B. E. Patt, Electronics for X-ray and Gamma Ray Spectrometers
M. Schieber, R. B. James, and T. E. Schlesinger, Summary and Remaining Issues for Room Temperature Radiation Spectrometers

Volume 44   II–IV Blue/Green Light Emitters: Device Physics and Epitaxial Growth

J. Han and R. L. Gunshor, MBE Growth and Electrical Properties of Wide Bandgap ZnSe-based II–VI Semiconductors

Shizuo Fujita and Shigeo Fujita, Growth and Characterization of ZnSe-based II–VI Semiconductors by MOVPE

Easen Ho and Leslie A. Kolodziejski, Gaseous Source UHV Epitaxy Technologies for Wide Bandgap II–VI Semiconductors

Chris G. Van de Walle, Doping of Wide-Band-Gap II–VI Compounds — Theory

Roberto Cingolani, Optical Properties of Excitons in ZnSe-Based Quantum Well Heterostructures

A. Ishibashi and A. V. Nurmikko, II–VI Diode Lasers: A Current View of Device Performance and Issues

Supratik Guha and John Petruzello, Defects and Degradation in Wide-Gap II–VI-based Structures and Light Emitting Devices

Volume 45   Effect of Disorder and Defects in Ion-Implanted Semiconductors: Electrical and Physiochemical Characterization

Heiner Ryssel, Ion Implantation into Semiconductors: Historical Perspectives

You-Nian Wang and Teng-Cai Ma, Electronic Stopping Power for Energetic Ions in Solids

Sachiko T. Nakagawa, Solid Effect on the Electronic Stopping of Crystalline Target and Application to Range Estimation

G. Müller, S. Kalbitzer and G. N. Greaves, Ion Beams in Amorphous Semiconductor Research

Jumana Boussey-Said, Sheet and Spreading Resistance Analysis of Ion Implanted and Annealed Semiconductors

M. L. Polignano and G. Queirolo, Studies of the Stripping Hall Effect in Ion-Implanted Silicon

J. Stoemenos, Transmission Electron Microscopy Analyses

Roberta Nipoti and Marco Servidori, Rutherford Backscattering Studies of Ion Implanted Semiconductors

P. Zaumseil, X-ray Diffraction Techniques

Volume 46   Effect of Disorder and Defects in Ion-Implanted Semiconductors: Optical and Photothermal Characterization

M. Fried, T. Lohner and J. Gyulai, Ellipsometric Analysis

Antonios Seas and Constantinos Christofides, Transmission and Reflection Spectroscopy on Ion Implanted Semiconductors

Andreas Othonos and Constantinos Christofides, Photoluminescence and Raman Scattering of Ion Implanted Semiconductors. Influence of Annealing

Constantinos Christofides, Photomodulated Thermoreflectance Investigation of Implanted Wafers. Annealing Kinetics of Defects

U. Zammit, Photothermal Deflection Spectroscopy Characterization of Ion-Implanted and Annealed Silicon Films

Andreas Mandelis, Arief Budiman and Miguel Vargas, Photothermal Deep-Level Transient Spectroscopy of Impurities and Defects in Semiconductors

R. Kalish and S. Charbonneau, Ion Implantation into Quantum-Well Structures

Alexandre M. Myasnikov and Nikolay N. Gerasimenko, Ion Implantation and Thermal Annealing of III-V Compound Semiconducting Systems: Some Problems of III-V Narrow Gap Semiconductors

Volume 47 Uncooled Infrared Imaging Arrays and Systems

R. G. Buser and M. P. Tompsett, Historical Overview
P. W. Kruse, Principles of Uncooled Infrared Focal Plane Arrays
R. A. Wood, Monolithic Silicon Microbolometer Arrays
C. M. Hanson, Hybrid Pyroelectric-Ferroelectric Bolometer Arrays
D. L. Polla and J. R. Choi, Monolithic Pyroelectric Bolometer Arrays
N. Teranishi, Thermoelectric Uncooled Infrared Focal Plane Arrays
M. F. Tompsett, Pyroelectric Vidicon
T. W. Kenny, Tunneling Infrared Sensors
J. R. Vig, R. L. Filler and Y. Kim, Application of Quartz Microresonators to Uncooled Infrared Imaging Arrays
P. W. Kruse, Application of Uncooled Monolithic Thermoelectric Linear Arrays to Imaging Radiometers

Volume 48 High Brightness Light Emitting Diodes

G. B. Stringfellow, Materials Issues in High-Brightness Light-Emitting Diodes
M. G. Craford, Overview of Device issues in High-Brightness Light-Emitting Diodes
F. M. Steranka, AlGaAs Red Light Emitting Diodes
C. H. Chen, S. A. Stockman, M. J. Peanasky, and C. P. Kuo, OMVPE Growth of AlGaInP for High Efficiency Visible Light-Emitting Diodes
F. A. Kish and R. M. Fletcher, AlGaInP Light-Emitting Diodes
M. W. Hodapp, Applications for High Brightness Light-Emitting Diodes
I. Akasaki and H. Amano, Organometallic Vapor Epitaxy of GaN for High Brightness Blue Light Emitting Diodes
S. Nakamura, Group III-V Nitride Based Ultraviolet-Blue-Green-Yellow Light-Emitting Diodes and Laser Diodes

ISBN 0-12-752157-7